Undergraduate Lecture Notes

CW01497480

For further volumes:
http://www.springer.com/series/8917

Undergraduate Lecture Notes in Physics (ULNP) publishes authoritative texts covering topics throughout pure and applied physics. Each title in the series is suitable as a basis for undergraduate instruction, typically containing practice problems, worked examples, chapter summaries, and suggestions for further reading.

ULNP titles must provide at least one of the following:

- An exceptionally clear and concise treatment of a standard undergraduate subject.
- A solid undergraduate-level introduction to a graduate, advanced, or non-standard subject.
- A novel perspective or an unusual approach to teaching a subject.

ULNP especially encourages new, original, and idiosyncratic approaches to physics teaching at the undergraduate level.

The purpose of ULNP is to provide intriguing, absorbing books that will continue to be the reader's preferred reference throughout their academic career.

Series Editors

Neil Ashby
Professor Emeritus, University of Colorado, Boulder, CO, USA

William Brantley
Professor, Furman University, Greenville, SC, USA

Michael Fowler
Professor, University of Virginia, Charlottesville, VA, USA

Michael Inglis
Professor, SUNY Suffolk County Community College, Selden, NY, USA

Heinz Klose
Oldenburg, Niedersachsen, Germany

Helmy Sherif
Professor Emeritus, University of Alberta, Edmonton, AB, Canada

Charles Keeton

Principles of Astrophysics

Using Gravity and Stellar Physics to Explore the Cosmos

 Springer

Charles Keeton
Department of Physics and Astronomy
Rutgers University
Piscataway, NJ, USA

ISSN 2192-4791   ISSN 2192-4805 (electronic)
ISBN 978-1-4614-9235-1   ISBN 978-1-4614-9236-8 (eBook)
DOI 10.1007/978-1-4614-9236-8
Springer New York Heidelberg Dordrecht London

Library of Congress Control Number: 2014935057

Printed on acid-free paper

Springer is part of Springer Science+Business Media (www.springer.com)

To my parents, who helped me find my path

Preface

This book is designed to show how physical principles can be used at the advanced undergraduate level to understand astronomical systems such as planets, stars, galaxies, and the universe as a whole. It emerges from a pair of courses at Rutgers University that attract not just astrophysics students but a broad audience of physics and engineering students. The organization is therefore "physics-first": we start with key principles of physics and then examine applications to astronomical systems.

At Rutgers, each half of the book constitutes a coherent semester-length course; while there is a little overlap (notably with cosmology in Chaps. 11 and 20), the two halves are largely independent and complementary. Part I focuses on gravity, because this is the dominant force in many astronomical systems and it governs many types of motions we observe. The goal of Chaps. 2–11 is to develop a progressively richer understanding of gravity and the way astrophysicists use gravitational motion to investigate mass.

Part II centers on one of the "big questions" we humans ask. *Why are we here?* is admittedly beyond the realm of physics, but a related question is within our reach: *How did we come to be here?* As the Sun was forming, various elements came together in the right combination to form a rocky planet with a tenuous atmosphere. On this planet Earth, the energy from the Sun and the gas in the atmosphere were just right to allow the emergence of life. The energy that sustains us originates deep inside our star, thanks to $E = mc^2$. The atoms that comprise our bodies were forged in previous generations of stars. Literally, we are star dust. The goal of Chaps. 12–20 is to understand the roles that electromagnetism as well as gas, atomic, and nuclear physics play in this remarkable story.

I hope this book will help you learn to think like an astrophysicist. Rather than memorizing facts about specific astronomical systems, you will learn to break the systems into pieces you can analyze and understand using material that should be familiar from introductory physics and vector calculus. (The necessary physics topics are reviewed as they arise; vital aspects of vector calculus are reviewed in Appendix A.) Then you will be equipped to investigate interesting systems that you encounter in the future, even if they are not addressed in this book. Astrophysics is a dynamic field of research—and one in which you can understand the physical principles that underlie even the newest discoveries. So let's have fun!

Piscataway   Chuck Keeton
December 2013

Acknowledgements

"No book is an island, entire of itself." That is not what John Donne actually wrote, but it could have been. It is certainly apt here. This book would not exist in its present form without the help of many people.

Arthur Kosowsky originally developed the structure for the astrophysics courses at Rutgers, which is reflected in the makeup of this book. Saurabh Jha, Eric Gawiser, and John Moustakas have taught from this material at various stages of development, and provided critical feedback. All contributed ideas for homework problems; and many rounds of students have (perhaps to their chagrin) field-tested a lot of the problems. The Rutgers Department of Physics and Astronomy, and in particular the astrophysics group, has provided an environment where excellence in research and teaching are both encouraged and supported.

Art Congdon, Allan Moser, Erik Nordgren, Barnaby Rowe, and Tim Jones have done yeomen's work with the manuscript. They provided extensive and insightful comments throughout the drafting process, catching everything from typos to muddled thinking. All remaining errors are my fault, not theirs!

Many researchers have graciously let me use images and figures to illustrate the material. They are too numerous to list here, but are credited in the figure captions.

A number of books have contributed to my own learning, but two in particular stand out. *An Introduction to Modern Astrophysics* by Bradley W. Carroll and Dale A. Ostlie is a monumental survey of astrophysics at the undergraduate level. The "big orange book" maintains a respected place on every astronomer's bookshelf. *Astrophysics in a Nutshell* by Dan Maoz is a more focused treatise that shares a lot of the spirit animating this book. Both have influenced my thinking about how to present this material, as indicated throughout the text.

Last but not least, my wife and son have not merely endured this absorbing project, but actively endorsed it. To Kelly: thank you for letting me dream. To Evan: if you can dream it, you can do it, but it might take more effort than you imagine.

This work has received financial support from the U.S. National Science Foundation through grant AST-0747311.

Contents

Part III Appendices

List of Symbols

Constants of Nature

Speed of light (in vacuum)	$c\ =\ 2.9979 \times 10^8\,\mathrm{m\,s^{-1}}$
Newton's gravitational constant	$G\ =\ 6.6738 \times 10^{-11}\,\mathrm{m^3\,kg^{-1}\,s^{-2}}$
Planck's constant	$h\ =\ 6.6261 \times 10^{-34}\,\mathrm{J\,s}$
	$=\ 4.1357 \times 10^{-15}\,\mathrm{eV\,s}$
	$\hbar\ =\ 1.0546 \times 10^{-34}\,\mathrm{J\,s}$
	$=\ 6.5821 \times 10^{-16}\,\mathrm{eV\,s}$
Electron charge[a]	$e\ =\ 1.5189 \times 10^{-14}\,\mathrm{kg^{1/2}\,m^{3/2}\,s^{-1}}$
Electron mass	$m_e\ =\ 9.1094 \times 10^{-31}\,\mathrm{kg}$
Proton mass	$m_p\ =\ 1.6726 \times 10^{-27}\,\mathrm{kg}$
Neutron mass	$m_n\ =\ 1.6749 \times 10^{-27}\,\mathrm{kg}$
Boltzmann's constant	$k_B\ =\ 1.3806 \times 10^{-23}\,\mathrm{J\,K^{-1}}$
	$=\ 8.6173 \times 10^{-5}\,\mathrm{eV\,K^{-1}}$
Stefan-Boltzmann constant	$\sigma\ =\ 5.6704 \times 10^{-8}\,\mathrm{kg\,s^{-3}\,K^{-4}}$

[a]Note: See Chap. 1 for remarks about the units of charge

Unit Conversions

Energy	$\mathrm{eV} = 1.6022 \times 10^{-19}\,\mathrm{J}$
Time	$\mathrm{yr} = 3.1557 \times 10^7\,\mathrm{s}$
Angle	$\mathrm{rad} = 2.0626 \times 10^5\,\mathrm{arcsec}$

Astrophysical Scales

Mass	Earth mass	$M_\oplus = 5.974 \times 10^{24}\,\text{kg}$
	Jupiter mass	$M_J = 1.899 \times 10^{27}\,\text{kg}$
	Solar mass	$M_\odot = 1.989 \times 10^{30}\,\text{kg}$
Length	Earth radius	$R_\oplus = 6.378 \times 10^6\,\text{m}$
	Jupiter radius	$R_J = 7.149 \times 10^7\,\text{m}$
	Solar radius	$R_\odot = 6.955 \times 10^8\,\text{m}$
	Astronomical unit	$\text{AU} = 1.496 \times 10^{11}\,\text{m}$
	light-year	$\text{ly} = 9.461 \times 10^{15}\,\text{m}$
	parsec	$\text{pc} = 3.086 \times 10^{16}\,\text{m}$
Luminosity	Solar luminosity	$L_\odot = 3.839 \times 10^{26}\,\text{J s}^{-1}$

Astrophysical Symbols

$\odot$ Sun
$\oplus$ Earth

Mathematical Symbols

Astrophysicists employ a variety of mathematical relations. One skill I hope you will develop is a sense of how and when they apply. This book uses what I think are conventional symbols to indicate the different relations:

$=$ strict equality
$\equiv$ strict equivalence (often used for definitions)
$\propto$ strict proportionality
$\approx$ close approximation (e.g., Taylor series expansion)
$\sim$ order-of-magnitude estimate

Here are symbols for comparative relations (with similar symbols for 'less than'):

$>$ strictly greater than
$\gtrsim$ greater than or approximately equal to
$\gg$ much (i.e., order of magnitude) greater than

We sometimes use symbols to indicate logical relationships:

$\Rightarrow$ implies
$\Leftrightarrow$ if and only if

We use two symbols to indicate a statistical average (choosing between them based on which notation is simpler in a given context):

$\langle x \rangle$ or $\bar{x}$ average of x

In Taylor series expansions, we indicate higher-order terms as follows:

$\mathcal{O}(x^n)$ a term proportional to x^n

Chapter 1
Introduction: Tools of the Trade

A concise way to state the scientific method in astrophysics is this: We use theory to make quantitative predictions that can be compared with observations. Sometimes we can solve the relevant equations with pencil and paper in a modest number of steps, but other times we cannot. How do we proceed? Often we can use physical insight and approximate calculations to understand the salient features of a system without sweating the details. Before diving into technical material, it is good to see how physical reasoning and estimation techniques (such as toy models, scaling relations, Taylor series approximations, and dimensional analysis) offer a potent approach to astrophysics.

1.1 What Is Gravity?

Understanding gravity opens the door to studying many fascinating systems, so it is a natural place to begin. Plus, it provides a nice way to illustrate the analytic tools that infuse our inquiry. You can probably recite Newton's law of gravity,

$$F = \frac{GMm}{r^2} \tag{1.1}$$

but where does it come from? Put yourself in Isaac Newton's shoes and imagine you are trying to understand the motion of planets. Johannes Kepler has combed through reams of observational data and distilled three laws of planetary motion:

I. Planets move in elliptical orbits, with the Sun at one focus.
II. A line that connects a planet to the Sun sweeps out equal areas in equal times.
III. A planet's orbital period P and average distance from the Sun a are related by

$$P^2 \propto a^3$$

C. Keeton, *Principles of Astrophysics: Using Gravity and Stellar Physics to Explore the Cosmos*, Undergraduate Lecture Notes in Physics, DOI 10.1007/978-1-4614-9236-8_1,
© Springer Science+Business Media New York 2014

These are examples of **empirical laws**; they are extracted from, and provide a powerful summary of, observational data, but they do not explain in any physical way *why* planets move as they do. Empirical laws can, however, offer clues that help us find physical explanations, if we know how to reason with them.

The first step is to recognize that Kepler's third law is an example of a **scaling relation**. It answers the question: If you move a planet farther from the Sun, will its orbital period increase or decrease, and by how much? The second step is to see if we can relate the scaling relation we know to something we want to learn. While I cannot say for certain, I imagine Newton's reasoning was something like this: Galileo famously demonstrated that objects of different mass fall at the same rate under the influence of gravity. Since a more massive object has more inertia, it must feel more gravity; the gravitational force should therefore be proportional to m. Then by Newton's third law of motion (equal and opposite reactions),[1] the force must be proportional to the product Mm. Surely gravity depends on the distance between two objects; intuitively it should decrease with distance, so let's postulate

$$F \propto \frac{Mm}{r^n}$$

where n is unknown. Let's call the constant of proportionality K and write

$$F = \frac{KMm}{r^n} \tag{1.2}$$

The third step is to connect the two scaling relations. Here we might introduce a **toy model** that is deliberately simple but (we hope) rich enough to capture the essential physics. To build a toy model for motion under the influence of gravity, we ignore Kepler's lesson about ellipses and just consider circles. From Newton's laws of motion, we know the force required to keep an object of mass m in a circular orbit of radius r and speed v is

$$F = \frac{mv^2}{r} = \frac{4\pi^2 m r}{P^2} \tag{1.3}$$

where we replace the orbital speed v with the period $P = 2\pi r/v$ in order to connect with Kepler III. We then equate the force we have available (1.2) with the force we need to explain the motion (1.3):

$$\frac{KMm}{r^n} = \frac{4\pi^2 m r}{P^2}$$

---

[1] Newton's laws of motion are independent of his law of gravity. We will discuss them later; for now we take them as given.

Rearranging yields

$$P^2 = \frac{4\pi^2}{KM} r^{n+1}$$

If we want to explain Kepler's third law ($P^2 \propto r^3$), we apparently need the gravitational force to follow an inverse square law ($n = 2$). This argument is only heuristic; it cannot be taken as *proof* of Eq. (1.1). But imagine you were Newton and had no one to tell you the law of gravity. An analysis like this would strongly suggest the hypothesis that gravity is described by an inverse square law.[2]

Once we know the gravitational force law, we might wonder how it affects our everyday experience on Earth. Strictly speaking, we already have everything we need to determine how gravity weakens with height (h) above the surface of Earth (indicated by the radius $R_\oplus$):

$$F = \frac{GM_\oplus m}{(R_\oplus + h)^2}$$

This formula can be a little unwieldy, though, if we just want to know what happens when we climb a mountain or fly in an airplane. Is there any way to simplify the analysis when h is much smaller than $R_\oplus$? Yes! Rewriting F slightly lets us make the following approximation:

$$F = \frac{GM_\oplus m}{R_\oplus^2}\left(1 + \frac{h}{R_\oplus}\right)^{-2} \approx \frac{GM_\oplus m}{R_\oplus^2}\left[1 - 2\frac{h}{R_\oplus} + \mathcal{O}\left(\left(\frac{h}{R_\oplus}\right)^2\right)\right] \quad (1.4)$$

If $h \ll R_\oplus$ then the second term in square brackets is much smaller than the first, and the third term is smaller still so we can neglect it without making a significant error. What we have done here is make a **Taylor series expansion** of F. This is a form of estimation that we will use from time to time when we encounter functions that are cumbersome, or we want to examine a function's behavior over some fairly narrow range. In Eq. (1.4), the Taylor series shows that at "lowest order" (i.e., in the first term) the force of gravity is independent of height above the surface of Earth. In elementary mechanics classes we often write this as $F = mg$ where[3]

$$g = \frac{GM_\oplus}{R_\oplus^2} = \frac{(6.67 \times 10^{-11}\,\text{m}^3\,\text{kg}^{-1}\,\text{s}^{-2}) \times (5.97 \times 10^{24}\,\text{kg})}{(6.38 \times 10^6\,\text{m})^2} = 9.80\,\text{m s}^{-2}$$

---

[2]See p. 57 of *Isaac Newton* by James Gleick [1] for more discussion.

[3]Notice how I write and keep track of all units when doing the calculation. I *strongly* encourage you to get in the habit of doing this; it will help you catch errors and remember to convert units when necessary.

The minus sign in the second term of Eq. (1.4) then says gravity weakens with height. While we knew that already, the approximation offers a simple way to quantify this effect. Suppose we ask how much gravity varies when you go up in a building or an airplane, or into the upper parts of Earth's atmosphere:

Example	h	$2h/R_\oplus$	$1 - 2h/R_\oplus$
Building	$\sim$6 m	2×10^{-6}	0.999998
Airplane	$\sim$6 km	2×10^{-3}	0.998
Upper atmosphere	$\sim$60 km	0.02	0.98

These numbers help us understand that you have to go pretty high (relative to the atmosphere) for any change to be significant.

To recap: we have combined an empirical scaling relation with a toy model to deduce the form of the gravitational force law. We did not do any complicated math; rather, we used careful physical reasoning. We also used a Taylor series expansion to examine how gravity varies with height. I hope this book will help you cultivate these types of analysis skills, which can be quite valuable throughout astrophysics and beyond.

1.2 Dimensions and Units

Most of the quantities we discuss in physics and astrophysics come as numbers with some **units** attached (such as meters or light-years). The units are crucial; the numbers are meaningless without them. That said, units themselves are merely conventions for how we express measurements. The more fundamental quantities are **dimensions** (such as length). The distinction may seem subtle, but it is important because units are fungible while dimensions are not. Analyzing the dimensions that matter for a particular problem can be a good first step, as we are about to see.

In this book we use a combination of SI and astrophysical units. While it may seem unnecessarily complicated to mix different sets of units, there can be some advantages. Using certain units can help build your intuition about the relevant scales for different problems (e.g., it is more enlightening to specify star masses in units of the mass of the Sun than in kilograms). Also, *knowing* that you may encounter different sets of units can make you more vigilant about checking them. As a general rule:

In calculations, always check dimensions and units!

1.2.1 Fundamental Dimensions

The three key dimensions we use in physics are length, mass, and time. Here are their units in the SI system:

	Dimension	Unit
Length	$[L]$	m
Mass	$[M]$	kg
Time	$[T]$	s

Other familiar quantities involve combinations of the fundamental dimensions:

velocity	$\mathbf{v} = \dfrac{d\mathbf{x}}{dt}$	$[LT^{-1}]$
acceleration	$\mathbf{a} = \dfrac{d^2\mathbf{x}}{dt^2}$	$[LT^{-2}]$
force	$\mathbf{F} = m\mathbf{a}$	$[MLT^{-2}]$
kinetic energy	$K = \dfrac{1}{2}mv^2$	$[ML^2T^{-2}]$
momentum	$\mathbf{p} = m\mathbf{v}$	$[MLT^{-1}]$
angular momentum	$\mathbf{L} = \mathbf{r} \times \mathbf{p}$	$[ML^2T^{-1}]$
pressure	$P = \dfrac{\text{force}}{\text{area}}$	$[ML^{-1}T^{-2}]$
number density	$n = \dfrac{\text{number}}{\text{volume}}$	$[L^{-3}]$
mass density	$\rho = \dfrac{\text{mass}}{\text{volume}}$	$[ML^{-3}]$

We sometimes invent special units to measure certain quantities. Some of the special units are clearly combinations of fundamental dimensions (and their associated units):

Force	Newton	$N = kg\,m\,s^{-2}$
Energy	Joule	$J = N\,m$
Energy	Electron volt	$eV = 1.60 \times 10^{-19}\,J$

Other special units might seem to be unique but turn out to be composites as well:

- **Temperature** is often measured on the Fahrenheit, Celsius, or Kelvin scale, but it is actually a measure of energy. We can always convert a temperature in Kelvins to an equivalent energy using $E = k_B T$ where

$$k_B = 1.38 \times 10^{-23}\,J\,K^{-1} = 8.62 \times 10^{-5}\,eV\,K^{-1}$$

is Boltzmann's constant.[4] Astronomers sometimes invoke the equivalence between temperature and energy by reporting the "temperature" of hot gas in keV.

• **Charge** has a special unit—the Coulomb—in the SI system of units, but it can actually be expressed in terms of the three fundamental dimensions. In the Gaussian system of units, the force between charges q_1 and q_2 separated by a distance r is written with no proportionality constant[5]:

$$F = \frac{q_1 q_2}{r^2}$$

With this convention, we can identify the dimensions of charge as follows:

$$q_1 q_2 = r^2 F$$
$$[Q^2] = [L^2 \times MLT^{-2}]$$
$$\Rightarrow \quad [Q] = [M^{1/2} L^{3/2} T^{-1}]$$

This is one case in which I favor the Gaussian system, because thinking of charge in terms of the three fundamental dimensions turns out to be very helpful for dimensional analysis (as we will see below). In centimeter-gram-second units the value of the electron charge is $e = 4.8032 \times 10^{-10}\, \mathrm{g}^{1/2}\, \mathrm{cm}^{3/2}\, \mathrm{s}^{-1}$. Converting to meter-kilogram-second units yields $e = 1.5189 \times 10^{-14}\, \mathrm{kg}^{1/2}\, \mathrm{m}^{3/2}\, \mathrm{s}^{-1}$.

1.2.2 Constants of Nature

There are some special, fundamental numbers in physics:

Speed of light (in vacuum)	c	$= 2.9979 \times 10^8\, \mathrm{m\,s^{-1}}$	$[LT^{-1}]$
Newton's grav. constant	G	$= 6.6738 \times 10^{-11}\, \mathrm{m^3\,kg^{-1}\,s^{-2}}$	$[M^{-1}L^3T^{-2}]$
Planck's constant	$\hbar$	$= 1.0546 \times 10^{-34}\, \mathrm{kg\,m^2\,s^{-1}}$	$[ML^2T^{-1}]$
Electron charge	e	$= 1.5189 \times 10^{-14}\, \mathrm{kg^{1/2}\,m^{3/2}\,s^{-1}}$	$[M^{1/2}L^{3/2}T^{-1}]$
Electron mass	m_e	$= 9.1094 \times 10^{-31}\, \mathrm{kg}$	$[M]$
Proton mass	m_p	$= 1.6726 \times 10^{-27}\, \mathrm{kg}$	$[M]$
Neutron mass	m_n	$= 1.6749 \times 10^{-27}\, \mathrm{kg}$	$[M]$

---

[4]We often drop the subscript B to simplify the notation. Any k that appears in conjunction with T is probably Boltzmann's constant.

[5]You might ask whether we could do something similar to redefine the dimensions of mass. The answer is no, because mass appears not only in $F = GMm/r^2$ but also in $F = ma$. We cannot eliminate proportionality constants from both relations at the same time.

These allow conversions between the fundamental dimensions:

- Time $\leftrightarrow$ length: $\ell = ct$ or $t = \ell/c$ (think of a "light-year")
- Mass $\leftrightarrow$ length: $\ell = GM/c^2$
- Energy $\leftrightarrow$ mass: $E = mc^2$
- Energy $\leftrightarrow$ time: $E = h\nu$ where ν is frequency (or inverse time)

Using these conversions, you could argue in principle that there is really one fundamental dimension: length. Theoretical studies of general relativity or quantum mechanics often do such conversions. We will stick with length, mass, and time, though, because they are familiar and keeping track of all three dimensions can help us check and interpret calculations.

1.2.3 Astrophysical Units

There are some numbers that are used so frequently in astrophysics that they act as a de facto set of units. Using astrophysical units can help us interpret quantities quickly; for example, it is easier to get an impression of an exoplanet's properties if we quote its mass and radius as $0.7 \, M_{\text{Jupiter}}$ and $1.6 \, R_{\text{Jupiter}}$ than if we specify them as 1.3×10^{27} kg and 1.1×10^8 m. We need to remember, though, that the quantities we take as reference values are not *fundamental*; they just happen to be quantities that are familiar in our corner of the universe. (Part of our goal as astrophysicists is to see if we can explain why these quantities have the values they do.) Here are some of the quantities we will use as astrophysical units:

Mass	Earth mass	$M_\oplus$	=	5.974×10^{24} kg
	Jupiter mass	M_J	=	1.899×10^{27} kg
	Solar mass	$M_\odot$	=	1.989×10^{30} kg
length	Earth radius	$R_\oplus$	=	6.378×10^6 m
	Jupiter radius	R_J	=	7.149×10^7 m
	Solar radius	$R_\odot$	=	6.955×10^8 m
	Astronomical unit	AU	=	1.496×10^{11} m
	Light-year	ly	=	9.461×10^{15} m
	Parsec	pc	=	3.086×10^{16} m

Our earlier discussion of Kepler's third law illustrates the value of picking good units. The proportionality means there is some constant K such that $P^2 = Ka^3$. When we study planets orbiting the Sun, we can eliminate K by taking a ratio with respect to Earth:

$$P^2 = Ka^3 \quad \text{and} \quad P_\oplus^2 = Ka_\oplus^3 \quad \Rightarrow \quad \left(\frac{P}{P_\oplus}\right)^2 = \left(\frac{a}{a_\oplus}\right)^3$$

Since $P_\oplus = 1$ yr and $a_\oplus = 1$ AU (by definition), we can write

$$\left(\frac{P}{1\,\text{yr}}\right)^2 = \left(\frac{a}{1\,\text{AU}}\right)^3 \tag{1.5}$$

If we measure a planet's distance from the Sun in AU and orbital period in years, we can write $P^2 = a^3$ without any additional constants.[6] Using appropriate units for a problem can simplify things quite a bit.

1.2.4 Dimensional Analysis

Thinking about dimensions can be a good way to begin analyzing a particular system. Before doing any detailed calculations, we might be able to make an "educated guess" about the properties of a system just by finding combinations of constants and scales that have the right dimensions. This approach cannot pin down numerical factors of order unity (e.g., 2, π, etc.), but those are rarely essential for conceptual understanding. Nor can it tell us what to do if we find several combinations of constants and scales that have the right dimensions. If that happens, we can use physical reasoning to choose among the possibilities. Let's see how this works in a few examples.

Planetary Motion

Consider a planet orbiting at distance r from a star of mass M, and suppose we want to determine the period of the orbit. To make a dimensional analysis estimate, we start by listing the scales or constants that are involved in the problem. We are given r and M, and we know gravity plays a role, so we write this list:

Distance	r	$[L]$
Mass	M	$[M]$
Gravity	G	$[M^{-1}L^3T^{-2}]$

If we want to form a combination that has dimensions of time, we clearly need to start with $G^{-1/2}$. Then we include $M^{-1/2}$ to eliminate mass, and $r^{3/2}$ to eliminate length. Thus, we guess that the expression for orbital period should look like

---

[6]This works only for objects orbiting the Sun, because K depends on the mass of the central object. Equation (1.5) implies that we can write $K = 1\,\text{yr}^2\,\text{AU}^{-3}$ for motion around the Sun.

$$P \sim \frac{r^{3/2}}{(GM)^{1/2}}$$

Does our guess make sense? Consider the scalings: as M increases, the gravitational force gets stronger, so things move faster and P decreases. Also, as r increases, P increases with the specific relation

$$P^2 \propto r^3$$

This is the scaling in Kepler's third law! In other words, we can recover Kepler III from dimensional analysis alone. The exact calculation for a circular orbit (which we did in Sect. 1.1) gives

$$P = 2\pi \frac{r^{3/2}}{(GM)^{1/2}}$$

Our dimensional analysis estimate was right up to a factor of $2\pi \approx 6$, which is not bad for such a simple analysis. Even more important is the fact that we got the scalings correct.

Black Hole

In Einstein's general theory of relativity, a point mass has an "event horizon" out of which no physical object can escape (see Sect. 10.6). What is the radius of the event horizon of a black hole with mass M? Again, we begin by listing the scales and constants we think are relevant:

Mass	M	$[M]$
Gravity	G	$[M^{-1}L^3T^{-2}]$
Relativity	c	$[LT^{-1}]$

The combination that has dimensions of length is

$$R \sim \frac{GM}{c^2}$$

The exact answer is the Schwarzschild radius of a black hole,

$$R_S = \frac{2GM}{c^2}$$

Here dimensional analysis comes within a factor of 2.

Suppose we had incorrectly invoked quantum mechanics rather than gravity. Then we would have used $\hbar = [ML^2T^{-1}]$ and constructed

$$R \sim \frac{\hbar}{Mc}$$

Think about this for a moment: it would imply that more massive objects have smaller event horizons. That would not make sense! There may be different combinations of scales and constants with the dimensions we are looking for, but considering the physical scalings can help us identify the best choice.

Atom

How big is an atom? The size is determined by electrons orbiting under the influence of the electric force from the nucleus. The force must involve the electric charge, while an electron's response to the force is affected by its mass. And the whole problem is quantum mechanical in nature. Thus, we have:

Quantum mechanics	$\hbar$	$[ML^2T^{-1}]$
Electric force	e	$[M^{1/2}L^{3/2}T^{-1}]$
Electron mass	m_e	$[M]$

The combination with dimensions of length is

$$\frac{\hbar^2}{m_e\, e^2}$$

The scalings with e and m_e make sense: increasing the charge would strengthen the electric force and pull the electrons closer, while increasing that mass would mean the electrons do not move as much (less acceleration for a given force). In fact, the combination we have found is the **Bohr radius** a_0, which is the radius of the lowest electron energy level in the Bohr model of the hydrogen atom (see Sect. 13.4.1). We take it as characteristic of the sizes of atoms.

1.3 Using the Tools

In Part II of this book we will encounter gases in various astrophysical contexts. Even before we study the details, we can use dimensional analysis to understand the key properties of the gases, and then deduce a few features—some straightforward, some unexpected—of different types of stars.

1.3.1 Phases of an Electron Gas

Our first goal is to uncover the **equation of state** relating the pressure of a gas to its other physical properties. In many settings we will study, the gas is ionized and most of the pressure comes from free electrons; hence we consider an electron gas. There are different scenarios depending on whether the behavior of the gas depends on quantum physics, relativity, both, or neither.

Ideal Gas

If quantum physics is not important, we can think of the gas as being made of point particles that hardly interact with one another; this is a classic "ideal" gas. Pressure is caused by particles bouncing off the walls of any container holding the gas. Dimensionally, pressure is force per unit area so

$$[P] = [ML^{-1}T^{-2}]$$

What quantities might influence the pressure of a classical ideal electron gas? The speed with which particles hit the wall depends on the temperature, and the rate at which that happens depends on how many particles there are. Are temperature (or equivalent energy) and number density enough?

Temperature	kT	$[ML^2T^{-2}]$
Number density	n	$[L^{-3}]$

In fact, simply multiplying these quantities gives dimensions of pressure, so we put

$$P \sim nkT$$

A detailed analysis reveals that there are no dimensionless factors, and we have actually recovered the famous **ideal gas law** (see Sect. 12.1.3).

We might wonder whether relativity is important for an ideal gas. That is the subject of Problem 1.5.

Classical Degenerate Gas

What happens when the density increases significantly? As the electrons squeeze closer together, the main contribution to pressure comes from the fact that different particles cannot occupy the same quantum state; in effect, the Pauli exclusion principle kicks in to create what is known as **electron degeneracy pressure**. This pressure would exist even if the temperature were zero, so the equation of state must not involve T. What does it depend on?

Number density	n	$[L^{-3}]$
Particle mass	m_e	$[M]$
Quantum mechanics	$\hbar$	$[ML^2T^{-1}]$

To this point we have built dimensional analysis estimates basically by trial and error. We can be more systematic, though. Let's postulate that the equation of state has the form

$$P \sim \hbar^\alpha \, m_e^\beta \, n^\gamma$$

where the exponents α, β, and γ are to be determined. Plugging in the dimensions, we obtain

$$[M \, L^{-1} \, T^{-2}] \sim [M^\alpha \, L^{2\alpha} \, T^{-\alpha} \times M^\beta \times L^{-3\gamma}]$$
$$\sim [M^{\alpha+\beta} \, L^{2\alpha-3\gamma} \, T^{-\alpha}]$$

To match the dimensions on the left- and right-hand sides, we need

$$1 = \alpha + \beta$$
$$-1 = 2\alpha - 3\gamma$$
$$-2 = -\alpha$$

This is a system of three equations in three unknowns, whose solution is $\alpha = 2$, $\beta = -1$, and $\gamma = 5/3$. Thus, our equation of state for a degenerate electron gas is

$$P \sim \frac{\hbar^2}{m_e} \, n^{5/3}$$

A complete analysis gives a dimensionless factor of $(3\pi^2)^{2/3}/5 = 1.91$ (see Sect. 17.1).

To find the transition between an ideal gas and a degenerate gas, we want to find the point at which the two systems have comparable pressures. This is equivalent to requiring

$$P_{\text{ideal}} \sim P_{\text{deg}} \quad \Rightarrow \quad kT \sim \frac{\hbar^2}{m_e} \, n^{2/3}$$

To find the transition between a classical and relativistic system, we can estimate a typical speed

$$v \sim \frac{\hbar}{m_e} \, n^{1/3}$$

and find that v becomes comparable to c when the density reaches

$$n \sim \left(\frac{m_e c}{\hbar}\right)^3 \sim 2 \times 10^{37} \, \text{m}^{-3}$$

Relativistic Degenerate Gas

Finally we come to the case of a degenerate gas in which the particles are moving near the speed of light. The energy of relativistic particles is dominated by motion rather than mass, so m_e presumably drops out of the equation of state and c enters. Thus, our list of ingredients becomes:

Number density	n	$[L^{-3}]$
Relativity	c	$[LT^{-1}]$
Quantum mechanics	$\hbar$	$[ML^2T^{-1}]$

As before, we put

$$P \sim \hbar^\alpha \, c^\beta \, n^\gamma$$

$$[ML^{-1}T^{-2}] = [M^\alpha \, L^{2\alpha+\beta-3\gamma} \, T^{-\alpha-\beta}]$$

and solve to find $\alpha = 1$, $\beta = 1$, $\gamma = 4/3$. This yields the equation of state

$$P \sim \hbar c \, n^{4/3}$$

Where is the transition between a relativistic ideal gas and a relativistic degenerate gas? That would correspond to

$$P_{\text{ideal}} \sim P_{\text{deg}} \quad \Rightarrow \quad kT \sim \hbar c \, n^{1/3} \quad \Rightarrow \quad n \sim \left(\frac{kT}{\hbar c}\right)^3$$

Phase Diagram

To recap, here are the equations of state we have estimated for the various scenarios we have considered:

Ideal gas	$P \sim n \, k \, T$
Classical degenerate gas	$P \sim \hbar^2 \, m_e^{-1} \, n^{5/3}$
Relativistic degenerate gas	$P \sim \hbar c \, n^{4/3}$

All of these expressions have dimensions of pressure, but our physical reasoning has let us understand which expression corresponds to which physical context.

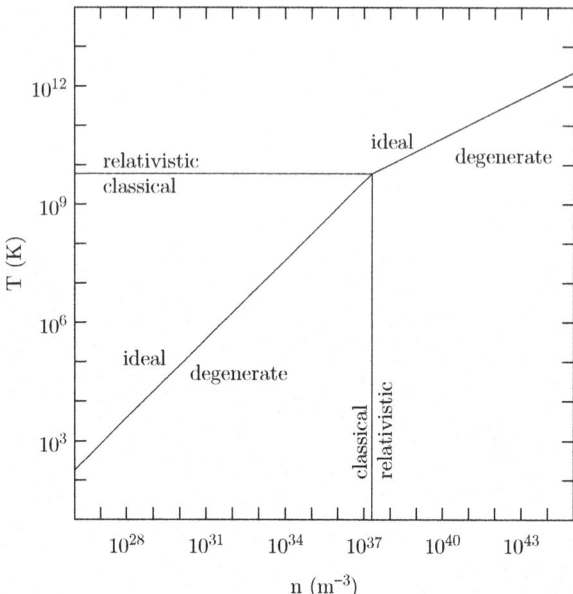

Fig. 1.1 Phase diagram for an electron gas, identifying the regimes discussed in the text: classical ideal gas, relativistic ideal gas, classical degenerate gas, and relativistic degenerate gas

We also found the transitions between different regimes, so we can sketch a phase diagram as shown in Fig. 1.1. The boundaries between the different regions are not sharp (because we have only done dimensional analysis, which is not exact). But this analysis does give a general picture of the type of gas we will encounter in different settings.

1.3.2 Stars, Familiar and Exotic

The preceding analysis may have seemed esoteric, but it proves to be very useful for understanding different kinds of stars. To make the connection, let's shift from microscopic quantities like density and pressure to the macroscopic quantities that we typically use to characterize an astrophysical object: mass M and radius R. We can relate them as follows:

$$\text{mass density} \quad \rho \sim \frac{M}{R^3}$$

$$\text{number density} \quad n \sim \frac{M}{m_p R^3}$$

$$\text{pressure} \quad P \sim \frac{G M^2 / R^2}{R^2} \sim \frac{G M^2}{R^4}$$

(Note that m_p appears in the number density because protons dominate the mass even if electrons dominate the pressure.) Now we can answer some interesting questions about stars.

Ideal Gas

What is the temperature of a normal star composed of ideal gas?

$$P \sim nkT$$

$$\Rightarrow \quad T \sim \frac{P}{nk} \sim \frac{GM^2 R^{-4}}{M m_p^{-1} R^{-3} k} \sim \frac{GM m_p}{kR}$$

For the Sun, plugging in numbers gives

$$T \sim \frac{(6.67 \times 10^{-11} \text{ m}^3 \text{ kg}^{-1} \text{ s}^{-2}) \times (1.99 \times 10^{30} \text{ kg}) \times (1.67 \times 10^{-27} \text{ kg})}{(1.38 \times 10^{-23} \text{ kg m}^2 \text{ s}^{-2} \text{ K}^{-1}) \times (6.96 \times 10^8 \text{ m})}$$

$$\sim 2 \times 10^7 \text{ K}$$

This estimate agrees surprisingly well with detailed stellar models (see Sect. 16.2.2).

Classical Degenerate Gas

What would a star composed of a degenerate electron gas be like?

$$P \sim \hbar^2 m_e^{-1} n^{5/3}$$

$$GM^2 R^{-4} \sim \hbar^2 m_e^{-1} (M m_p^{-1} R^{-3})^{5/3}$$

$$\Rightarrow \quad R \sim \frac{\hbar^2}{G m_e m_p^{5/3} M^{1/3}}$$

The scaling $R \propto M^{-1/3}$ implies that more massive stars are *smaller*. While this may seem counterintuitive, it is confirmed by more detailed calculations (see Sect. 17.2). Consider a **white dwarf** with $M \sim M_\odot$:

$$R \sim \frac{(1.05 \times 10^{-34} \text{ kg m}^2 \text{ s}^{-1})^2}{(9.11 \times 10^{-31} \text{ kg}) \times (1.67 \times 10^{-27} \text{ kg})^{5/3}}$$

$$\times \frac{1}{(6.67 \times 10^{-11} \text{ m}^3 \text{ kg}^{-1} \text{ s}^{-2}) \times (1.99 \times 10^{30} \text{ kg})^{1/3}}$$

$$\sim 6 \times 10^6 \text{ m}$$

A white dwarf is roughly the size of Earth.

Relativistic Degenerate Gas

Now let's actually consider a *neutron* gas. What would a star composed of a relativistic degenerate neutron gas be like?

$$P \sim \hbar c n^{4/3}$$

$$\frac{GM^2}{R^4} \sim \hbar c \left(\frac{M}{m_n R^3} \right)^{4/3}$$

$$\Rightarrow \quad M \sim \frac{1}{m_n^2} \left(\frac{\hbar c}{G} \right)^{3/2} \sim 4 \times 10^{30} \, \text{kg} \sim 2 \, M_\odot$$

All stars composed of a (highly) relativistic degenerate neutron gas have roughly the same mass. In order for them to be relativistic, we need:

$$n \gtrsim \left(\frac{m_n c}{\hbar} \right)^3$$

$$\Rightarrow \quad R \lesssim \frac{1}{m_n^2} \left(\frac{\hbar^3}{G c} \right)^{1/2} \sim 3 \, \text{km}$$

Such a star would be a little more massive that the Sun, but only as big as a city. In fact, we observe this kind of object as a **neutron star**. (Real neutron stars are probably not ultra-relativistic, but this analysis still gives a useful sense of the physics. See Chap. 17 for more discussion.)

Problems

1.1. Use dimensional analysis to derive a relationship between the total mass M of a gravitationally bound system, its size R, and the typical speed v of its components. Then use it to answer the following questions.

(a) At what speed does the Earth orbit the Sun?
(b) Globular clusters typically contain $\sim 10^6$ stars moving at speeds of $\sim 10 \, \text{km s}^{-1}$. How big are they?
(c) Spiral galaxies are typically about 10 kpc in size and rotate such that the stars move at $\sim 200 \, \text{km s}^{-1}$. Estimate the mass of a spiral galaxy (in $M_\odot$).

1.2. Type Ia supernovae are exploding stars that have played an important role in observational cosmology (see Chap. 18).

(a) The exploding stars are white dwarfs that have a mass of about 1.4 $M_\odot$ and a radius of about 5,000 km. Use dimensional analysis to estimate the gravitational binding energy of such a star.

(b) The explosion is powered by nuclear fusion. How much mass must be converted to energy ($E = mc^2$) in order to overcome the binding energy and explode the star?

1.3. Explosions such as Type Ia supernovae produce blast waves.

(a) Use dimensional analysis to estimate the size R of a blast wave at time t after an explosion with energy E, propagating into a medium of ambient density ρ. (Hint: these are all the quantities you need; gravity is not directly relevant here.)
(b) Information about the first atomic bomb tests was kept secret, but the physicist Geoffrey Taylor estimated the energy of one test from published photographs showing a fireball expanding through the air [2]. If the blast wave reached 100 m just 0.02 s after the explosion, what was the energy? What mass was converted into energy? (Hint: you will need to look up the density of air.)
(c) How large would the remnant of a supernova ($E \sim 10^{44}$ J) be 1,000 years after the explosion, as it expands into the interstellar medium with a typical density of 10^6 hydrogen atom per cubic meter?

1.4. The universe is believed to be about 14 billion years old. Use dimensional analysis to estimate the average density of the universe. About how many hydrogen atoms are there in 1 m^3 of "empty" space?

1.5. Can we treat the center of the Sun as a classical ideal gas? Let's find out.

(a) Consider a gas at temperature T composed of particles of mass m. Use dimensional analysis to estimate the typical speed of the particles.
(b) Recall our estimate of the Sun's central temperature, $T \sim 2 \times 10^7$ K. This is hot enough to ionize atoms, so electrons and nuclei move independently. What is the typical speed of electrons? Of hydrogen nuclei? Are they relativistic?
(c) At roughly what temperature does an electron gas become relativistic ($v \sim c$)?

1.6. Light carries momentum, so it creates pressure when it shines on something. This has led people to propose using "solar sails" on interplanetary or interstellar spacecraft.

(a) Use dimensional analysis to estimate the light pressure at a distance d from a star with luminosity L (energy per unit time).
(b) Estimate the force on a solar sail with an area of 1 km^2 that is 1 AU from the Sun.
(c) Suppose that sail is pulling a 10 ton spacecraft. How long would it take to reach Jupiter's orbit (5.2 AU from the Sun)? For simplicity, assume the acceleration remains constant even though it actually varies with distance from the star.

References

1. J. Gleick, *Isaac Newton* (Vintage Books, New York, 2004)
2. G. Taylor, Proc. R. Soc. Lond. Ser. A. Math. Phys. Sci. **201**(1065), 175 (1950)

Part I
Using Gravity and Motion
to Measure Mass

Chapter 2
Celestial Mechanics

Patterns of motion in the sky played a significant role in the historical development of mechanics. Briefly reviewing the history lets us see how physical concepts and models emerged from the empirical facts.

2.1 Motions in the Sky

Science often begins when people notice patterns in nature and try to understand what causes them. One well-known pattern is the daily rising and setting of the Sun, Moon, and stars. As the stars move across the sky each night, they look for all the world like points of light on some kind of crystalline sphere rotating around Earth. The Sun seems to move around Earth as well, although the relative positions of the Sun and stars vary throughout the year (the collection of visible stars changes with the season) so there must be two different crystalline spheres. The Moon is a little more complicated because its position and phase both change throughout the month, but both effects can be explained by placing the Moon on a sphere of its own. In other words, most of the obvious motions in the sky can be explained with the intuitive notion that Earth is fixed and objects in the sky move around us. This is the classic **geocentric model** of the universe.

Problems arise, though, when we notice another set of motions in the sky: planets are points of light that seem to "wander" among the stars.[1] Ancient societies knew of five planets (the discovery of others had to await the invention of the telescope). Mercury and Venus always stay fairly close to the Sun, appearing either in the west after sunset or in the east before sunrise. Jupiter and Saturn can be seen across a much wider range of positions, moving from west to east relative to the stars from one night to the next. Mars is a bit like Jupiter and Saturn, but with a twist. Most of

---

[1]The term "planet" comes from the ancient Greek term *aster planetes*, or "wandering star."

C. Keeton, *Principles of Astrophysics: Using Gravity and Stellar Physics to Explore the Cosmos*, Undergraduate Lecture Notes in Physics, DOI 10.1007/978-1-4614-9236-8_2, © Springer Science+Business Media New York 2014

Fig. 2.1 Pictures taken across several months have been combined to illustrate Mars's retrograde motion. Relative to the background stars, Mars usually moves from right to left. However, from November 15, 2007 to January 30, 2008 the planet moved from left to right, producing the loop pattern shown here. In other cases retrograde motion can create a zigzag pattern. (Credit: Tunç Tezel (TWAN), reproduced by permission)

the time it moves from west to east, but every once in a while Mars appears to stop, turn around and go from east to west for a few weeks, then turn around again and resume its "normal" motion (relative to the stars). Today we can see this **retrograde motion** very clearly in composite photographs, as shown in Fig. 2.1.

When scholars in ancient Greece tried to explain the apparent motions of planets, they started with the assumption that the intrinsic motions involve circles. Apollonius (c. 200 BC) constructed a model in which a planet moves on a small circle (called an "epicycle") that itself moves along a larger circle (called the "deferent"). As shown in Fig. 2.2, the composite motion can allow the planet to move backward at certain points in its orbit (depending on the relative sizes and speeds of the epicycle and deferent; see Problem 2.1). As the measurements became more precise, Ptolemy (c. 100 AD) refined the model by shifting the center of the deferent away from Earth and introducing yet a different point (called the "equant") around which the angular speed was defined.[2]

While Ptolemy's model was admittedly complex, its quantitative success kept it successful well into the Renaissance. Nicolaus Copernicus (1473–1543) introduced the first mathematically detailed alternative with the Sun at the center of motion.[3]

---

[2]There is a common misconception that Ptolemy and his successors added more and more epicycles. They couldn't; even one was hard enough to compute. See Chap. 4 of *The Book Nobody Read* by Owen Gingerich [1].

[3]The geocentric model had been questioned much earlier by Aristarchus (c. 300 BC), but without a fully developed alternative.

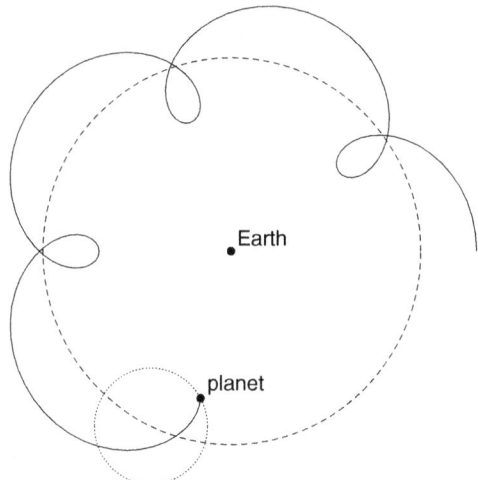

Fig. 2.2 In the geocentric model, a planet moves on an epicycle (*dotted*) whose center moves along a curve called the deferent (*dashed*). The combined motion (*solid*) can cause the planet to move backward as viewed from Earth. In the full Ptolemaic model, the deferent was not perfectly centered on Earth

In this **heliocentric model**, retrograde motion is an illusion that occurs when fast-moving Earth overtakes a slower-moving outer planet (see Problem 2.2); planets never actually move backward in space. Offering a simple explanation of retrograde motion is not all that Copernicus's model had going for it. The heliocentric model also explained why the observed planets fall into two categories: Mercury and Venus are never seen far from the Sun because their orbits are smaller than Earth's; while Mars, Jupiter, and Saturn can be seen near the Sun, on the opposite side of the sky, or anywhere in between because their orbits are larger than Earth's. Last but not least, Copernicus's model revealed a simple pattern in the quantitative relation between a planet's distance from the Sun and its orbital period. To Copernicus, this was a striking success: "In no other way," he wrote, "do we find a wonderful commensurability and a sure harmonious connection between the size of the orbit and the planet's period" (quoted by Gingerich [1, p. 54]).

That said, the original heliocentric model was not without fault. Like the Greeks, Copernicus assumed that planetary motion involved circles. While he was able to eliminate equants and large epicycles, he still needed small epicycles to make the model fit the data. That made Copernicus's model about as mathematically complex as Ptolemy's, even if it was conceptually simpler.

Copernicus's model made an important prediction: Earth moves in space. If that is true, then our perspective on the stars should change as Earth travels from one side of its orbit to the other (e.g., from January to July). Tycho Brahe (1546–1601), who was perhaps the world's greatest naked-eye astronomer, set out to test this prediction. He amassed years' worth of careful measurements of planet and star

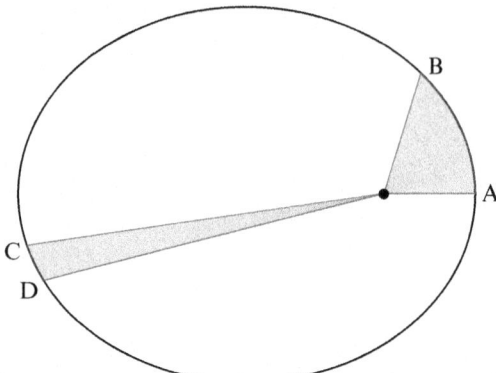

Fig. 2.3 Illustration of Kepler's first and second laws of planetary motion. I. The ellipse indicates the orbit, and the *dot* indicates the Sun at one focus. II. The time it takes the planet to travel from *A* to *B* is the same as the time to travel from *C* to *D*, so the areas of the *two shaded regions* are the same. This is a rather extreme example; the orbits of planets in our Solar System are much less elongated

positions in an attempt to measure **parallax**, or small shifts in the apparent positions of stars that should arise when we look from different sides of Earth's orbit. Tycho failed to find clear evidence for parallax, although now we know that stars are so far away that parallax can only be detected with a good telescope. Tycho's efforts did ultimately provide support for the heliocentric model, although not in the way he expected.

Shortly before he died, Tycho hired Johannes Kepler (1571–1630) as an assistant. Kepler combed through Tycho's measurements of planet positions and tried to find a geometric model to explain the motion. He initially adopted Copernicus's heliocentric model with circular orbits modified by epicycles. Kepler found, though, that the model could not quite reproduce Tycho's high-quality data, notably for Mars. Once he considered more general forms of motion, Kepler discovered that he could fit the data using elliptical orbits. Working through the details, he eventually extracted three **laws of planetary motion**:

I. Planets move in elliptical orbits, with the Sun at one focus.
II. A line that connects a planet to the Sun sweeps out equal areas in equal times (see Fig. 2.3).
III. A planet's orbital period P (in years) and average distance from the Sun a (in AU) are related by $P^2 = a^3$.

Suddenly the heliocentric model had an attractive and powerful quantitative framework. Still, people continued to struggle with the notion of a moving Earth.

That situation finally began to change thanks to the work of Galileo Galilei (1564–1642), who was arguably the first great experimental physicist. Using the newly-invented telescope, Galileo made two key discoveries related to planetary motion. First, he observed that Venus has phases just like the Moon. In the geocentric model, Venus would always stay between Earth and the Sun so it could

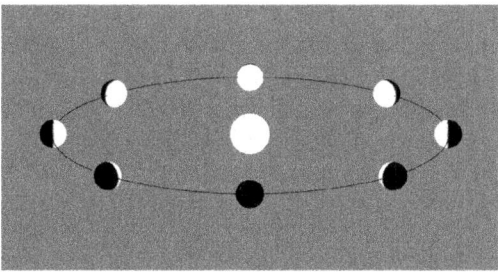

Fig. 2.4 Phases of Venus in the heliocentric model (not to scale). Full and gibbous phases can occur only if Venus travels to the far side of the Sun

only have new and crescent phases. Galileo saw that Venus has quarter and gibbous phases as well, which implies that Venus can go "behind" the Sun (as seen from Earth; see Fig. 2.4). In other words, if we know that Venus is closer to the Sun than Earth is, and the planet has a full cycle of phases, then it must orbit the Sun. Second, Galileo discovered four moons orbiting Jupiter. While this did not directly prove that planets orbit the Sun, it did demonstrate that objects can orbit something other than Earth. On the basis of his evidence, Galileo argued strongly in favor of the heliocentric model, most famously in his book *Dialogue on the Two Chief World Systems*. The work violated dictates from the Catholic Church, causing the book to be banned by the Roman Inquisition and Galileo to be placed under house arrest. More than three and a half centuries later, Pope John Paul II renounced the Church's condemnation of Galileo.

2.2 Laws of Motion

All of those ideas set the stage for Isaac Newton (1642–1727) to devise the fields we now know as theoretical physics and calculus (among other accomplishments). In 1665, Newton graduated from Cambridge but the university then closed because of the plague. He went home and, working alone, entered a period of remarkable intellectual creativity.[4] Newton started with mathematics, inventing the idea of

---

[4]Historical aside: In 1665–1666 Newton solved the problems of motion and gravity to his satisfaction, keeping a detailed notebook but not publishing his work. In 1684, Edmund Halley visited Newton to pose the question: If gravity has an inverse square force law, what curve will a planet follow? Newton knew the answer was an ellipse (see Sect. 3.1), but only after battling Robert Hooke for some time did he finally decide to write his famous work *Philosophiae Naturalis Principia Mathematica*, or "Mathematical Principles of Natural Philosophy." Newton's introduction of *mathematical* principles was profoundly important for the further development of physics and astrophysics. See *Isaac Newton* by James Gleick [2] for more about the life and work of this fascinating figure.

plotting solutions of equations as curves (a topic now known as algebraic geometry). He developed calculus so he could analyze curves, using derivatives to represent tangent lines and integrals to compute areas. Then Newton began to think about curves representing trajectories of objects in motion. Before he could apply his mathematical tools to motion, though, Newton had to introduce some new physical concepts that became his famous **laws of motion**:

I.   **Inertia.** An object will remain at rest or in uniform motion in a straight line unless acted on by an unbalanced force.
II. **Force and acceleration.** A net force acting on an object produces an acceleration in the same direction as the applied force. The acceleration and force are related by

$$\mathbf{F} = m\mathbf{a} = m\,\frac{d\mathbf{v}}{dt} \tag{2.1}$$

III. **Equal/opposite reaction.** If object #1 exerts a force on object #2, then object #2 exerts an equal and opposite force back on object #1: $\mathbf{F}_{12} = -\mathbf{F}_{21}$.

These laws are general; they are not specific to planets. In fact, to explain planetary motion Newton had to add one more law specifying the force. We will come to the law of gravity in Sect. 2.3.

While they are often introduced as above, Newton's laws of motion can be restated in terms of quantities that do not change with time. Think of a rod: the (x, y, z) coordinates of the endpoints depend on whether the rod is moving or rotating, but the *distance* between the two endpoints is always the same. A quantity that is "conserved" is usually thought to represent some fundamental property of a system (such as the length of the rod). Stating physical theories in terms of **conservation laws** can often help us find the simplest expressions of those theories. Let's see a few examples that are probably familiar but nonetheless valuable.

Momentum is defined by

$$\mathbf{p} \equiv m\mathbf{v}$$

We can use this to rewrite Eq. (2.1) as

$$\mathbf{F} = \frac{d\mathbf{p}}{dt}$$

While this might seem trivial, it is actually a nice generalization of Newton's second law. It helps us see that when there is no net force, momentum does not change. Thus, Newton's first law is fundamentally a statement of conservation of momentum.

Angular momentum is defined by

$$\mathbf{L} \equiv \mathbf{r} \times \mathbf{p} = m(\mathbf{r} \times \mathbf{v}) \tag{2.2}$$

We will sometimes use the **specific angular momentum**, defined to be the angular momentum per unit mass:

$$\boldsymbol{\ell} \equiv \frac{\mathbf{L}}{m} = \mathbf{r} \times \mathbf{v} \tag{2.3}$$

Let's take the derivative of angular momentum with respect to time:

$$\frac{d\mathbf{L}}{dt} = \frac{d\mathbf{r}}{dt} \times \mathbf{p} + \mathbf{r} \times \frac{d\mathbf{p}}{dt}$$
$$= \mathbf{v} \times (m\,\mathbf{v}) + \mathbf{r} \times \mathbf{F}$$
$$= \mathbf{r} \times \mathbf{F}$$

(The cross product of a vector with itself is zero, so the first term vanishes.) Clearly if there is no net force then angular momentum is conserved. More interesting is a situation in which the force is purely radial, $\mathbf{F} = F(r)\,\hat{\mathbf{r}}$. In this case,

$$\frac{d\mathbf{L}}{dt} = \mathbf{r} \times [F(r)\,\hat{\mathbf{r}}] = 0$$

We see that if a force is applied but there is no angular component to the force, then angular momentum is conserved.

Energy. If a force acts on an object, it takes "work" to move the object against the force. The amount of work required to go from some initial position $\mathbf{r}_i$ to final position $\mathbf{r}_f$ can be calculated as

$$\Delta U = - \int_{\mathbf{r}_i}^{\mathbf{r}_f} \mathbf{F} \cdot d\mathbf{r} \tag{2.4}$$

We call this **potential energy** because it is energy that would be released if the object were to move back to the initial position. We include a minus sign because the work acts against the force $\mathbf{F}$, and we write ΔU to emphasize that this is an energy difference. If desired, we can pick a reference point at which the potential is defined to be zero and thus obtain a potential energy function $U(\mathbf{r})$. Then Eq. (2.4) can be inverted to say the force is obtained by differentiating the potential energy:

$$\mathbf{F} = -\nabla U \tag{2.5}$$

(This is independent of the choice of zeropoint because any additive constant vanishes in the derivative.) Now let's return to Eq. (2.4) and use Newton's second law along with $\mathbf{v} = d\mathbf{r}/dt$ to see what we can learn:

$$\Delta U = - \int_{t_i}^{t_f} \left(m\,\frac{d\mathbf{v}}{dt} \right) \cdot (\mathbf{v}\,dt)$$

$$= -\int_{t_i}^{t_f} m \, \frac{d}{dt} \left(\frac{1}{2} |\mathbf{v}|^2 \right) dt$$

$$= -\left(\frac{1}{2} m \, v_f^2 - \frac{1}{2} m \, v_i^2 \right)$$

$$= -(K_f - K_i)$$

$$= -\Delta K$$

Toward the end we identify $K = (1/2) \, m \, v^2$ as the **kinetic energy**, or energy of motion. Trivially rewriting the final equation gives

$$\Delta U + \Delta K = 0$$

or

$$\Delta E_{\text{tot}} = 0 \qquad \text{where} \qquad E_{\text{tot}} = U + K$$

This is the statement of conservation of energy. Note that potential and kinetic energy are not separately conserved; in fact, one can be traded for the other. But the combination—the total energy—is conserved. This is true for any force, at least in the context of Newtonian physics.

2.3 Law of Gravity

In order to apply his general laws of motion to planets, Newton had to specify the force that acts on planets to generate their motion. We saw in Chap. 1 how he used Kepler's third law to motivate the inverse square law form. To give a precise formulation, let's suppose that an object of mass M exerts a gravitational force on a second object of mass m whose position relative to the first object is given by the vector $\mathbf{r}$. If the objects are both point masses, Newton's law of gravity in vector form reads

$$\mathbf{F}_{\text{grav}}(\mathbf{r}) = -\frac{GMm}{r^2} \, \hat{\mathbf{r}} \tag{2.6}$$

where $\hat{\mathbf{r}}$ reminds us that the force is radial, and the minus signs indicates that gravity is an attractive force.

What if the two objects are not point masses? One of Newton's triumphs was to show that the gravitational force outside a spherically symmetric object of mass M is the *same* as that from a point mass M at the center of the object. Also, the gravitational force inside a spherical shell is zero. To understand these results, consider the setup in Fig. 2.5. Let's use spherical coordinates[5] but modify them

---

[5] See Sect. A.2 for a review.

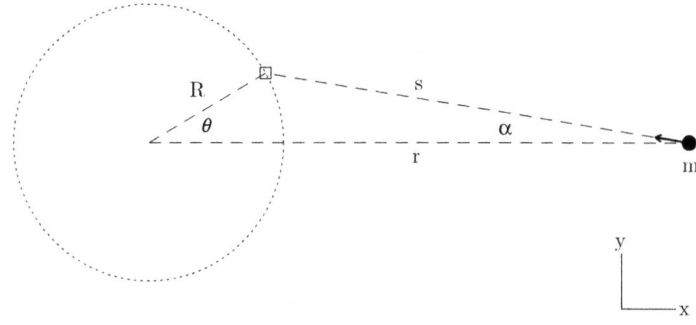

Fig. 2.5 Setup for computing the gravitational force from an extended spherical object

so θ is measured from the x-axis while ϕ is in the direction perpendicular to the page. Then complete the triangle by defining the side s and angle α as shown. By symmetry, the net force on m is in the x-direction. The contribution to F_x from a small volume element dV at r and θ is

$$dF_x = -\frac{Gm\rho\,dV}{s^2}\cos\alpha \tag{2.7}$$

We would like to rewrite this in terms of R and θ. From the law of cosines,

$$s^2 = r^2 + R^2 - 2rR\cos\theta \tag{2.8}$$

and from the law of sines,

$$\frac{\sin\alpha}{R} = \frac{\sin\theta}{s} \quad\Rightarrow\quad \sin\alpha = \frac{R\sin\theta}{(r^2 + R^2 - 2rR\cos\theta)^{1/2}}$$

Then the familiar trigonometric identity $\cos^2\alpha + \sin^2\alpha = 1$ yields

$$\cos\alpha = \frac{r - R\cos\theta}{(r^2 + R^2 - 2rR\cos\theta)^{1/2}}$$

Putting the pieces together, we can write Eq. (2.7) as

$$dF_x = -Gm\frac{r - R\cos\theta}{(r^2 + R^2 - 2rR\cos\theta)^{3/2}}\,\rho\,dV$$

We obtain the net force by integrating, using the spherical volume element $dV = R^2\sin\theta\,dr\,d\theta\,d\phi$:

$$F_x = -Gm\int dR\,R^2\rho(R)\int_0^\pi d\theta\,\sin\theta\int_0^{2\pi}d\phi\,\frac{r - R\cos\theta}{(r^2 + R^2 - 2rR\cos\theta)^{3/2}}$$

(We discuss the limits for the R integral below.) The ϕ integral gives 2π. To evaluate the θ integral, change integration variables to s using Eq. (2.8). This yields

$$
\begin{aligned}
F_x &= -2\pi Gm \int dR\, R^2 \rho(R) \int_{|r-R|}^{r+R} ds\, \frac{r^2 - R^2 + s^2}{2Rr^2 s^2} \\
&= -\frac{\pi Gm}{r^2} \int dR\, R\, \rho(R) \left[-\frac{r^2 - R^2}{s} + s \right]_{s=|r-R|}^{s=r+R}
\end{aligned}
\tag{2.9}
$$

Because of the absolute value, the value of the quantity in square brackets depends on whether $r - R$ is positive or negative:

$$
R < r \quad \Rightarrow \quad \left[-\frac{r^2 - R^2}{s} + s \right]_{s=r-R}^{s=r+R} = 4R
$$

$$
R > r \quad \Rightarrow \quad \left[-\frac{r^2 - R^2}{s} + s \right]_{s=R-r}^{s=r+R} = 0
$$

The second result says there is no contribution to the integral in Eq. (2.9) from the region with $R > r$. In other words, mass outside of r does not contribute to the gravitational force at r (given spherical symmetry). Using the first result in Eq. (2.9) lets us write

$$
F_x = -\frac{Gm}{r^2} \int_0^r 4\pi R^2 \rho(R)\, dR
\tag{2.10}
$$

This integral gives $M(r)$, or the total mass enclosed within radius r, Thus, we can write the gravitational force from an extended, spherically-symmetric object (now in vector form) as

$$
\mathbf{F}_{\text{grav}}(\mathbf{r}) = -\frac{GM(r)m}{r^2}\, \hat{\mathbf{r}}
\tag{2.11}
$$

Using Eq. (2.4), we can now determine the **gravitational potential energy** for point masses:

$$
\begin{aligned}
\Delta U_{\text{grav}} &= -\int_{\mathbf{r}_i}^{\mathbf{r}_f} \mathbf{F}_{\text{grav}} \cdot d\mathbf{r} \\
&= GMm \int_{\mathbf{r}_i}^{\mathbf{r}_f} \frac{1}{r^2} \hat{\mathbf{r}} \cdot d\mathbf{r} \\
&= GMm \int_{r_i}^{r_f} \frac{1}{r^2}\, dr \\
&= -GMm \left(\frac{1}{r_f} - \frac{1}{r_i} \right)
\end{aligned}
$$

This is also the potential energy outside any spherical object with total mass M. As noted above, we must pick a reference point in order to define the full potential energy function. The most common choice in astrophysics is to put the reference point at infinity and define the potential energy to be zero there. This yields

$$U(r) = -\frac{GMm}{r} \tag{2.12}$$

It can be valuable to factor out m:

$$\Phi(r) = \frac{U(r)}{m} = -\frac{GM}{r} \tag{2.13}$$

This function is independent of m, so it describes the gravitational field around M in a general way. We call it the **gravitational potential** of M. To see its utility, consider:

$$m\,\mathbf{a} = \mathbf{F} = -\nabla U = -m\nabla\Phi \quad \Rightarrow \quad \mathbf{a} = -\nabla\Phi$$

All objects at a given position in the gravitational field of M experience the same acceleration, regardless of their mass.

If we focus attention near the surface of Earth (as in introductory physics courses), it may be convenient to adopt a different convention and let the reference point for the potential be Earth's surface. Then the potential energy at a height h above the surface is written as

$$U(h) = -GM_\oplus m \left(\frac{1}{R_\oplus + h} - \frac{1}{R_\oplus} \right)$$

If $h \ll R_\oplus$, we can make a Taylor series expansion and find

$$U(h) \approx mgh \qquad \text{where} \qquad g = \frac{GM_\oplus}{R_\oplus^2} = 9.80\,\mathrm{m\,s^{-2}}$$

Remember that this is valid only near the surface of Earth.

Application: Escape

In the next chapter we will see how Newton's laws of motion and gravity come together to explain Kepler's laws. First, though, it is useful to do a short example that illustrates how conservation laws can help us analyze certain problems quickly and easily.

"What goes up must come down," according to the common saying, but Newton begged to differ. He discerned that the force causing an apple to fall from a tree is the same force keeping the Moon in orbit around Earth; the key difference is that the Moon's forward motion keeps it from crashing into the ground. In principle, if we could throw an apple hard enough we could give it enough motion to go up

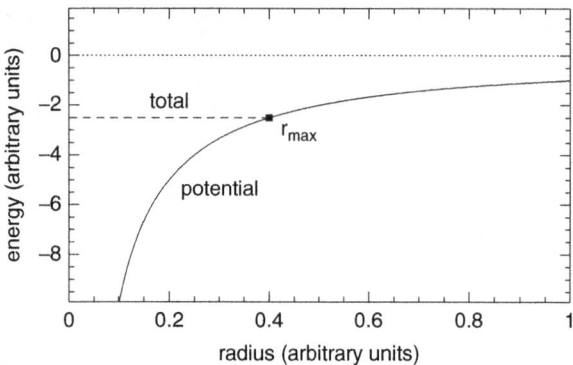

Fig. 2.6 The *solid curve* shows the gravitational potential energy; the *dashed horizontal line* shows the total energy (which is conserved); and the difference between the two gives the kinetic energy. Since kinetic energy cannot be negative, the object can never go beyond r_{max}

and never come back down. (This works better with rockets than apples.) How hard would we have to throw it?

To find out, suppose an object with mass m is at radius r and moving with speed v in the gravitational field around mass M. Is there any limit on how far the small object can go? If so, what is the maximum radius (r_{max}) it can reach? How fast do we need to make the object move if we want it to escape?

If we wanted to work with the original version of Newton's laws of motion, we would have to solve the differential equation $d^2\mathbf{r}/dt^2 = -(GM/r^2)\hat{\mathbf{r}}$ for all trajectories that originate at radius r with speed v, and then we would have to search among those trajectories to find r_{max}. That does not sound like a simple task. But the analysis gets *much* easier if we turn to conservation of energy. At any given r, the total energy is the sum of the potential and kinetic terms,

$$E = -\frac{GMm}{r} + \frac{1}{2}mv^2 \tag{2.14}$$

We can think about this in terms of an energy diagram as in Fig. 2.6. The total energy must be independent of radius. Since the kinetic term is non-negative, the potential energy can never exceed the total energy. The maximum allowed radius is the place where the kinetic energy vanishes and the potential energy equals the total energy,

$$E = -\frac{GMm}{r_{max}} \tag{2.15}$$

Equating (2.14) and (2.15) lets us solve for r_{max}:

$$r_{max} = \left(\frac{1}{r} - \frac{v^2}{2GM}\right)^{-1} \tag{2.16}$$

Notice that we reached this answer in just three lines of algebra; we did not have to specify the direction of motion or examine specific trajectories, or deal with vectors and differential equations at all. Applying conservation of energy is a powerful approach to this problem.

We can now ask how fast the object would have to be moving when it is at radius r in order to escape the gravitational field altogether. This is the speed that allows r_{max} to become infinity, and it is given by the value of v that causes the quantity in parentheses in Eq. (2.16) to vanish:

$$v_{esc} = \left(\frac{2GM}{r}\right)^{1/2} \tag{2.17}$$

We call this the **escape velocity** at a distance r from an object of mass M.

Problems

2.1. Consider a geocentric model for retrograde motion. Suppose the deferent has radius R and angular speed Ω, while the epicycle has radius $a < R$ and angular speed ω (about its center). Find the velocity vector in polar coordinates centered on Earth. By analyzing the tangential velocity at the innermost points, show that the condition to have retrograde motion is $a\omega > R\Omega$.

2.2. Here is a way to understand retrograde motion in the heliocentric model using geometric reasoning (no equations required). Consider a system with two planets orbiting the Sun along circles in the same plane. Suppose the outer planet takes twice as long as the inner planet to orbit the star. Let $t = 0$ be the time when the two planets are lined up on one side of the star.

(a) Sketch the orbits, and add some distant stars. Suppose both planets orbit and spin in the counterclockwise direction. Indicate the directions in the star field that an observer on the inner planet would identify as "east" and "west."
(b) Sketch the positions of the planets a little before and after $t = 0$. In which direction across the sky does the outer planet appear to move, as viewed from the inner planet?
(c) Repeat part (b) at times when the planets are not lined up (for example, when the inner planet has completed 1/4 or 1/2 of its orbit).

2.3. To practice/review working with vectors, compute the specific angular momentum for straight line motion $\mathbf{r}(t) = vt\,\hat{\mathbf{x}} + b\,\hat{\mathbf{y}}$. Is angular momentum conserved? Should it be?

2.4. Consider conservation of energy and angular momentum as applied to an elliptical orbit.

(a) At what point in an elliptical orbit does a planet move fastest? Slowest?

(b) Sketch the kinetic and potential energies as a function of time for a planet in an elliptical orbit.

2.5. Consider a uniform sphere with mass M and radius R. Compute the gravitational force on a particle of mass m at any radius $0 < r < \infty$. Then compute the corresponding gravitational potential. You make take the potential to be zero at the center of the sphere.

2.6. Consider a particle of mass m released from rest at a distance r_0 from a point mass M (and assume $M \gg m$ so M does not move). Use conservation of energy to find the speed v, which is also dr/dt. Then compute the acceleration and show that the motion satisfies Newton's laws.

2.7. For a sufficiently small object, compute the radius at which the escape velocity equals the speed of light. Since nothing can go faster than the speed of light, this is the "Schwarzschild radius" for the event horizon of a black hole. What is the Schwarzschild radius of a black hole the mass of Earth? Of the Sun?

2.8. Could you jump off an asteroid? Let's find out.

(a) Estimate the velocity you achieve when you jump straight up on Earth. Hint: use the height you reach to estimate the change in your potential energy, and then use conservation of energy to estimate your initial kinetic energy.
(b) Now estimate the size of the largest asteroid you could escape from by jumping. You will need to make an assumption about the asteroid's density; just explain your reasoning.

References

1. O. Gingerich, *The Book Nobody Read: Chasing the Revolutions of Nicolaus Copernicus* (Walker, New York, 2004)
2. J. Gleick, *Isaac Newton* (Vintage Books, New York, 2004)

Chapter 3
Gravitational One-Body Problem

Newton's laws of motion and gravity come together to explain the motion of planets around the Sun, plus a wide range of other astrophysical systems. In this chapter we study systems in which the source of gravity (e.g., the Sun) is stationary and a single object (e.g., a planet) is in motion. While Newton's third law tells us that a planet's gravitational pull must also cause the Sun to move, the Sun is so much more massive than any of its planets that its motion can be neglected as a first approximation. In Chap. 4 we will generalize to the case in which both objects move.

3.1 Deriving Kepler's Laws

Kepler's laws provide a great way to analyze orbital motion, since they are already focused on relevant properties of orbits, but in their initial form they were purely empirical and limited to motion around the Sun. If we can use Newton's laws to justify and generalize Kepler's, then we can use the latter to analyze orbital motion in a wide range of settings.

Since Kepler taught us to work with ellipses, we begin by reviewing their geometry. An ellipse is specified mathematically as the solution of the equation

$$\frac{x^2}{a^2} + \frac{y^2}{b^2} = 1 \tag{3.1}$$

We can assume $a > b$ without loss of generality, so Eq. (3.1) is written in a coordinate system where the long or "major" axis of the ellipse is along the x-axis, while the short or "minor" axis is along the y-axis. There are two special points inside the ellipse called **foci** (plural of **focus**) that are a distance $c = \sqrt{a^2 - b^2}$ from the center along the major axis. They are special because the combined distance to the two foci is constant along the ellipse. We define the **eccentricity** of an ellipse to be the dimensionless ratio $e = c/a$, such that a circle has $e = 0$ and more elongated

C. Keeton, *Principles of Astrophysics: Using Gravity and Stellar Physics to Explore the Cosmos*, Undergraduate Lecture Notes in Physics, DOI 10.1007/978-1-4614-9236-8_3, © Springer Science+Business Media New York 2014

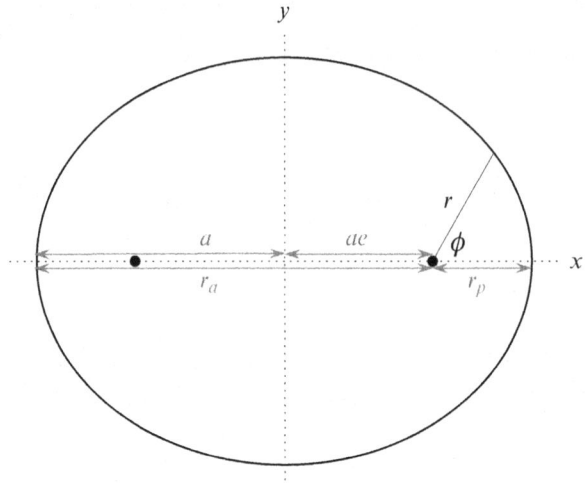

Fig. 3.1 An ellipse with eccentricity $e = 0.6$. The distance from the center to the curve is a along the major axis, and the foci (points) are located a distance ae from the center. The area of the ellipse is $A = \pi a^2 (1 - e^2)^{1/2}$. In the text we use polar coordinates (r, ϕ) centered on one focus. The pericenter and apocenter distances are indicated: $r_p = a(1 - e)$ and $r_a = a(1 + e)$

ellipses have higher eccentricities up to the limit $e = 1$. Using the eccentricity we can rewrite $b = a\sqrt{1 - e^2}$ and then specify the size and shape of an ellipse using (a, e) instead of (a, b).

Kepler also taught us that the Sun is at one focus of an elliptical orbit, so if we introduce polar coordinates (r, ϕ) centered on the Sun then we have (see Fig. 3.1)

$$x = ae + r \cos \phi \qquad y = r \sin \phi$$

Plugging into Eq. (3.1) yields

$$\frac{(ae + r \cos \phi)^2}{a^2} + \frac{(r \sin \phi)^2}{a^2 (1 - e^2)} = 1$$

Rearranging, we can write this as

$$\frac{1 - e^2 \cos^2 \phi}{1 - e^2} \frac{r^2}{a^2} + 2e \cos \phi \frac{r}{a} - 1 + e^2 = 0$$

This is a quadratic equation for r, so it has two solutions. Taking the positive solution (since radius must be positive), we obtain

$$r = \frac{a(1 - e^2)}{1 + e \cos \phi} \qquad\qquad (3.2)$$

This is the equation for an ellipse in polar coordinates centered on a focus. The points on the ellipse that are closest and farthest from the star have $\phi = 0$ and $\phi = \pi$, respectively; these are known as **pericenter** and **apocenter**.[1] Their radii are

$$\text{pericenter, } r_p = a(1-e) \qquad \text{apocenter, } r_a = a(1+e) \qquad (3.3)$$

Our goal now is to connect the geometry to the physical principles represented by Newton's laws. Since the gravitational force is radial it makes sense to use spherical polar coordinates in which the acceleration vector has the form (see Sect. A.2)

$$
\begin{aligned}
\mathbf{a} = &\left[\frac{d^2 r}{dt^2} - r\left(\frac{d\theta}{dt}\right)^2 - r\sin^2\theta \left(\frac{d\phi}{dt}\right)^2 \right] \hat{\mathbf{r}} \\
&+ \left[r\frac{d^2\theta}{dt^2} + 2\frac{dr}{dt}\frac{d\theta}{dt} - r\sin\theta\cos\theta\left(\frac{d\phi}{dt}\right)^2 \right] \hat{\boldsymbol{\theta}} \\
&+ \left[r\sin\theta\frac{d^2\phi}{dt^2} + 2\sin\theta\frac{dr}{dt}\frac{d\phi}{dt} + 2r\cos\theta\frac{d\theta}{dt}\frac{d\phi}{dt} \right] \hat{\boldsymbol{\phi}}
\end{aligned}
$$

Newton's second law gives $\mathbf{a} = \mathbf{F}/m$, which yields the three component equations

$$\frac{d^2 r}{dt^2} - r\left(\frac{d\theta}{dt}\right)^2 - r\sin^2\theta\left(\frac{d\phi}{dt}\right)^2 = -\frac{GM}{r^2} \qquad (3.4a)$$

$$r\frac{d^2\theta}{dt^2} + 2\frac{dr}{dt}\frac{d\theta}{dt} - r\sin\theta\cos\theta\left(\frac{d\phi}{dt}\right)^2 = 0 \qquad (3.4b)$$

$$r\sin\theta\frac{d^2\phi}{dt^2} + 2\sin\theta\frac{dr}{dt}\frac{d\phi}{dt} + 2r\cos\theta\frac{d\theta}{dt}\frac{d\phi}{dt} = 0 \qquad (3.4c)$$

We can solve Eq. (3.4b) if θ is fixed to $\pi/2$, so the motion is confined to a plane (which we are taking to be the equatorial plane). Then Eq. (3.4c) simplifies to

$$\frac{1}{r}\frac{d}{dt}\left(r^2\frac{d\phi}{dt}\right) = 0 \qquad (3.5)$$

If we recall the specific angular momentum from Eq. (2.3),

$$\boldsymbol{\ell} = |\boldsymbol{\ell}| = |\mathbf{r}\times\mathbf{v}| = \left| (r\,\hat{\mathbf{r}})\times\left(\frac{dr}{dt}\hat{\mathbf{r}} + r\frac{d\phi}{dt}\hat{\boldsymbol{\phi}}\right) \right| = r^2\frac{d\phi}{dt} \qquad (3.6)$$

---

[1] Special versions of these terms are used for certain situations: *perigee/apogee* for an orbit around Earth, and *perihelion/aphelion* for an orbit around the Sun.

then we see that Eq. (3.5) says angular momentum is conserved (also see Sect. 2.2). This, finally, lets us rewrite the radial equation (3.4a) as

$$\frac{d^2 r}{dt^2} - \frac{\ell^2}{r^3} = -\frac{GM}{r^2} \tag{3.7}$$

To solve this equation, let's shift from t to ϕ as the independent variable and also make the substitution $r = 1/u$. The derivative becomes

$$\frac{dr}{dt} = \frac{d(1/u)}{d\phi}\frac{d\phi}{dt} = -\frac{1}{u^2}\frac{du}{d\phi}\ell u^2 = -\ell\frac{du}{d\phi}$$

In the second step we use the chain rule of derivatives, and in the third step we use $d\phi/dt = \ell/r^2$. By a similar analysis, the second derivative is

$$\frac{d^2 r}{dt^2} = \frac{d}{d\phi}\left(-\ell\frac{du}{d\phi}\right)\ell u^2 = -\ell^2 u^2\frac{d^2 u}{d\phi^2}$$

Plugging this into Eq. (3.7) and simplifying yields

$$\frac{d^2 u}{d\phi^2} + u = \frac{GM}{\ell^2} \tag{3.8}$$

If the right-hand side were zero, this would be the equation for a simple harmonic oscillator and the solution would have the form $u_0 = B\cos\phi$ where B is some constant. To deal with the constant on the right-hand side, we just need to add GM/ℓ^2 to u_0 (which works because the constant does not affect the derivative term). In other words, the solution has the form $u = B\cos\phi + GM/\ell^2$. Without loss of generality, we can define a new constant e such that $B = GMe/\ell^2$ and our final solution is

$$\frac{1}{r(\phi)} = u(\phi) = \frac{GM}{\ell^2}(1 + e\cos\phi) \tag{3.9}$$

Comparing with Eq. (3.2), we see that this curve describes an ellipse, and the constant e we have defined here is nothing more than the eccentricity of the ellipse.

To examine Kepler's second law, we need to consider the area dA swept out by a planet's motion in some small time interval dt. From the geometry shown in Fig. 3.2, the area is

$$dA = \frac{1}{2}\left|\mathbf{r} \times \mathbf{v}\,dt\right| = \frac{1}{2}|\boldsymbol{\ell}|\,dt \quad\Rightarrow\quad \frac{dA}{dt} = \frac{\ell}{2} \tag{3.10}$$

This is constant because angular momentum is conserved. Thus, Kepler's second law is a direct consequence of the fact that gravity is a central force.

Fig. 3.2 A particle at
position **r** moving with
velocity **v** for an infinitesimal
time interval d*t* sweeps out a
small triangle. By the
properties of the cross
product, the area of the
triangle is
$dA = (1/2)|\mathbf{r} \times \mathbf{v}\, dt|$

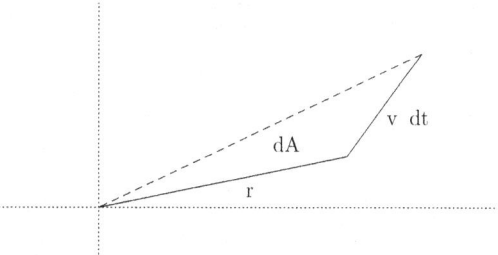

Now we come to Kepler's third law. By comparing Eqs. (3.2) and (3.9), we can express the specific angular momentum in terms of the orbital elements as

$$\ell = \left[GMa(1 - e^2)\right]^{1/2} \tag{3.11}$$

Then from Eq. (3.10) the rate at which area is swept out is

$$\frac{dA}{dt} = \frac{1}{2}\left[GMa(1 - e^2)\right]^{1/2}$$

Since this is constant, the area swept out in one period is

$$A = \frac{dA}{dt} \times P = \frac{P}{2}\left[GMa(1 - e^2)\right]^{1/2} \tag{3.12}$$

But this has to equal the area of the ellipse, which is

$$A = \pi\, a\, b = \pi\, a^2 \left(1 - e^2\right)^{1/2} \tag{3.13}$$

Equating (3.12) and (3.13) and solving for P yields

$$P^2 = \frac{4\pi^2}{GM}\, a^3 \tag{3.14}$$

This is Kepler's third law, but now in a general form that explicitly shows the proportionality factor between P^2 and a^3, which depends on the mass of the central object.

To recap, here again are Kepler's empirical laws of planetary motion, along with Newton's physical explanation of them:

I. The orbit is an ellipse because that shape is the solution of Newton's laws of motion under the influence of an inverse square gravitational force.

II. The rate at which area is swept out is constant because of conservation of angular momentum, which holds because gravity is a central force.

III. The relation $P^2 \propto a^3$ holds because gravity has an inverse square force law.

3.2 Using Kepler III: Motion → Mass

With Newton's generalization, Kepler's third law becomes a powerful principle for astrophysics. Rearranging Eq. (3.14), we can write

$$M = \frac{4\pi^2 a^3}{GP^2} \qquad (3.15)$$

This form is notable because the right-hand side involves quantities we can measure—the size and period of an orbit—while the left-hand side is something we may want to know—the mass of an astrophysical object. As we explore applications, we will encounter a number of practical challenges (mostly related to measuring a accurately), but the fundamental principle remains valid: if we can observe motion and interpret it using Newton's laws, we can infer mass. Mass is a key property of astronomical systems that is difficult to measure directly, so the motion→mass principle is valuable in a wide range of contexts.

3.2.1 The Black Hole at the Center of the Milky Way

At the center of the Milky Way galaxy is a compact source known as **Sagittarius A*** (often abbreviated as Sgr A*) that emits light across the electromagnetic spectrum. At radio wavelengths, high-resolution observations have constrained the size to be $\lesssim 0.3$ AU [1]. At X-ray wavelengths corresponding to photon energies[2] between 2 and 10 keV, the luminosity is greater than 10^{26} J s^{-1} [2]. What could be so energetic yet compact?

Beginning in the 1990s, powerful telescopes and clever observational techniques made it possible to resolve individual stars in the Galactic Center, as shown in Fig. 3.3. Dedicated observers discovered that the stars are moving, mapped the motions, and ultimately found that the orbits appear to be ellipses with Sgr A* as a common focus. In other words, the stars orbiting Sgr A* form a Keplerian system that is directly analogous to the planets orbiting the Sun.

We can therefore use the motion → mass principle to measure the mass of Sgr A*. Stars #2, 16, and 19 (labeled in Fig. 3.3) are particularly important because they have been tracked long enough to pass pericenter, so their orbits are well constrained. Fitting ellipses to the motion yields the following orbital parameters (taken from Ghez et al. [3]; see Gillessen et al. [4] for updated data):

Star	P (yr)	a (AU)	r_p (AU)
2	14.53	919	122
16	36	1,680	45
19	37.3	1,720	287

---

[2]X-ray astronomers often quote energy rather than frequency or wavelength using the quantum relation $E = h\nu = hc/\lambda$.

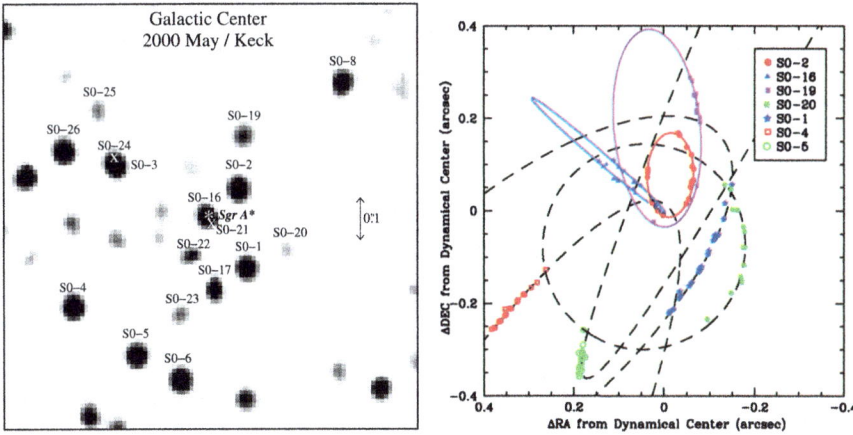

Fig. 3.3 Stars near the Galactic Center. The *left panel* shows a snapshot from May 2000, while the *right panel* shows some of the orbits traced over time (plotted on a different scale) (Credit: Ghez et al. [3]. Reproduced by permission of the AAS)

Applying Eq. (3.15) to star #2 yields

$$M = \frac{4\pi^2 \times (919 \times 1.50 \times 10^{11}\,\text{m})^3}{(6.67 \times 10^{-11}\,\text{m}^3\,\text{kg}^{-1}\,\text{s}^{-2}) \times (14.53 \times 3.16 \times 10^7\,\text{s})^2}$$

$$= 7.3 \times 10^{36}\,\text{kg}$$

$$= 3.7 \times 10^6\,M_\odot$$

Repeating the analysis for other stars gives consistent results. In other words, from the motions of stars we conclude that there is an object with nearly four million times the mass of the Sun lying at the center of the Milky Way. From the radio and X-ray observations, and the pericenter distances, we know this object is luminous and compact. What could it be? The only plausible answer is a black hole—indeed, a **supermassive black hole (SMBH)**.

At this point you may have some questions:

- **Why did we treat this as a one-body problem?**
  The black hole is even more massive relative to the stars than the Sun is compared to the planets, so its reflex motion is negligible.
- **Could Sgr A* be anything other than a black hole?**
  Could it be a single star? No: in Chap. 16 we will see that there is no way for a single star to be anywhere near this massive. Could it be a cluster of stars? Again, no: in Sect. 3.3.2 below we will see that such a massive and compact star cluster would "evaporate" due to stellar dynamical effects.

- **If it is a black hole, why haven't we used relativity?**

  As we will see in Chap. 10, relativistic effects become important on scales comparable to the Schwarzschild radius of a black hole. For Sgr A* this is

$$R_S = \frac{2GM}{c^2} = 1.1 \times 10^{10}\,\text{m} = 0.07\,\text{AU}$$

  Even star #16 stays far enough from the black hole that Newtonian gravity gives a reasonable approximation to the motion.

- **Can we see the event horizon?**

  The Galactic Center is about $R_{GC} \approx 8\,\text{kpc}$ away, so the angle subtended by the black hole's event horizon is (using the small-angle approximation)

$$\phi \approx \frac{R_S}{R_{GC}}$$
$$= \frac{1.1 \times 10^{10}\,\text{m}}{8 \times 3.09 \times 10^{19}\,\text{m}}$$
$$= 4.5 \times 10^{-11}\,\text{rad} \times \frac{180\,\text{deg}}{\pi\,\text{rad}} \times \frac{3,600\,\text{arcsec}}{1\,\text{deg}}$$
$$= 9.3 \times 10^{-6}\,\text{arcsec}$$

At optical wavelengths, the best resolution that can be achieved today is 0.05–0.1 arcsec (with the Hubble Space Telescope, or adaptive optics from the ground). At radio wavelengths, it is possible to use an array of telescopes with a technique called interferometry to achieve a resolution of 10^{-4} arcsec or better. While observations have not directly revealed the event horizon, they do seem to be on the verge of resolving some of the interesting structure in Sgr A* [1].

3.2.2 Supermassive Black Holes in Other Galaxies

Our galaxy is not the only one with a supermassive black hole at the center; evidence is growing that every massive galaxy hosts such an object. In most cases we cannot study the black holes in as much detail as Sgr A*, but we can still use the motion→mass principle to infer their masses.

NGC 4258

After the Milky Way, the galaxy with the best constraints on a supermassive black hole is NGC 4258. (The name refers to the galaxy's entry in the *New General Catalogue of Nebulae and Clusters of Stars* [5].) Radio observations reveal water

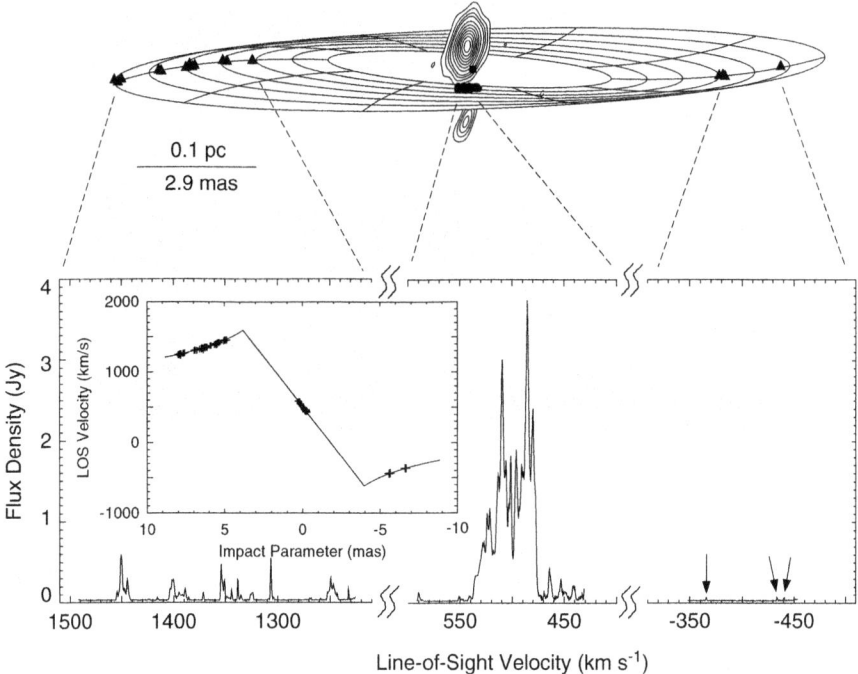

Fig. 3.4 The *top panel* shows a sketch of the disk of gas orbiting the black hole at the center of NGC 4258, with some maser positions indicated. The *bottom panel* shows the radio spectrum. The inset shows the line-of-sight velocity as a function of position, along with a Keplerian rotation curve. (The middle part of the Keplerian curve corresponds to "sideways" motion in the front part of the rotating disk) (Reprinted by permission from Macmillan Publishers Ltd: Herrnstein et al. [6], © 1999)

masers[3] orbiting the center of the galaxy. While the orbital period is too long for us to see the masers shift position, we can still measure motion. Masers emit light at very specific wavelengths, but if they are moving toward or away from us the emission is shifted to shorter or longer wavelengths by the **Doppler effect**. For non-relativistic motion, the shift in wavelength is $\Delta\lambda/\lambda_e = v/c$ where λ_e is the emitted wavelength, and v is the component of velocity along the line of sight with the convention that $v > 0$ if the object is moving away from us and $v < 0$ if it is moving toward us. (See Sect. 10.2.4 for a full discussion of the relativistic Doppler effect.) Figure 3.4 shows that masers closer to the center of NGC 4258 move faster, and the motion is consistent with orbits around an object with mass $(3.9\pm0.1)\times10^7\,M_\odot$ [6].

---

[3]Maser originally stood for "microwave amplification by stimulated emission of radiation," although "microwave" is now sometimes replaced by "molecular." A laser is similar to a maser except that it operates in the visible portion of the electromagnetic spectrum (the "l" stands for "light," specifically meaning visible light).

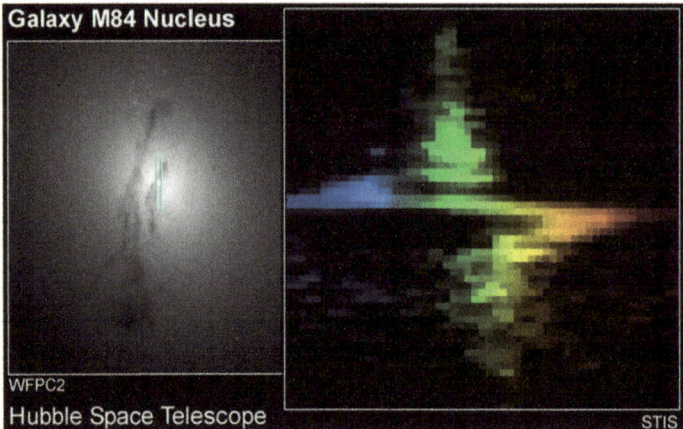

Fig. 3.5 On the *left* is an image of the galaxy NGC 4374 (also known as M84), taken with the Wide Field and Planetary Camera 2 on the Hubble Space Telescope. The *small box* shows the region whose spectrum was recorded with the Space Telescope Imaging Spectrograph, as shown on the right. The *zigzag pattern* is created by the Doppler shift of light from stars and gas orbiting a supermassive black hole at the center of the galaxy (Credit: Gary Bower, Richard Green (NOAO), the STIS Instrument Definition Team, and NASA)

The current upper limit on the size of the object is 0.16 pc, so the size constraint is not nearly as strong as for Sgr A*. Nevertheless, astronomers believe the central object is a black hole.

NGC 4374

At present there are no other galaxies where we can observe *individual* objects moving around the center of the galaxy. Still, we can measure *collections* of stars or gas moving around in the centers of many galaxies. As an example, Fig. 3.5 shows an optical spectrum of the galaxy NGC 4374. The light from stars and gas on one side of the galaxy center is shifted toward bluer wavelengths by the Doppler effect, while the light from stars and gas on the other side of the center is shifted toward redder wavelengths. Also, objects closer to the center move faster. The motion again reveals a central massive object, this time with a mass of nearly $9 \times 10^8 \, M_\odot$ [7].

A Supermassive Black Hole in Every Galaxy?

Similar observations in other galaxies have shown that whenever we can make good measurements we find evidence for supermassive black holes. Astronomers now suspect that every massive galaxy harbors a central black hole, and the black hole masses range from a few million to more than a billion times the mass of

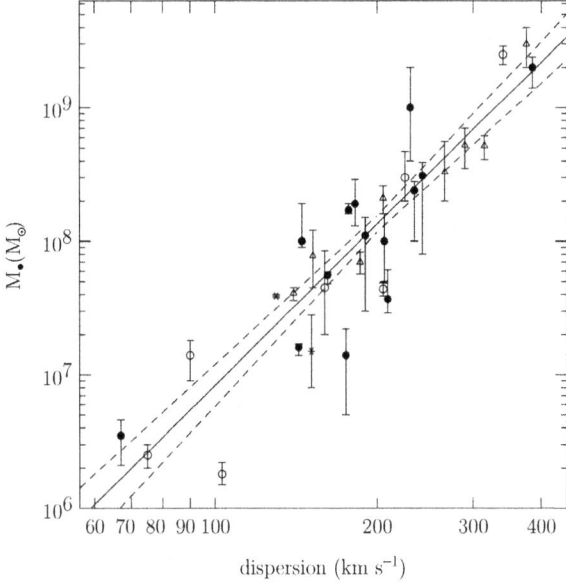

Fig. 3.6 Relation between *black hole mass* (indicated here by $M_\bullet$) and galaxy velocity dispersion. The *solid line* shows the best fit to the data, which scales as $M \propto \sigma^{4.02}$. The *dashed lines* show uncertainties in the fit (Credit: Tremaine et al. [8]. Reproduced by permission of the AAS. (See [9] for an updated version of this relation))

the Sun. What's more, the mass of the black hole appears to be closely related to the properties of the galaxy in which it resides.

We will study galaxies later (in Chaps. 7 and 8), but for now we note that most galaxies can be described in terms of two types of structures: a flat *disk* in which the star orbits lie mostly in a plane; and a rounder *spheroid* in which the star orbits have random orientations. Spiral galaxies usually have large disks surrounding smaller spheroids known as bulges, while elliptical galaxies are pure spheroids. Since the motion in spheroids is random, we characterize it by examining the distribution of star velocities (strictly speaking, the component along the line of sight) and computing the statistical standard deviation, which we call the **velocity dispersion**.

A striking discovery about supermassive black holes is that the black hole mass is correlated with the velocity dispersion of the spheroidal component of its host galaxy, as shown in Fig. 3.6. You may wonder: why should it be remarkable that motion (σ) is closely related to mass (M)? Most of the stars used to measure the velocity dispersion lie far enough from the black hole that they should hardly notice its gravity.[4] Yet the stars seem to know how much the black hole weighs— or, conversely, the black hole knows how fast the stars move. Astronomers are still

---

[4]We quantify this idea in terms of a gravitational "sphere of influence" in Sect. 3.3.1.

trying to understand how this came to be: observers are trying to see whether the M-σ relation was the same in the past, while theorists are trying to understand whether the processes by which black holes and spheroids grow might be related to one another. The final answers are not known, but the discovery of the M-σ relation has sparked a lot of new research.

3.2.3 Active Galactic Nuclei

Direct motion-based measurements of black hole masses can be made only in relatively nearby galaxies, where we can resolve the motion on small scales. Nevertheless, strong indirect evidence suggests that supermassive black holes are common in galaxies throughout the universe.

The evidence comes from Active Galactic Nuclei (AGN)—an umbrella term for galaxies that emit huge amounts of energy from their centers. There are many types of AGN but for our purposes there are two key features. First, these objects can be very luminous, reaching $L \sim 10^{12} L_\odot$. Second, AGN can vary on time scales as short as $\Delta t \sim 1\,\mathrm{h}$. The variability lets us place an upper limit on the size, because a source can change coherently only if information about the physical conditions can travel across the source. If we imagine that something changes in the middle of the source, the time it would take for that information to reach the edge is $\Delta t \gtrsim R/c$ (and perhaps much longer if the information propagates at less than the speed of light). If $\Delta t \sim 1\,\mathrm{h}$ then we can infer

$$R \lesssim c\,\Delta t \sim 3.0 \times 10^8\,\mathrm{m\,s^{-1}} \times 3{,}600\,\mathrm{s} \sim 10^{12}\,\mathrm{m} \sim 7\,\mathrm{AU}$$

In other words, an AGN can be *as bright as a galaxy, but smaller than the Solar System!* What might be so energetic? A supermassive black hole.

You may ask: Aren't black holes supposed to be black? How can they emit so much energy? While nothing can escape a black hole once it has fallen in, a lot of energy can be emitted *as matter approaches a black hole*. Imagine mass falling in at a rate $\dot{M} = \mathrm{d}M/\mathrm{d}t$. In time $\mathrm{d}t$, an amount $\dot{M}\,\mathrm{d}t$ falls into the black hole, and as it falls from infinity to the event horizon it releases potential energy

$$\mathrm{d}U \sim -\frac{GM}{R_s}\,\dot{M}\,\mathrm{d}t$$

(We will use Newtonian gravity for this simple estimate.) As atoms fall in, their kinetic energy must increase to conserve energy. As they speed up, they bump into one another more and more often, causing the gas to heat up and radiate. If all the potential energy that was liberated gets converted to light, the total luminosity (light energy per unit time) could be as large as

$$L \sim \left|\frac{\mathrm{d}U}{\mathrm{d}t}\right| \sim \frac{GM}{R_s}\,\dot{M} \sim \frac{GM}{2GM/c^2}\,\dot{M} \sim \frac{1}{2}\,\dot{M}\,c^2$$

As mass falls into a black hole, a significant fraction of its rest mass energy could be converted into light.

There are some caveats. The energy release is probably gradual; it does not all happen at the event horizon. Some of the energy might even vanish into the black hole. Also, a proper analysis should account for relativity. Detailed analyses indicate that the energy release has the form (e.g., [10])

$$L \approx \varepsilon \, \dot{M} \, c^2$$

where the "efficiency" is $\varepsilon \sim 0.06 - 0.42$ and $\varepsilon \approx 0.1$ is a typical value. Even so, it is fair to say that black holes are the most efficient machines in the universe for converting mass into energy.[5]

3.3 Related Concepts

Let us briefly step away from the main story to address two topics that arose in Sect. 3.2. The notion of a gravitational sphere of influence is important for interpreting the M-σ relation, and it is an interesting variant of the one-body problem. The concept of stellar dynamical evaporation is important for interpreting constraints on supermassive black holes (particularly Sgr A*), and it provides a nice application of dimensional analysis.

3.3.1 Sphere of Influence

In Sect. 3.2.2 we mentioned that astronomers were surprised to find a tight relation between the masses of supermassive black holes and the velocity dispersions of the spheroids in which they are embedded. Why was that a surprise? To find out, let's estimate the size of the region in which a black hole has a significant influence on the motions of stars. To be more specific, let's define a black hole's "sphere of influence" to be the region where the gravity from the black hole is stronger than the gravity from the rest of the matter in the galaxy. At radius r, the strength of the gravitational force from the black hole is

$$F_{bh}(r) = \frac{GM_{bh}m}{r^2}$$

What about the force from the galaxy? For simplicity, let's assume the galaxy is spherically symmetric. From Eq. (2.11), the force is then

---

[5]For comparison, the energy released by fusion in stars corresponds to an efficiency $\varepsilon = 0.007$ (see Sect. 15.2).

$$F_{\text{gal}}(r) = \frac{GM_{\text{gal}}(r)\,m}{r^2}$$

where $M_{\text{gal}}(r)$ is the mass enclosed by a sphere of radius r. In Chaps. 7 and 8 we will see that a simple model for a galaxy with velocity dispersion σ is the isothermal sphere, which has density

$$\rho_{\text{gal}}(r) = \frac{\sigma^2}{2\pi G r^2}$$

The mass enclosed by radius r is

$$M_{\text{gal}}(r) = \int_0^r \frac{\sigma^2}{2\pi G(r')^2}\, 4\pi (r')^2 \, \mathrm{d}r' = \frac{2\sigma^2}{G}\, r$$

so the gravitational force from the galaxy is

$$F_{\text{gal}}(r) = \frac{Gm}{r^2} \frac{2\sigma^2 r}{G} = \frac{2\sigma^2 m}{r}$$

In order to have the force from the black hole exceed the force from the rest of the mass in the galaxy, we need

$$\frac{GM_{\text{bh}}m}{r^2} > \frac{2\sigma^2 m}{r}$$

Thus, the black hole's sphere of influence is the region with $r < R_0$ where

$$R_0 \equiv \frac{GM_{\text{bh}}}{2\sigma^2}$$

From the observed M-σ relation, a galaxy with $\sigma \approx 200\,\text{km s}^{-1}$ hosts a black hole of about $M_{\text{bh}} \approx 10^8\,M_\odot$. By our estimate, the black hole's sphere of influence is then

$$R_0 = \frac{(6.67 \times 10^{-11}\,\text{m}^3\,\text{kg}^{-1}\,\text{s}^{-2}) \times (10^8 \times 1.99 \times 10^{30}\,\text{kg})}{2 \times (2 \times 10^5\,\text{m s}^{-1})^2} = 1.7 \times 10^{17}\,\text{m} = 5.4\,\text{pc}$$

For a massive galaxy with $\sigma \approx 330\,\text{km s}^{-1}$ that hosts a huge black hole with $M_{\text{bh}} \approx 10^9\,M_\odot$, we get

$$R_0 = 6.1 \times 10^{17}\,\text{m} = 20\,\text{pc}$$

These distances are very small compared with the size of a galaxy (which is typically measured in kpc). In other words, even a supermassive black hole does not have enough mass compared with its galaxy to have a strong effect on the entire galaxy.

The M-σ relation must arise from some indirect connection between the way galaxies form and the way supermassive black holes grow inside galaxies.

3.3.2 Stellar Dynamical Evaporation

In Sect. 3.2.1 we learned that stellar motions reveal Sgr A* to be massive and compact, but they do not definitively prove it to be a black hole, so we should consider alternatives. We said it cannot be a single star, but could it be a cluster of stars?

If millions of stars are confined to a small space, they will occasionally pass very close to each other. Since gravity gets strong when separations get small, close interactions can impart enough force to eject one of the stars from the cluster. Let's use dimensional analysis to estimate the time it would take for a star cluster to "evaporate" in this way.[6] Suppose there are N stars of mass m (so the total mass is $M = Nm$), in a region of size R. For dimensional analysis, what do we have to work with?

Cluster mass	M	$[M]$
Star mass	m	$[M]$
Number of stars	N	—
Cluster size	R	$[L]$
Gravity	G	$[M^{-1}L^3T^{-2}]$

We need $G^{-1/2}$ to get a time, and then we need $R^{3/2}$ to eliminate length. We have a choice of mass: M or m. Since the evaporation interactions involve individual stars, I think the key mass is m. There may also be some factor that depends on the number of stars N; we will come back to that in a moment. To this point, our analysis of dimensions gives a guess of the form

$$t_{evap} \sim \frac{R^{3/2}}{(Gm)^{1/2}}$$

Now let's consider the number of stars. I imagine that there are two places where N enters. First, since stars are ejected one by one the time it takes to evaporate the cluster should have a factor of N. Second, if we pack more stars into a fixed space, gravity will be stronger, and the stars will move faster. That will cause interactions to happen more quickly, decreasing the evaporation time. In Problem 1.1 you used dimensional analysis to estimate the typical velocity of stars in a gravitationally bound system; the upshot is that speed scales as

---

[6]See Sect. 3.2 of *Galaxies in the Universe* by Sparke and Gallagher [11] for a complementary analysis of evaporation.

$v \propto N^{1/2}$, which suggests that the evaporation time should have a factor of $N^{-1/2}$. Incorporating both of these factors yields

$$t_{\text{evap}} \sim \frac{R^{3/2}}{(Gm)^{1/2}} \frac{N}{N^{1/2}} \sim \left(\frac{NR^3}{Gm}\right)^{1/2} \sim \left(\frac{MR^3}{Gm^2}\right)^{1/2}$$

Let's plug in numbers. Our mass estimate for Sgr A* is $M = 3.7 \times 10^6 \, M_\odot$. Strictly speaking, all we know from the motion is that the mass is confined within a region $R < 45 \, \text{AU}$. If we assume all stars are like the Sun, we have $m \sim M_\odot$. Then:

$$t_{\text{evap}} \sim \left[\frac{(3.7 \times 10^6 \times 1.99 \times 10^{30} \, \text{kg}) \times (45 \times 1.50 \times 10^{11} \, \text{m})^3}{(6.67 \times 10^{-11} \, \text{m}^3 \, \text{kg}^{-1} \, \text{s}^{-2}) \times (1.99 \times 10^{30} \, \text{kg})^2}\right]^{1/2}$$

$$\sim 2.9 \times 10^{12} \, \text{s}$$

$$\sim 90{,}000 \, \text{yr}$$

While this estimate from dimensional analysis may be fairly crude, it certainly indicates that if Sgr A* were a cluster of normal stars it would have evaporated long ago.

We are left with the conclusion that Sgr A* is probably a black hole. Even though we have not yet detected the event horizon—that is the holy grail of black hole studies—we have assembled a strong case in which the Kepler's laws and the motion $\rightarrow$ mass principle have played a key role.

Problems

3.1. Sketch the orbital speed v as a function of orbital size r for a planet in a circular orbit about the Sun. This is known as a *Keplerian rotation curve*.

3.2. Consider a rocket orbiting Earth in an orbit that is initially circular.

(a) If the rocket fires a short burst from its engine to apply a force in the *same* direction as its motion, what happens to the shape of the orbit? Sketch the before and after orbits. Hint: think about the kinetic and potential energies just before and just after the burst, and refer back to Problem 2.4.
(b) Repeat part (a) with the engine firing in the *opposite* direction.
(c) How would a rocket have to fire its engine if it wanted to move to an orbit that is larger but still circular?

3.3. Suppose a comet orbits the Sun with a period of 27 years, and the closest it gets to the Sun is 3 AU. At the point in its orbit when it is moving slowest, how far is the comet from the Sun?

3.4. If the Moon orbited above Earth's equator at a distance of 42,200 km from Earth's center how would it appear to an observer on Earth? Describe the cycle of phases the observer would see.

3.5. Use the orbital data for Jupiter's Galilean moons to compute Jupiter's mass. Verify that all four moons give consistent results.

	P (days)	a (10^3 km)
Io	1.769	421.7
Europa	3.551	670.9
Ganymede	7.155	1,070.4
Callisto	16.689	1,882.7

3.6. Derive expressions for the orbital speeds at pericenter and apocenter of an elliptical orbit. Then consider the stars observed to orbit the black hole at the center of the Milky Way. Which star moves fastest at pericenter? (Be quantitative.)

3.7. Suppose you discover an extrasolar planet orbiting a star of mass $2M_\odot$ with an orbital period of 3 months. What is the semimajor axis of the planet's orbit?

3.8. The black hole in NGC 4374 has been studied using the Doppler shift of light with a wavelength of about 6,600 Å. What is the wavelength shift of light emitted from gas that orbits the black hole at a distance of 30 pc?

3.9. Revisit the analysis of a black hole's sphere of influence (Sect. 3.3.1) assuming a uniform density of stars. Express your answer in terms of ρ_{stars} and equivalently in terms of the mass and radius of a spherical galaxy with uniform density.

References

1. S.S. Doeleman et al., Nature **455**, 78 (2008)
2. F.K. Baganoff et al., Astrophys. J. **591**, 891 (2003)
3. A.M. Ghez, S. Salim, S.D. Hornstein, A. Tanner, J.R. Lu, M. Morris, E.E. Becklin, G. Duchêne, Astrophys. J. **620**, 744 (2005)
4. S. Gillessen, F. Eisenhauer, S. Trippe, T. Alexander, R. Genzel, F. Martins, T. Ott, Astrophys. J. **692**, 1075 (2009)
5. J. Dreyer, R. Sinnott, *NGC 2000.0: The Complete New General Catalogue and Index Catalogues of Nebulae and Star Clusters by J.L.E. Dreyer* (Cambridge University Press, New York, 1988)
6. J.R. Herrnstein, J.M. Moran, L.J. Greenhill, P.J. Diamond, M. Inoue, N. Nakai, M. Miyoshi, C. Henkel, A. Riess, Nature **400**, 539 (1999)
7. J.L. Walsh, A.J. Barth, M. Sarzi, Astrophys. J. **721**, 762 (2010)
8. S. Tremaine et al., Astrophys. J. **574**, 740 (2002)
9. J. Kormendy, L.C. Ho, Annu. Rev. Astron. Astrophys. **51**, 511 (2013)
10. S.L. Shapiro, S.A. Teukolsky, *Black Holes, White Dwarfs and Neutron Stars: The Physics of Compact Objects* (Wiley, New York, 1986)
11. L.S. Sparke, J.S. Gallagher, III, *Galaxies in the Universe: An Introduction* (Cambridge University Press, Cambridge, 2007)

Chapter 4
Gravitational Two-Body Problem

Now we are ready to study what happens when two objects interact via gravity and both are free to move. As we will see, there is a deep connection between the one-body and two-body problems that provides a powerful opportunity to understand binary star systems and extrasolar planets.

4.1 Equivalent One-Body Problem

Our first task is to solve the equations of motion and find the orbits in the two-body problem. We can do this by uncovering a mathematical equivalence with the one-body problem, which we have already solved.

4.1.1 Setup

Consider the gravitational interaction between mass m_1 at position $\mathbf{r}_1$ and mass m_2 at position $\mathbf{r}_2$, as sketched in Fig. 4.1. Introducing a few new quantities will clarify our analysis. Define the **separation vector**,

$$\mathbf{r} = \mathbf{r}_2 - \mathbf{r}_1 \tag{4.1}$$

and the **center of mass position**,

$$\mathbf{R} = \frac{m_1 \mathbf{r}_1 + m_2 \mathbf{r}_2}{m_1 + m_2} \tag{4.2}$$

Also define the **total mass**,

$$M = m_1 + m_2 \tag{4.3}$$

C. Keeton, *Principles of Astrophysics: Using Gravity and Stellar Physics to Explore the Cosmos*, Undergraduate Lecture Notes in Physics, DOI 10.1007/978-1-4614-9236-8_4, © Springer Science+Business Media New York 2014

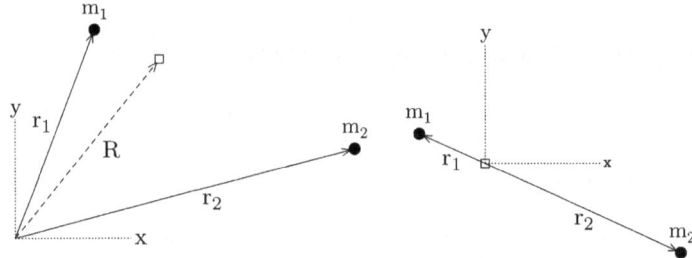

Fig. 4.1 Geometry of the two-body problem. The *left panel* shows a general reference frame and indicates the vectors to the two objects ($\mathbf{r}_1$ and $\mathbf{r}_2$) along with the vector to the center of mass ($\mathbf{R}$). The *right panel* shows the reference frame with the center of mass at the origin

and the **reduced mass**,

$$\mu = \frac{m_1 m_2}{m_1 + m_2} \quad \Leftrightarrow \quad \frac{1}{\mu} = \frac{1}{m_1} + \frac{1}{m_2} \tag{4.4}$$

As defined, the total and reduced masses obey the product relation

$$M\mu = m_1 m_2 \tag{4.5}$$

With these definitions, we can rewrite the positions as

$$\mathbf{r}_1 = \mathbf{R} - \frac{\mu}{m_1}\mathbf{r} \quad \text{and} \quad \mathbf{r}_2 = \mathbf{R} + \frac{\mu}{m_2}\mathbf{r} \tag{4.6}$$

Notice that the two objects are *always on opposite sides of the center of mass*. While this should be apparent from the term "center," it is a good point to keep in mind when visualizing motion in the two-body problem.

Intuitively, the gravitational force on object #1 should point toward object #2, which means the force vector $\mathbf{F}_1$ is parallel to the separation vector $\mathbf{r}$. The force on object #2 points in the opposite direction, so $\mathbf{F}_2$ has the opposite sign. Newton's law of gravity tells us that both forces have strength $Gm_1 m_2 / r^2$. Putting these pieces together, we can write the forces as

$$\text{force on \#1:} \quad \mathbf{F}_1 = +\frac{Gm_1 m_2}{r^2}\,\hat{\mathbf{r}} \tag{4.7a}$$

$$\text{force on \#2:} \quad \mathbf{F}_2 = -\frac{Gm_1 m_2}{r^2}\,\hat{\mathbf{r}} \tag{4.7b}$$

4.1.2 Motion

Let's first consider the acceleration of the center of mass:

$$\frac{d^2\mathbf{R}}{dt^2} = \frac{1}{m_1 + m_2}\left(m_1 \frac{d^2\mathbf{r}_1}{dt^2} + m_2 \frac{d^2\mathbf{r}_2}{dt^2}\right) = \frac{1}{m_1 + m_2}\left(\mathbf{F}_1 + \mathbf{F}_2\right) = 0$$

In the first step we replace $\mathbf{R}$ using Eq. (4.2). In the second step we use Newton's second law to put $m_i\, d^2\mathbf{r}_i/dt^2 = \mathbf{F}_i$, and in the third step we use Newton's third law (in the form of 4.7). We learn that the center of mass does not accelerate.

Therefore we can define an inertial reference with the center of mass at the origin, so $\mathbf{R} = 0$. Shifting to this center of mass frame for the remainder of the analysis, we can write

$$\mathbf{r}_1 = -\frac{\mu}{m_1}\mathbf{r} \quad \text{and} \quad \mathbf{r}_2 = \frac{\mu}{m_2}\mathbf{r} \tag{4.8}$$

Note that when we deal with vectors, $\mathbf{r}_1$ and $\mathbf{r}_2$ have opposite signs, and the separation vector still includes a minus sign: $\mathbf{r} = \mathbf{r}_2 - \mathbf{r}_1$. But if we just consider the *lengths* of vectors, we know the length of the separation vector is (not surprisingly) the sum of the lengths of $\mathbf{r}_1$ and $\mathbf{r}_2$:

$$|\mathbf{r}| = |\mathbf{r}_1| + |\mathbf{r}_2|$$

The ratio of lengths is interesting:

$$\frac{|\mathbf{r}_2|}{|\mathbf{r}_1|} = \frac{m_1}{m_2} \tag{4.9}$$

Even before we fully characterize the motion, we realize that the orbits of the two objects are scaled versions of one another, with the scaling given by the (inverse) mass ratio.

To analyze the motion in detail, consider the equation of motion for object #1:

$$m_1 \frac{d^2\mathbf{r}_1}{dt^2} = \mathbf{F}_1$$

$$-\mu \frac{d^2\mathbf{r}}{dt^2} = \frac{Gm_1m_2}{r^2}\hat{\mathbf{r}}$$

$$\Rightarrow \quad \frac{d^2\mathbf{r}}{dt^2} = -\frac{GM}{r^2}\hat{\mathbf{r}} \tag{4.10}$$

We first use Eq. (4.8) for $\mathbf{r}_1$ and Eq. (4.7a) for $\mathbf{F}_1$, and then use Eq. (4.5) to replace $m_1 m_2$. Considering object #2 yields the same equation. This equation should look familiar: it is the equation of motion for the gravitational one-body problem. The key lesson is that *a two-body problem with masses m_1 and m_2 is mathematically equivalent to a one-body problem with mass $M = m_1 + m_2$.*

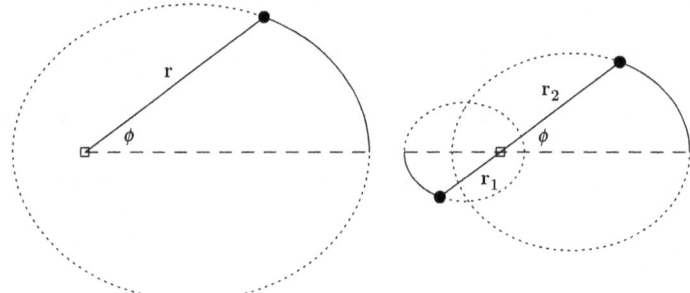

Fig. 4.2 Sample two-body problem with a 2:1 mass ratio and eccentricity $e = 0.6$ (*right*) and the equivalent one-body problem (*left*). The two-body orbits are scaled down versions of the one-body ellipse, with the same eccentricity. They share a common focus at the center of mass of the system (denoted by $\square$). As the separation vector sweeps around, it is pinned at the center of mass

We know from Eq. (3.9) that the solution to the one-body problem has the form

$$r = \frac{a\left(1 - e^2\right)}{1 + e \cos \phi} \tag{4.11}$$

We can then use Eq. (4.8) to say that the orbits for the two-body problem are smaller ellipses with semimajor axes

$$a_1 = \frac{\mu}{m_1} a \quad \text{and} \quad a_2 = \frac{\mu}{m_2} a \tag{4.12}$$

The orbits are arranged so the two ellipses share a common focus (at the center of mass) and the two objects always lie on opposite sides of the center of mass. The association between a two-body problem and its equivalent one-body analog is illustrated in Fig. 4.2.

As we use the one-body analogy, we need to keep in mind that it is a *mathematical* connection more than a *physical* one. It is not correct to say that a problem with masses m_1 and m_2 orbiting each other is physically equivalent to a problem with masses M and μ orbiting each other. The issue is that a physical scenario with masses M and μ would itself be a two-body problem so both objects would move, but the mathematical equivalence is to a one-body problem in which M is *stationary*. The analogy between the two-body and one-body problems is powerful, but it must be used with some care.

4.1.3 Energy and Angular Momentum

We have seen the analogy with the equation of motion, but does it extend to energy and angular momentum? Let's start with kinetic energy. Equation (4.8) implies that the velocity vectors are related by

$$\mathbf{v}_1 = -\frac{\mu}{m_1}\mathbf{v} \quad \text{and} \quad \mathbf{v}_2 = \frac{\mu}{m_2}\mathbf{v} \tag{4.13}$$

where $\mathbf{v} = d\mathbf{r}/dt$ is the time derivative of the separation vector. (We are still working in the center of mass frame.) The kinetic energy of each object is then

$$K_i = \frac{1}{2}m_i|\mathbf{v}_i|^2 = \frac{1}{2}\frac{\mu^2}{m_i}|\mathbf{v}|^2$$

The gravitational potential energy between the two objects is

$$U = -\frac{Gm_1m_2}{r} = -\frac{GM\mu}{r}$$

where we use Eq. (4.5). The total energy can therefore be written as

$$\begin{aligned} E &= \frac{1}{2}m_1|\mathbf{v}_1|^2 + \frac{1}{2}m_2|\mathbf{v}_2|^2 - \frac{Gm_1m_2}{r} \\ &= \frac{1}{2}\left(\frac{1}{m_1} + \frac{1}{m_2}\right)\mu^2|\mathbf{v}|^2 - \frac{GM\mu}{r} \\ &= \frac{1}{2}\mu|\mathbf{v}|^2 - \frac{GM\mu}{r} \end{aligned} \tag{4.14}$$

where we use Eq. (4.4) to simplify the first term. A similar analysis of the angular momentum yields

$$\begin{aligned} \mathbf{L} &= m_1\mathbf{r}_1 \times \mathbf{v}_1 + m_2\mathbf{r}_2 \times \mathbf{v}_2 \\ &= (\mu\mathbf{r}) \times \left(\frac{\mu\mathbf{v}}{m_1}\right) + (\mu\mathbf{r}) \times \left(\frac{\mu\mathbf{v}}{m_2}\right) \\ &= \left(\frac{1}{m_1} + \frac{1}{m_2}\right)\mu^2\,\mathbf{r} \times \mathbf{v} \\ &= \mu\,\mathbf{r} \times \mathbf{v} \end{aligned} \tag{4.15}$$

The analogy continues to be useful: the final expressions for both energy and angular momentum have forms appropriate for an object of mass μ orbiting a (stationary) object of mass M in a one-body problem.

4.1.4 Velocity Curve

Equation (4.13) gives general relations for the velocity, but it is worthwhile to dig into the details because a lot of what we can learn about binary stars and exoplanets comes from analyzing velocities. We focus here on **v** for the one-body problem, since $\mathbf{v}_1$ and $\mathbf{v}_2$ can be obtained from it. To begin, we find the components of **v** in polar coordinates. The angular component is

$$v_\phi = r \frac{d\phi}{dt} = \frac{\ell}{r} = \frac{\ell(1 + e \cos \phi)}{a(1 - e^2)} \tag{4.16}$$

where we recall that the specific angular momentum $\ell = r^2 \, d\phi/dt$ is constant, and we use Eq. (4.11) for r. The radial component of velocity is

$$v_r = \frac{dr}{dt} = \frac{dr}{d\phi} \frac{d\phi}{dt} = \frac{a(1 - e^2)e \sin \phi}{(1 + e \cos \phi)^2} \frac{d\phi}{dt} = \frac{\ell e \sin \phi}{a(1 - e^2)} \tag{4.17}$$

We use the chain rule to rewrite the derivative, then evaluate $dr/d\phi$ from Eq. (4.11), and finally substitute for $d\phi/dt$ using Eq. (4.16). We can convert to Cartesian coordinates as follows:

$$\begin{bmatrix} v_x \\ v_y \end{bmatrix} = \begin{bmatrix} \cos(\phi + \phi_0) & -\sin(\phi + \phi_0) \\ \sin(\phi + \phi_0) & \cos(\phi + \phi_0) \end{bmatrix} \begin{bmatrix} v_r \\ v_\phi \end{bmatrix}$$

where we now allow a general coordinate system in which the major axis of the ellipse lies at angle ϕ_0. Carrying out the matrix multiplication yields

$$v_x = -\frac{\ell[e \sin \phi_0 + \sin(\phi + \phi_0)]}{a(1 - e^2)} \quad \text{and} \quad v_y = \frac{\ell[e \cos \phi_0 + \cos(\phi + \phi_0)]}{a(1 - e^2)} \tag{4.18}$$

To this point we have mainly characterized the orbit as a function of ϕ, and we have not discussed $\phi(t)$ in much detail. It turns out to be easier to keep ϕ as the independent variable and compute the time dependence as $t(\phi)$. Recall from Eq. (3.10) that area in the ellipse is swept out at the rate $dA/dt = \ell/2$ where $\ell = \sqrt{GMa(1 - e^2)}$ is the specific angular momentum. If we rewrite this as $dt = (2/\ell) \, dA$ and use $dA = (1/2)r^2 \, d\phi$ in polar coordinates, we can integrate to obtain

$$t = \frac{1}{\ell} \int r(\phi)^2 \, d\phi$$

Using $r(\phi)$ from Eq. (4.11) yields[1]

---

[1] With help from Mathematica [1].

$$t = \frac{a^2(1-e^2)^2}{\ell} \int \frac{d\phi}{(1+e\cos\phi)^2}$$

$$= \frac{a^2(1-e^2)^2}{\ell} \left\{ \frac{2}{(1-e^2)^{3/2}} \tan^{-1}\left[\left(\frac{1-e}{1+e}\right)^{1/2} \tan\frac{\phi}{2} \right] - \frac{e\sin\phi}{(1-e^2)(1+e\cos\phi)} \right\}$$

(We choose the constant of integration so $t = 0$ at $\phi = 0$.) It is convenient to deal with the factor involving a and ℓ by expressing t in units of the orbital period,

$$\frac{t}{P} = \frac{1}{2\pi} \left\{ 2\tan^{-1}\left[\left(\frac{1-e}{1+e}\right)^{1/2} \tan\frac{\phi}{2} \right] - \frac{e(1-e^2)^{1/2}\sin\phi}{1+e\cos\phi} \right\} \qquad (4.19)$$

Note that a circular orbit has $e = 0$ and hence $t/P = \phi/2\pi$, which makes sense.

Now we have the ingredients to understand the shapes of orbits and velocity curves for the two-body problem. Figure 4.3 shows examples with different eccentricities. Recall that the orbits must share a common focus at the center of mass, and the two objects must always lie on opposite sides of this point. If the eccentricity is zero, the orbits are circular and concentric, and the velocity we would measure with the Doppler effect is a sinusoidal function (because it is a projection of circular motion). If the eccentricity is nonzero, the orbit centers are offset from one another, and the velocity curve is less regular. These two effects give us the ability to determine the eccentricity from the shape of the orbits or velocity curves.

4.1.5 Application to the Solar System

Let's see how the two-body theory applies to the Solar System and consider whether it was reasonable for Kepler to neglect the Sun's motion. We just want to get a sense of the numbers, so we examine the Sun's interaction with one planet at a time and assume circular orbits for simplicity. For the Sun/Earth system, here are the key quantities:

$$a = 1\,\text{AU} = 1.50 \times 10^{11}\,\text{m}$$
$$P = 1\,\text{yr} = 3.16 \times 10^7\,\text{s}$$
$$m_1 = 1.99 \times 10^{30}\,\text{kg}$$
$$m_2 = 5.97 \times 10^{24}\,\text{kg}$$

The corresponding reduced mass is

$$\mu = \frac{m_1 m_2}{m_1 + m_2} = 5.97 \times 10^{24}\,\text{kg}$$

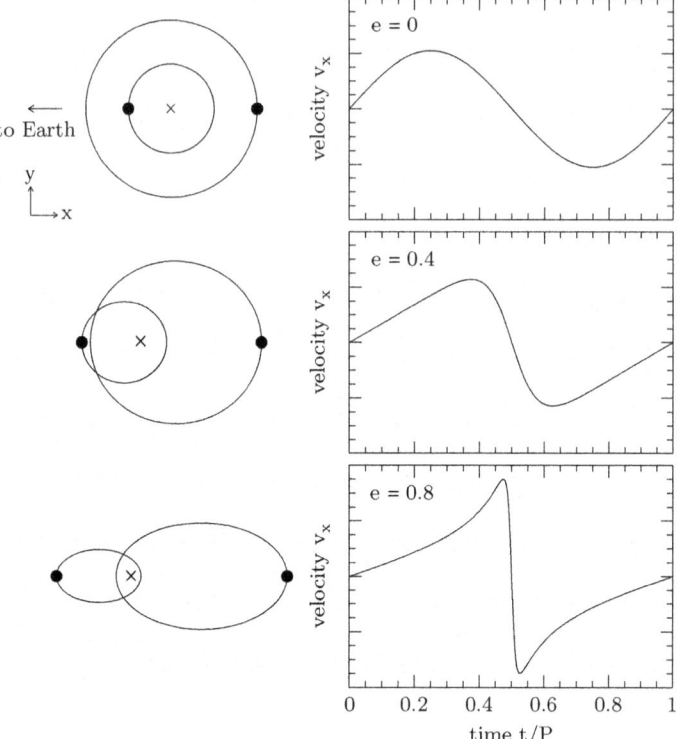

Fig. 4.3 Examples of two-body orbits and velocity curves. The three rows correspond to different eccentricities. The *left column* shows the orbital configuration, with × indicating the center of mass and • indicating the object positions at $t = 0$. The *right column* shows the Doppler velocity we would measure if Earth were off to the left (The observed velocity curve depends on how the orbit is oriented with respect to our line of sight; see Eq. 4.18)

(Note that $\mu \approx m_2$ when $m_2 \ll m_1$.) From Eq. (4.12), the amplitude of the Sun's motion induced by Earth is

$$a_1 = \frac{5.97 \times 10^{24}\,\text{kg}}{1.99 \times 10^{30}\,\text{kg}} \times 1.50 \times 10^{11}\,\text{m} = 4.49 \times 10^5\,\text{m} = 6.5 \times 10^{-4}\,R_\odot$$

The speed of this motion is

$$v_1 = \frac{2\pi a_1}{P} = \frac{2\pi \times (4.49 \times 10^5\,\text{m})}{3.16 \times 10^7\,\text{s}} = 0.089\,\text{m s}^{-1}$$

Since Jupiter is the most massive planet, let's consider it as well:

$$a = 5.20\,\text{AU} = 7.78 \times 10^{11}\,\text{m}$$

$$P = 11.86 \,\text{yr} = 3.74 \times 10^8 \,\text{s}$$
$$m_1 = 1.99 \times 10^{30} \,\text{kg}$$
$$m_2 = 1.90 \times 10^{27} \,\text{kg}$$

The corresponding reduced mass is

$$\mu = \frac{m_1 m_2}{m_1 + m_2} = 1.90 \times 10^{27} \,\text{kg}$$

The amplitude of the Sun's motion induced by Jupiter is

$$a_1 = \frac{1.90 \times 10^{27} \,\text{kg}}{1.99 \times 10^{30} \,\text{kg}} \times (7.78 \times 10^{11} \,\text{m}) = 7.42 \times 10^8 \,\text{m} = 1.07 \, R_\odot$$

and the speed of this motion is

$$v_1 = \frac{2\pi \times (7.42 \times 10^8 \,\text{m})}{3.74 \times 10^8 \,\text{s}} = 12.45 \,\text{m s}^{-1}$$

Jupiter affects the Sun more than Earth does, because its larger mass more than compensates for its greater distance.

The Sun's actual motion is more complicated than we have accounted for here, because it is influenced by all objects in the Solar System at once. Even so, the lesson is that the Sun's position changes only by an amount comparable to its size, and its speed is around a dozen meters per second. Such motion was too small for Kepler to detect, which is why he and then Newton could treat planetary motion as a one-body problem.

4.1.6 Kepler III Revisited

To conclude our discussion of the theory, let's see how the motion→mass principle applies to the two-body problem. We know from Sect. 3.1 that the equation of motion (4.10) leads to an expression for the orbital period of the form

$$P^2 = \frac{4\pi^2 a^3}{GM}$$

Each object in the two-body system has this same orbital period (they have to stay on opposite sides of the center of mass, after all). Using $M = m_1 + m_2$ from Eq. (4.3) and $a = a_1 + a_2$ from Eq. (4.12), we can now write the generalized version of Kepler's third law for two-body problems:

$$P^2 = \frac{4\pi^2(a_1 + a_2)^3}{G(m_1 + m_2)} \tag{4.20}$$

We can still use motion to measure mass in binary systems, but we must understand that what Kepler's third law gives is *total* mass. In the applications below we will consider if and when it is possible to decompose the total mass into the two individual components.

4.2 Binary Stars

Binary systems provide an opportunity to measure accurate masses for stars using two-body theory. We identify three classes of binaries based on what we are able to observe. In a **visual binary**, we can watch the stars move on the sky and follow their orbits. In a **spectroscopic binary**, we can detect absorption lines[2] in the stars' spectra and use the Doppler effect to measure the velocities along the line of sight. In an **eclipsing binary**, the orbit is nearly edge-on and the stars periodically pass in front of each other. These categories are complementary; any given system may fall into one, two, or all three of them. The way we measure motion is different in each case, so let's take them one by one and see what we can learn about mass.

4.2.1 Background: Inclination

Before we proceed, there is one bit of technical background to discuss. We can observe two dimensions of position projected onto the "plane of the sky,"[3] but the third dimension of distance is often difficult to determine. Even when it can be found, the distance is not precise enough to reveal changes in position along the line of sight. The quantity we can measure along the line of sight is velocity, using the Doppler effect.

This is an issue for binary stars because the orbital plane can have an arbitrary orientation with respect to the line of sight. We define the inclination angle i to be the angle between the orbital plane and the plane of the sky, as shown in Fig. 4.4. To be more precise, let $\hat{\mathbf{n}}$ be a unit vector perpendicular to the orbital plane, which we call the **normal vector**. The inclination is the angle between the normal vector and the line of sight; this is the same as the polar angle θ if we express $\hat{\mathbf{n}}$ in spherical

---

[2]In Chap. 14 we study spectral lines created by atoms and molecules in the outer layers of stars.

[3]Strictly speaking, we measure angles on the spherical sky. If the angular extent of a system is small, we can project onto a plane tangent to the sphere to obtain Euclidean coordinates without making a significant error.

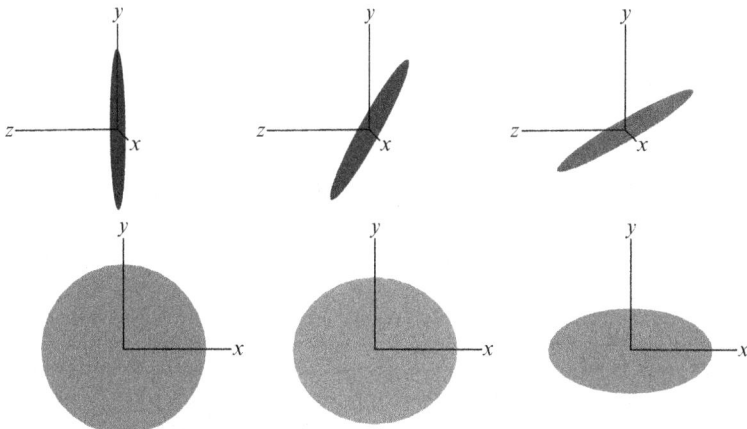

Fig. 4.4 Illustration of inclination. The *top row* shows nearly side-on views, while the *bottom row* shows the corresponding face-on views (looking down the z-axis). The columns display different inclinations: $i = 0°$ (*left*), $i = 30°$ (*middle*), and $i = 60°$ (*right*)

coordinates. With this definition, a face-on orbit has $i = 0°$ while an edge-on orbit has $i = 90°$.

To specify what we can measure, let $(x_{int}, y_{int}, z_{int})$ be the intrinsic coordinate system in which the orbital motion is in the (x_{int}, y_{int})-plane, while $(x_{obs}, y_{obs}, z_{obs})$ is the observed coordinate system in which we are looking along the z_{obs}-axis. The two frames are rotated with respect to one another by the angle i. Let's choose coordinates so the x-axes line up and the rotation applies to the y- and z-directions. Then the observed position is related to the intrinsic position by

$$x_{obs} = x_{int} \quad \text{and} \quad y_{obs} = y_{int} \cos i \qquad (4.21)$$

(Recall that the intrinsic orbital motion has $z_{int} = 0$.) The measured velocity along the line of sight is

$$v_{z,obs} = v_{y,int} \sin i \qquad (4.22)$$

The factors of $\cos i$ and $\sin i$ will be important in what follows. For each type of binary system, we need to consider whether the inclination can be determined, and how it affects our analysis.

Inclination can run between $0°$ and $90°$, but the values are not all equally likely. If orientations are random in space, the normal vector will be distributed uniformly over the unit sphere. Figure 4.5 shows that there is more area on the sphere with a larger value of i, and less area with a smaller value of i. In fact, the area is such that the probability distribution for inclination is

$$p(i) = \sin i \qquad (4.23)$$

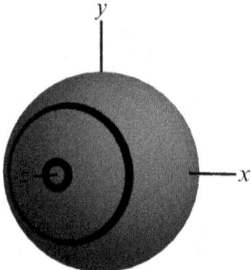

Fig. 4.5 If orbital orientations are random, the normal vector will be distributed uniformly over the unit sphere. The *small black band* indicates the set of normal vectors that correspond to inclinations in the range $5° < i < 10°$, while the large band has $40° < i < 45°$. Larger inclinations have a higher probability of being seen, with $p(i) = \sin i$

This factor of $\sin i$ is the same as the factor of $\sin \theta$ in the spherical coordinate volume element, $dV = r^2 \sin \theta \, dr \, d\theta \, d\phi$.

4.2.2 Visual Binary

If we can see both stars and watch them move, we can measure the period and trace the orbits directly. Can we determine the inclination? This might seem tricky at first because inclination causes an orbit to look squashed (due to the $\cos i$ factor in Eq. 4.21): a circle can look like an ellipse, or an ellipse can look more flattened than it truly is. There is, however, an important distinction between the configuration of orbits in a system with inclined circular orbits and a system with face-on elliptical orbits, as you can understand through Problem 4.1. The analysis is a little more subtle when the orbits are both elliptical and inclined, but the key idea is that the *true* orbits must satisfy Kepler's laws while the *projected* orbits may not. This principle makes it possible to deduce the true orbits and hence determine the inclination.

The challenge with visual binaries is that we can only measure the *angular* size of the orbits. If α_1 and α_2 are the angles subtended by the semimajor axes of the orbits, the corresponding physical lengths are

$$a_i = D \tan \alpha_i \approx D\alpha_i$$

where D is the distance to the binary system, and we are using the small-angle approximation $\tan \alpha_i \approx \alpha_i$. We can still find the mass ratio using Eq. (4.9): $m_2/m_1 = a_1/a_2 = \alpha_1/\alpha_2$. But if we want to find the actual masses using Eq. (4.20), we need to know the distance:

$$m_1 + m_2 = \frac{4\pi^2 D^3 (\alpha_1 + \alpha_2)^3}{GP^2}$$

Inclination is not a problem for visual binaries, but distance is.

4.2.3 Spectroscopic Binary

If a binary system is too distant and/or small, we may not be able to resolve the two stars on the sky. We can still analyze the motion, though, by using spectroscopy. As the stars move in their orbits, the Doppler effect causes each star's spectral lines to shift to shorter wavelengths when the star is moving toward us, and to longer wavelengths when the star is moving away.

Double-Line System

If we see distinct spectral lines from both stars, we can measure both of the Doppler velocity curves. The amplitude of the velocity curve for star #1 can be found by using Eqs. (4.13) and (4.18) for the intrinsic velocity and including a factor of $\sin i$ from projection (Eq. 4.22):

$$k_1 = \frac{\mu}{m_1} \frac{\ell}{a(1-e^2)} \sin i = \frac{\mu}{m_1} \frac{2\pi a}{P(1-e^2)^{1/2}} \sin i$$

where we simplify using Eqs. (3.11) and (4.20). The expression for k_2 is similar, with m_2 replacing m_1. If we measure both velocity amplitudes and take the ratio, most of the factors drop out,

$$\frac{k_2}{k_1} = \frac{m_1}{m_2} \tag{4.24}$$

and we can determine the ratio of masses directly from the measurements. Also, if we add the velocity amplitudes we find:

$$k_1 + k_2 = \mu \left(\frac{1}{m_1} + \frac{1}{m_2} \right) \frac{2\pi a}{P(1-e^2)^{1/2}} \sin i = \frac{2\pi a}{P(1-e^2)^{1/2}} \sin i$$

where we use Eq. (4.4) to simplify. Thus, we can write the semimajor axis in terms of the measurable[4] quantities k_1, k_2, P, and e as

$$a = \frac{P(1-e^2)^{1/2}}{2\pi} \frac{k_1 + k_2}{\sin i} \tag{4.25}$$

Using this in Kepler's third law gives the total mass as

$$m_1 + m_2 = \frac{P(1-e^2)^{3/2}}{2\pi G} \left(\frac{k_1 + k_2}{\sin i} \right)^3 \tag{4.26}$$

---

[4]Recall from Sect. 4.1.4 that we can determine e from the shape of the velocity curves.

For a spectroscopic binary, we can measure the absolute masses only if we know i. That makes spectroscopic binaries the opposite of visual binaries in the sense that distance is not a problem, but inclination is. If the inclination is unknown, the observables determine only the products $m_1 \sin^3 i$ and $m_2 \sin^3 i$.

Single-Line System

If one object (say, star #2) is faint, we may not be able to detect its absorption lines in the spectrum. We can still use the wavelength oscillations of the lines we do see to deduce that star #1 is in a binary orbit, and to measure its velocity amplitude k_1 as well as the orbital period P and eccentricity e. Now what can we do? Let's go back to Eq. (4.26) and use Eq. (4.24) to eliminate k_2, since it is not measurable:

$$m_1 + m_2 = \frac{P(1-e^2)^{3/2}}{2\pi G} \left(\frac{k_1 + k_1 m_1/m_2}{\sin i} \right)^3 = \frac{P(1-e^2)^{3/2}}{2\pi G} \left(\frac{k_1}{\sin i} \frac{m_2 + m_1}{m_2} \right)^3$$

Rearranging yields

$$\frac{m_2 \sin i}{(m_1 + m_2)^{2/3}} = \left(\frac{P}{2\pi G} \right)^{1/3} (1 - e^2)^{1/2} k_1 \qquad (4.27)$$

In other words, we can use the observables to infer a funny combination of masses, along with the usual inclination factor.

What good is this? Let's make two assumptions. First, suppose $m_2 \ll m_1$ so the left-hand side is approximately

$$\frac{m_2 \sin i}{m_1^{2/3}}$$

Second, suppose we have some way to estimate m_1 (perhaps from other properties of the star, such as its brightness and color). Then we can move m_1 to the right-hand side in Eq. (4.27) and write

$$m_2 \sin i = \left(\frac{m_1^2 P}{2\pi G} \right)^{1/3} (1 - e^2)^{1/2} k_1 \qquad (4.28)$$

As we will see in Sect. 4.3, these two assumptions are reasonable for extrasolar planets, so measuring Doppler velocities of stars lets us determine $m_2 \sin i$ for planets orbiting those stars.

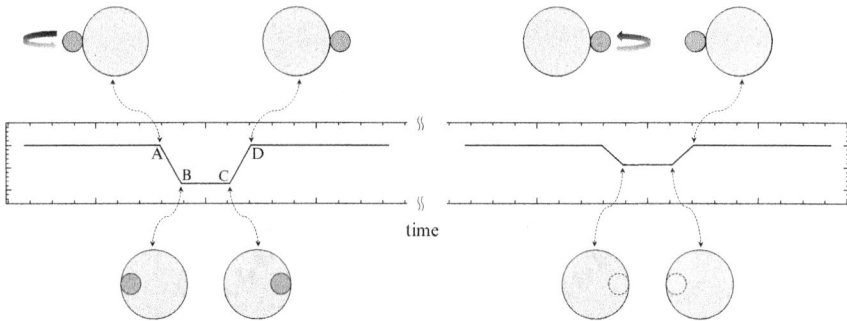

Fig. 4.6 Schematic light curve for an eclipsing binary, and the corresponding star configurations. (*Left*) During the primary eclipse, the small star is coming out of the page and moving left-to-right in front of the large star. (*Right*) During the secondary eclipse, the small star is going into the page and moving right-to-left behind the large star. In this example, the large star has a higher surface brightness (luminosity per unit area) than the small star

4.2.4 Eclipsing Binary

If a binary system is very close to edge-on, one star can fully or partially eclipse the other. The **light curve**, or brightness as a function of time, will dip during the eclipse events as shown in Fig. 4.6. Eclipses can occur only if $i \approx 90°$ (or $\sin i \approx 1$), so seeing them solves the inclination problem in spectroscopic binaries and lets us determine the absolute masses of the two stars.

Eclipses contain information about the *sizes* of stars as well. In Sect. 4.3.2, we will see that eclipse depth alone can reveal the relative sizes of the stars (or, in the case below, a star and a planet). If we combine eclipses with Doppler velocities, we can go a step further and determine the absolute sizes. For example, the time between points A and B in Fig. 4.6 is the time it takes for the stars to move (relative to one another) by the diameter of the small star. Since the stars are moving in opposite directions, their relative speed is $v_1 + v_2$. The radius of the smaller star is therefore

$$R_{\text{small}} = \frac{1}{2}(v_1 + v_2)(t_B - t_A)$$

(How would you determine the radius of the larger star?)

4.3 Extrasolar Planets

Since 1995, hundreds of planets have been discovered around other stars using the techniques we just discussed. A star+planet system acts as a single-line spectroscopic binary, while an edge-on system acts as an eclipsing binary. The effects

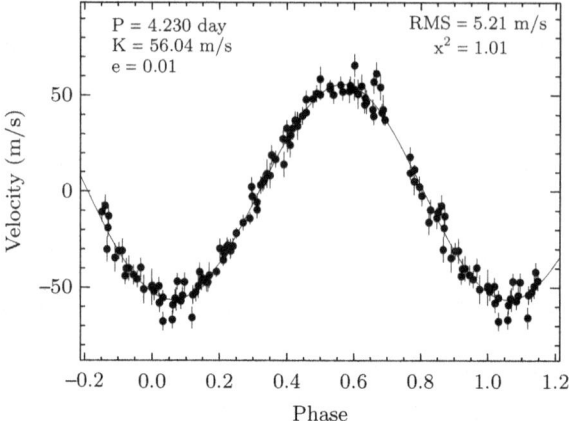

Fig. 4.7 Radial velocity curve for the star 51 Peg, where ϕ denotes orbital phase (Credit: Marcy et al. [2]. Reproduced by permission of the AAS (Also see Mayor et al. [3]))

are generally small—speeds are typically tens of meters per second or smaller, and eclipse depths are at the percent level or smaller—but they can now be measured routinely. Systems in which we can measure both motion and eclipses are particularly valuable, as we will see.

4.3.1 Doppler Planets

A star with a planet is a prime example of a single-line spectroscopic binary; the planet contributes very little light to the spectrum, so it does not introduce detectable absorption lines, but its gravity causes the star to "wobble" so the spectral lines oscillate in wavelength. As we saw in Eq. (4.28), if the planet is much less massive than the star then we can estimate $m_2 \sin i$, but we need to know the mass of the star. This can be often inferred from the star's visible properties; as we will see in Chap. 16, there are good relations between the mass, luminosity, color, and spectroscopic properties of stars.

The first extrasolar planet discovered orbits the star 51 Peg [3]. Figure 4.7 shows that the star's velocity curve is nearly sinusoidal, indicating that the orbit is close to circular. The measured period, eccentricity, and velocity amplitude are [4]

$$P = 4.23 \, \text{day} = 3.65 \times 10^5 \, \text{s}$$

$$e = 0.013$$

$$k_1 = 55.9 \, \text{m s}^{-1}$$

The mass of the star is estimated to be $m_1 = 1.05\,M_\odot = 2.09 \times 10^{30}\,\mathrm{kg}$. Using these values in Eq. (4.28) yields for the planet:

$$
m_2 \sin i = \left[\frac{(2.09 \times 10^{30}\,\mathrm{kg})^2 \times (3.65 \times 10^5\,\mathrm{s})}{2\pi \times (6.67 \times 10^{-11}\,\mathrm{m^3\,kg^{-1}\,s^{-2}})} \right]^{1/3} (1 - 0.013^2)^{1/2} \times 55.9\,\mathrm{m\,s^{-1}}
$$

$$
= 8.73 \times 10^{26}\,\mathrm{kg}
$$

$$
= 0.46\,M_J
$$

So $m_2 \sin i$ is comparable to the mass of Jupiter and much smaller than the mass of a typical star. Does that automatically imply that m_2 itself is small, i.e., that the second object is a planet? The alternative is that i is small, i.e., that the second object is a star but the orbits are very close to face-on. The early phase of exoplanet studies faced this key question: do low values of $m_2 \sin i$ indicate planets or just binary star systems in nearly face-on orbits?

One way to proceed is to make a statistical argument and point out that only a small fraction of orbits are nearly face-on. If observed $m_2 \sin i$ values are small because m_2 is large but i is small, then we would expect there to be many other systems where i and hence $m_2 \sin i$ are larger. How many? In order for us to misinterpret a stellar companion with mass M_s as a planet less massive than Jupiter, we would need

$$
M_s \sin i \le M_J \quad \Rightarrow \quad i \le \sin^{-1}\left(\frac{M_J}{M_s}\right)
$$

The probability for this to occur is

$$
\mathrm{Pr} = \int_0^{\sin^{-1}(M_J/M_s)} p(i)\,\mathrm{d}i
$$

where $p(i) = \sin i$ from Eq. (4.23). If there are N_{tot} systems overall, and N_J systems in which we think the companion is a planet less massive than Jupiter, then N_J/N_{tot} is given by this probability. Therefore we can compute

$$
\frac{N_J}{N_{\mathrm{tot}}} = \int_0^{\sin^{-1}(M_J/M_s)} \sin i\,\mathrm{d}i
$$

$$
= 1 - \cos\left(\sin^{-1}\frac{M_J}{M_s}\right)
$$

$$
= 1 - \sqrt{1 - \left(\frac{M_J}{M_s}\right)^2}
$$

$$
\approx \frac{M_J^2}{2M_s^2}
$$

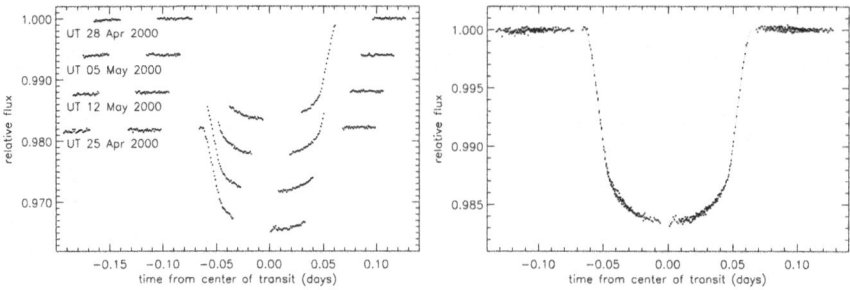

Fig. 4.8 Transit light curve for HD 209458, from Hubble Space Telescope observations. The *left panel* shows four individual eclipses (vertically offset for clarity), while the *right panel* shows all events superimposed (Credit: Brown et al. [7]. Reproduced by permission of the AAS)

where in the last step we make a Taylor series expansion assuming $M_J \ll M_s$. If the true companion mass were $M_s = M_\odot$, then we would expect $N_{tot}/N_J = 2.2 \times 10^6$, or more than a million times as many "stellar" companions as "planetary" companions. Even if the true companion mass were as low as $M_s = 0.08\, M_\odot$ (which is the smallest mass we consider to be a star; see Chap. 16), we would still expect $N_{tot}/N_J = 14{,}000$. In other words, if systems like 51 Peg were really stellar binaries seen nearly face-on, there ought to be many more systems seen at moderate inclinations with larger values of $m_2 \sin i$. The statistics suggested otherwise, but the argument was indirect and did not actually prove that the objects are planets.

4.3.2 Transiting Planets

Strong confirmation that some companions are in fact planets came with the discovery of planets that cross in front of their stars and produce eclipsing binaries. As we noted in Sect. 4.2.4, seeing a transit proves that a system is very close to edge-on, so $\sin i \approx 1$ and $m_2 \sin i$ accurately represents the companion's mass.

The first transiting planet found orbits a star called HD 209458 [5,6]. The eclipse light curve, shown in Fig. 4.8, is more complicated than the simple flat-bottomed curve sketched in Fig. 4.6. Previously we assumed the star was a flat, uniformly-bright disk, but in fact it is a sphere emitting light isotropically and we receive more light from the part of the surface that faces us and less light from the limbs. This

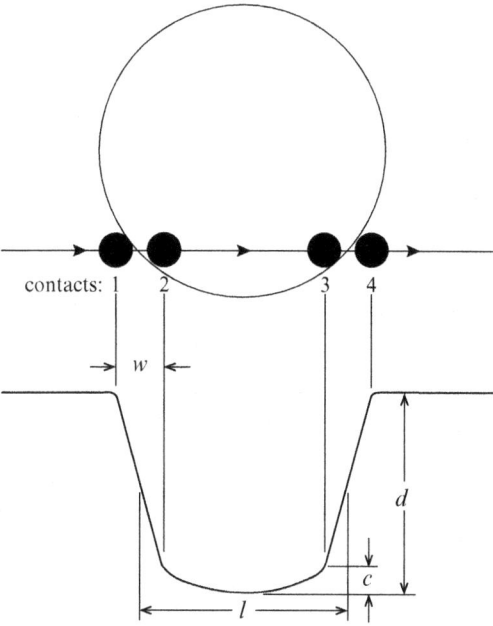

Fig. 4.9 Schematic diagram of the HD 209458 eclipse (Credit: Brown et al. [7]. Reproduced by permission of the AAS)

"limb darkening" effect can be incorporated into detailed models of the eclipse, leading to the picture shown in Fig. 4.9.

Transits reveal the size of the planet, with the simplest analysis using just the depth of the eclipse. The planet blocks a fraction of the star's visible area given by

$$ f_{\text{ecl}} = \frac{\pi R_2^2}{\pi R_1^2} = \left(\frac{R_2}{R_1} \right)^2 $$

where R_1 and R_2 are the radii of the star and planet, respectively. If we assume the star is a uniform disk (again, not correct but reasonable for a simple estimate), then f_{ecl} is also the fraction of the star's light that is blocked during the eclipse. Once we see an eclipse, we can use the depth to measure the size of the planet in relation to the size of the star. Then with an independent estimate of the star's size we can determine the planet's actual size, which we can finally combine with the mass to estimate the density. This is a big step toward understanding the physical properties and compositions of exoplanets.

Application to HD 209458

Let's examine the numbers for HD 209458b[5] [8]. This is a system with both Doppler and transit information, so we can use a joint analysis to learn a lot about the planet. The star's mass and radius are estimated to be

$$m_1 = 1.13 \, M_\odot = 2.25 \times 10^{30} \, \text{kg}$$

$$R_1 = 1.16 \, R_\odot = 8.07 \times 10^8 \, \text{m}$$

The orbital period and velocity amplitude for the star's motion are

$$P = 3.52 \, \text{day} = 3.04 \times 10^5 \, \text{s}$$

$$k_1 = 84.7 \, \text{m s}^{-1}$$

(The orbital eccentricity is small and assumed to be 0.) From the motion we can compute the mass of the companion:

$$m_2 = \left[\frac{(2.25 \times 10^{30} \, \text{kg})^2 \times (3.04 \times 10^5 \, \text{s})}{2\pi \times (6.67 \times 10^{-11} \, \text{m}^3 \, \text{kg}^{-1} \, \text{s}^{-2})} \right]^{1/3} \times 84.7 \, \text{m s}^{-1}$$

$$= 1.31 \times 10^{27} \, \text{kg}$$

$$= 0.69 \, M_J$$

where we use $\sin i \approx 1$. Also, rearranging Kepler's third law and approximating $m_1 + m_2 \approx m_1$ lets us find the semimajor axis, which is the distance of the planet from the star:

$$a \approx \left(\frac{G m_1 P^2}{4\pi^2} \right)^{1/3}$$

$$\approx \left[\frac{(6.67 \times 10^{-11} \, \text{m}^3 \, \text{kg}^{-1} \, \text{s}^{-2}) \times (2.25 \times 10^{30} \, \text{kg}) \times (3.04 \times 10^5 \, \text{s})^2}{4\pi^2} \right]^{1/3}$$

$$\approx 7.06 \times 10^9 \, \text{m}$$

$$\approx 0.047 \, \text{AU}$$

---

[5]By convention, planets are named by appending letters starting with "b" to the name of the star. For example, HD 209458b is a planet orbiting the star HD 209458.

The eclipse depth is 1.46 %, so the planet's size relative to the star is estimated to be

$$\frac{R_2}{R_1} = (0.0146)^{1/2} = 0.12$$

Factoring in the star's size yields for the planet:

$$R_2 = 9.8 \times 10^7 \, \text{m} = 1.4 \, R_J$$

Now combining the mass and radius lets us compute the mean density[6]

$$\rho_2 = \frac{3m_2}{4\pi R_2^3} = 340 \, \text{kg m}^{-3} = 0.34 \, \text{g cm}^{-3}$$

There are many things to say:

- The planet is roughly the mass and size of Jupiter, but is very close to its star.
- The density is much less than that of water, so the planet must be gaseous (as opposed to a rocky world like Earth).
- The planet is less massive but larger than Jupiter. It appears to be "puffed up" compared to Jupiter, presumably by heat from its star.

The discovery of large, massive planets very close to their stars—planets now called **hot Jupiters**—came as an enormous surprise and posed a significant challenge to theories of planet formation. In the traditional picture, which we will examine in Sect. 19.4.2, planets close to a star are expected to be rocky (like the terrestrial planets Mercury, Venus, Earth, and Mars in our Solar System) because it was too hot near the star for planetesimals to accumulate much gas or ice. Only planets forming farther from the star were able to collect volatile elements and grow much bigger. It seems difficult to change that picture, so the idea has emerged that hot Jupiters formed much farther from their stars than they are now, and then **migrated** inwards. Understanding how this migration occurred is a hot topic (pardon the pun) in planet formation theory.

4.3.3   Status of Exoplanet Research

Studies of exoplanets are advancing at an amazing rate. As of December 2013, more than 400 planets have been detected by the Doppler technique. With sensitive spectrographs it is now possible to measure star velocities as small as $0.25 \, \text{m s}^{-1}$

---

[6]We follow common practice and quote planet densities in CGS rather than MKS units because densities are of order unity in g cm^{-3}. For example, water has a density of $1 \, \text{g cm}^{-3}$ at standard temperature and pressure on Earth, while rocks and metals have densities of several g cm^{-3}. Earth's average density is about $5.5 \, \text{g cm}^{-3}$.

and thus to find planets with $m \sin i$ values comparable to the mass of Earth [9]. Well-measured velocity curves can reveal complicated motion caused by multiple planets; the most populous Doppler system found so far has at least five and perhaps as many as seven planets [10]. At the same time, more than 250 planets have been detected by the transit technique, along with some 2,500 more candidates from the Kepler mission. Kepler's precise transit measurements make it possible to discover planets as small as Mercury [11], systems with as many as six planets [12], and even planets orbiting binary stars [13,14]. (Another technique for finding planets is based on gravitational microlensing, which we will discuss in Sect. 9.2.4.)

After finding planets, the next step is to characterize their physical properties. As we saw with HD 209458b, measuring both mass and radius lets us use the mean density to investigate the bulk composition. There seems to be a lot of diversity: for example, the planet Kepler-10b has a mass of 4.6 $M_\oplus$ and a density of 8.8 g cm^{-3}, suggesting that it is made of rock and metals [15], while Kepler-11e has a mass of 8.0 $M_\oplus$ and a density of 0.58 g cm^{-3}, suggesting that it has a significant amount of light gas such as hydrogen and helium [12].

With transiting hot Jupiters we can investigate planetary atmospheres in some detail.[7] For example, spectra taking during a transit can reveal absorption by atoms and molecules when the star's light passes through the planet's atmosphere [16]. Infrared observations are sensitive to light *emitted* by hot planets. Most of time we receive light from both the star and planet, but during the secondary eclipse (when the planet goes behind the star; see Fig. 4.6) we receive light only from the star; we can use the difference to determine the brightness and temperature of the planet. We can even measure differences between daytime and nighttime temperatures and then investigate how effectively winds and clouds distribute heat across the planet [17, 18]. (For a more comprehensive review of work on exoplanet atmospheres, see [19].)

There is broad interest in finding planets similar to Earth. We could think about similarity in terms of mass, size, composition, etc., but perhaps the most tantalizing aspect is the ability to host life. On Earth it seems that liquid water is important for life, so we typically define the "habitable zone" around a star to be the region in which water could exist in liquid form (see Chap. 13, especially Problem 13.7). Kepler has found several planets that lie in the habitable zone and are between 40 and 140 % larger than Earth [20–22]. Their compositions are not known so it remains to be seen whether these planets are like Earth, Venus, or something altogether different. Nevertheless, it is remarkable to see how far exoplanet research has advanced in less than two decades since 51 Peg b was discovered—and to think that it all rests on the foundation of the gravitational two-body problem.

---

[7]We defer our own study of atmospheric physics to Chaps. 12 and 13; here we briefly summarize recent work on exoplanets.

Problems

4.1. Imagine that we see two visual binary systems with the orbits shown below (× denotes the center of mass). One represents a system with elongated orbits viewed face-on, while the other represents a system with circular orbits that are inclined to our line of sight. How can you determine which is which?

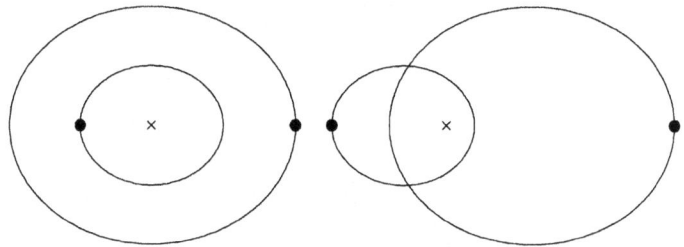

4.2. Here are the orbits of two stars in a binary system, along with Doppler velocity curves measured by an observer off to the left and in the plane of the orbits. (The velocity units are not important for this question.)

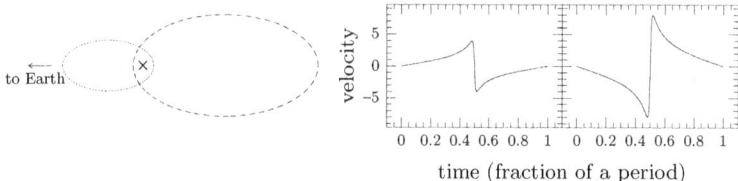

(a) Which orbit corresponds to the more massive star? How do you know?
(b) Which velocity curve belongs to which star? How do you know?
(c) Consider the points on the velocity curve marked below. Sketch the corresponding locations of the two stars on the orbits. Briefly explain your reasoning.

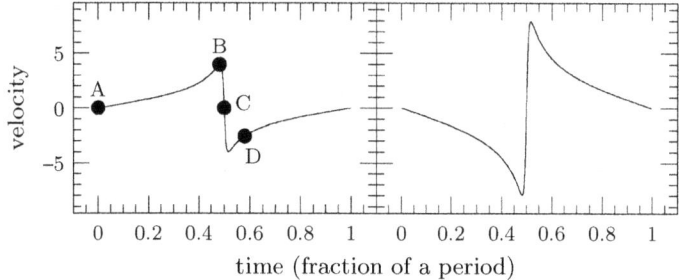

4.3. Consider the following eclipse light curve for a star. How many planets does the star have? What can you deduce about the relative sizes and orbital radii of the planets?

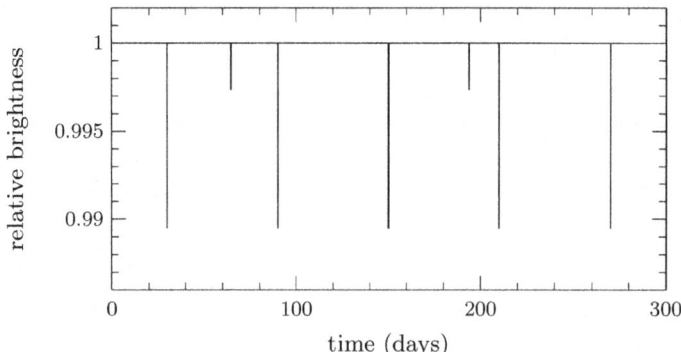

4.4. In a visual binary, we need to know the period (P), angular extent of the semimajor axis (α), and distance (d) to determine the star masses (see Sect. 4.2.2). Often the main source of uncertainty is d. If our measurement of the distance is $d \pm \sigma_d$, the fractional uncertainty is $f_d = \sigma_d / d$.

(a) If the fractional uncertainty in the distance is f_d, what is the corresponding fractional uncertainty in the total mass of the binary system?

(b) The brightest star in our night sky, Sirius, is a visual binary system. The brighter star has $\alpha_A = 2.5''$, the fainter star has $\alpha_B = 5.0''$, and the orbital period is 50.05 yr [23]. The distance to the Sirius system is 2.64 ± 0.01 pc. What are the masses of Sirius A and B? What are the uncertainties in the masses?

(c) When we analyze stars orbiting the black hole at the center of the Milky Way (Sect. 3.2.1), we are essentially studying a visual binary with one really massive component. If our estimate of the distance to the center of the Milky Way is 8.33 ± 0.35 kpc [24], what is the fractional uncertainty in our estimate of the black hole mass?

4.5. The binary system J0737−3039 has two pulsars orbiting with period $P = 0.102$ day and eccentricity $e = 0.088$ [25, 26]. It is nearly edge-on, and the velocity amplitudes are $k_1 = 302.9\,\mathrm{km\,s^{-1}}$ and $k_2 = 324.5\,\mathrm{km\,s^{-1}}$. What are the masses of the two pulsars? What is the distance between the pulsars?

4.6. Imagine that an alien astronomer observes Jupiter transiting the Sun. For this problem, you may take Jupiter's orbit to be circular and assume that Jupiter crosses the center of the star and does not emit any light itself. Define time $t = 0$ to be the middle of the eclipse.

(a) Plot the radial velocity curve the alien astronomer would measure, spanning at least one period. Be quantitative; label the axes with appropriate units.

(b) Plot the eclipse light curve. Make sure to identify all phases of the eclipse and quantify when each phase starts and ends. Also specify the depth of the eclipse (as a fraction of the uneclipsed brightness of the Sun).

4.7. Show that the geometric probability for having a system oriented so that we see a transiting planet is $P \approx R_*/a$ where R_* is the radius of the star and a is the orbital separation. Hint: use a geometric argument similar to the one in Sect. 4.3.1.

4.8. The Kepler space mission is searching for Earth-like planets using the transit technique.

(a) Kepler is observing about 100,000 stars. If every one is just like the Sun, with an Earth orbiting at 1 AU, how many would show transits? Use the probability from Problem 4.7

(b) Imagine that Kepler discovers a system that is an exact analog to our Solar System: "New Earth" orbiting "New Sol." How deep is the transit? How long does each transit last? Assume the planet crosses the center of the star.

(c) The reactionary group Just One Earth disputes the notion that "New Earth" is a planet and argues that it is a white dwarf instead. A white dwarf is about the same size as Earth but much more massive ($M_{WD} \approx 0.6\, M_\odot$). Calculate New Sol's radial velocity amplitude for the cases in which New Earth is (i) a planet, or (ii) a white dwarf. (Keep the orbital period the same.)

(d) We can now make radial velocity measurements with uncertainties of about $40\,\mathrm{cm\,s^{-1}}$. Could we tell whether New Earth is a planet or a white dwarf?

4.9. Kepler has found some planets that orbit binary star systems. The presence of two stars complicates the motion (see Chap. 6), but not too much if the planetary orbit is large compared with the stellar orbits. (In this problem, assume the stars and planet all move in the same plane.)

(a) Consider a coordinate system with the binary center of mass at the origin and the two stars on the x-axis. Let the semimajor axis of the binary orbit be a_{star}. Use a Taylor series expansion to show that the gravitational potential far from the stars can be written in polar coordinates (r, ϕ) as

$$\Phi \approx -\frac{G(M_1 + M_2)}{r} - \frac{GM_1 M_2}{M_1 + M_2}\frac{a_{\mathrm{star}}^2}{r^3}\frac{1 + 3\cos 2\phi}{4} + \mathcal{O}\left(\frac{1}{r^4}\right) \quad (4.29)$$

(b) Equation (4.29) indicates that a circumbinary orbit will be approximately Keplerian, with deviations that scale with the ratio $(a_{\mathrm{star}}/a_{\mathrm{planet}})^2$ where a_{planet} is the semimajor axis of the planetary orbit (in the Keplerian approximation). Compute this ratio for the Kepler-16 system [13]. The two stars have velocity amplitudes 13.7 and $46.5\,\mathrm{km\,s^{-1}}$, and their orbit has period 41.1 day and eccentricity 0.16. The planet has a nearly circular orbit with period 228.8 day.

References

1. Wolfram Research, Inc., *Mathematica*, 8th edn. (Wolfram Research, Champaign, 2010)
2. G.W. Marcy, R.P. Butler, E. Williams, L. Bildsten, J.R. Graham, A.M. Ghez, J.G. Jernigan, Astrophys. J. **481**, 926 (1997)
3. M. Mayor, D. Queloz, Nature **378**, 355 (1995)
4. R.P. Butler, J.T. Wright, G.W. Marcy, D.A. Fischer, S.S. Vogt, C.G. Tinney, H.R.A. Jones, B.D. Carter, J.A. Johnson, C. McCarthy, A.J. Penny, Astrophys. J. **646**, 505 (2006)

5. G.W. Henry, G.W. Marcy, R.P. Butler, S.S. Vogt, Astrophys. J. Lett. **529**, L41 (2000)
6. D. Charbonneau, T.M. Brown, D.W. Latham, M. Mayor, Astrophys. J. Lett. **529**, L45 (2000)
7. T.M. Brown, D. Charbonneau, R.L. Gilliland, R.W. Noyes, A. Burrows, Astrophys. J. **552**, 699 (2001)
8. G. Torres, J.N. Winn, M.J. Holman, Astrophys. J. **677**, 1324 (2008)
9. M. Mayor, D. Queloz, New Astron. Rev. **56**, 19 (2012)
10. C. Lovis et al., Astron. Astrophys. **528**, A112 (2011)
11. T. Barclay et al., Nature **494**, 452 (2013)
12. J.J. Lissauer et al., Nature **470**, 53 (2011)
13. L.R. Doyle et al., Science **333**, 1602 (2011)
14. W.F. Welsh et al., Nature **481**, 475 (2012)
15. N.M. Batalha et al., Astrophys. J. **729**, 27 (2011)
16. D. Charbonneau, T.M. Brown, R.W. Noyes, R.L. Gilliland, Astrophys. J. **568**, 377 (2002)
17. J. Harrington, B.M. Hansen, S.H. Luszcz, S. Seager, D. Deming, K. Menou, J.Y.K. Cho, L.J. Richardson, Science **314**, 623 (2006)
18. H.A. Knutson, D. Charbonneau, L.E. Allen, J.J. Fortney, E. Agol, N.B. Cowan, A.P. Showman, C.S. Cooper, S.T. Megeath, Nature **447**, 183 (2007)
19. S. Seager, D. Deming, Annu. Rev. Astron. Astrophys. **48**, 631 (2010)
20. W.J. Borucki et al., Astrophys. J. **745**, 120 (2012)
21. W.J. Borucki et al., Science **340**, 587 (2013)
22. T. Barclay et al., Astrophys. J. **768**, 101 (2013)
23. D. Benest, J.L. Duvent, Astron. Astrophys. **299**, 621 (1995)
24. S. Gillessen, F. Eisenhauer, S. Trippe, T. Alexander, R. Genzel, F. Martins, T. Ott, Astrophys. J. **692**, 1075 (2009)
25. M. Burgay et al., Nature **426**, 531 (2003)
26. M. Kramer et al., Science **314**, 97 (2006)

Chapter 5
Tidal Forces

Most of our analysis so far has used point masses. Now we ask whether the sizes of objects affect their gravitational interaction. For the *source* of gravity, size does not matter if the object is spherically symmetric (see Sect. 2.3). For the *target* of gravity, however, size does matter because gravity pulls harder on one side of the object than on the other. Newton studied this problem and realized that variations in the Moon's gravity across Earth's surface would "squeeze" the oceans and create the tides. This phenomenon is therefore known as the **tidal force**, and it has a variety of interesting consequences.

5.1 Derivation of the Tidal Force

Consider the gravitational force on an object of radius R from an object of mass M a distance r away (see Fig. 5.1). To use specific terminology, let's say the target of gravity is a planet and the source of gravity is a moon (later we will reverse the picture). Let's also say the moon lies in the planet's equatorial plane; while this is not quite correct for the Earth/Moon system, it allows us to use familiar geographic terms like equator, poles, and latitude. In this analysis we work in the plane containing the moon as well as the center and poles of the planet; everything else can be obtained by rotating around the line between the planet and moon.

Since the force of gravity scales as $1/r^2$, the side of planet that faces the moon feels a stronger force than the side of the planet away from the moon. The force on the center of planet causes the planet as a whole to move (orbiting the center of mass of the planet/moon system), so what creates the tidal force is the difference between the force at the surface and the force at the center. This differential force pulls "up" (relative to the planet's surface) near the equator and "down" near the poles.

Consider a small object of mass m on the surface at latitude θ. We draw the triangle shown in Fig. 5.1, and call s the length of the third side while α is the other

C. Keeton, *Principles of Astrophysics: Using Gravity and Stellar Physics to Explore the Cosmos*, Undergraduate Lecture Notes in Physics, DOI 10.1007/978-1-4614-9236-8_5, © Springer Science+Business Media New York 2014

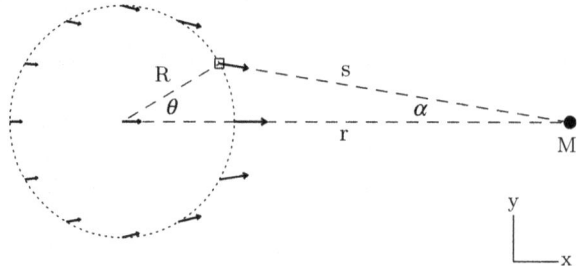

Fig. 5.1 The strength and direction of the gravitational force from the source on the right are denoted with *arrows*. Gravity varies across the surface of the planet, leading to a tidal force. The *dashed lines* indicate the geometry we use to analyze the tidal force at the position indicated by □

angle. These are important because s and α determine the strength and direction of the gravitational force, respectively. Specifically, the gravitational force from the moon on the small object m is

$$\mathbf{F}(\theta) = \frac{GMm}{s^2}\ (\cos\alpha\ \hat{\mathbf{x}} - \sin\alpha\ \hat{\mathbf{y}})$$

We would like to rewrite this in terms of coordinates on the planet (i.e., R and θ). As we saw with a similar analysis in Sect. 2.3, trigonometric identities let us write

$$s^2 = r^2\left(1 + \xi^2 - 2\xi\cos\theta\right)$$

$$\sin\alpha = \frac{\xi\sin\theta}{(1 + \xi^2 - 2\xi\cos\theta)^{1/2}}$$

$$\cos\alpha = \frac{1 - \xi\cos\theta}{(1 + \xi^2 - 2\xi\cos\theta)^{1/2}}$$

where we introduce $\xi = R/r$. We can then write the force as

$$\mathbf{F}(\theta) = \frac{GMm}{r^2}\ \frac{(1 - \xi\cos\theta)\hat{\mathbf{x}} - \xi\sin\theta\ \hat{\mathbf{y}}}{(1 + \xi^2 - 2\xi\cos\theta)^{3/2}}$$

$$\approx \frac{GMm}{r^2}\left[(1 + 2\xi\cos\theta)\hat{\mathbf{x}} - \xi\sin\theta\ \hat{\mathbf{y}} + \mathcal{O}\left(\xi^2\right)\right]$$

where we do a Taylor series expansion in ξ because we expect this ratio to be small for many planet/moon systems. We have found the force at the *surface* of the planet. For comparison, the force at the center of the planet is

$$\mathbf{F}_0 = \frac{GMm}{r^2}\ \hat{\mathbf{x}}$$

Fig. 5.2 The *arrows* indicate
the direction and amplitude of
the tidal force $\Delta \mathbf{F}$

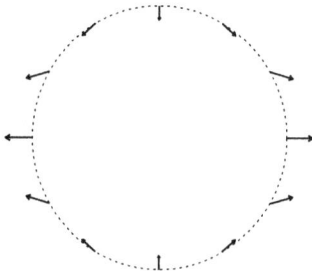

The tidal force is the difference,

$$\Delta \mathbf{F}(\theta) = \mathbf{F}(\theta) - \mathbf{F}_0 \approx \frac{GMm}{r^2} \left[2\xi \cos\theta \, \hat{\mathbf{x}} - \xi \sin\theta \, \hat{\mathbf{y}} + \mathcal{O}\left(\xi^2\right) \right]$$

Since the $\hat{\mathbf{x}}$ and $\hat{\mathbf{y}}$ components are both linear in ξ, we can pull this factor out front
and write

$$\Delta \mathbf{F}(\theta) \approx \frac{GMmR}{r^3} \left[2\cos\theta \, \hat{\mathbf{x}} - \sin\theta \, \hat{\mathbf{y}} + \mathcal{O}\left(\frac{R}{r}\right) \right] \qquad (5.1)$$

This is the general form of the tidal force on an object of size R that is a distance r
away from the source of gravity (in the approximation $R/r \ll 1$). The geometry is
shown in Fig. 5.2.

It is useful to consider the components of $\Delta \mathbf{F}$ relative to directions on the planet.
The component perpendicular to the surface ("vertical") is

$$\begin{aligned}
\Delta F_{\text{vert}} &= \Delta \mathbf{F} \cdot \hat{\mathbf{R}} \\
&= \frac{GMmR}{r^3} \left(2\cos\theta \, \hat{\mathbf{x}} - \sin\theta \, \hat{\mathbf{y}} \right) \cdot \left(\cos\theta \, \hat{\mathbf{x}} + \sin\theta \, \hat{\mathbf{y}} \right) \\
&= \frac{GMmR}{r^3} \left(2\cos^2\theta - \sin^2\theta \right) \\
&= \frac{GMmR}{r^3} \left(3\cos^2\theta - 1 \right) \qquad (5.2)
\end{aligned}$$

The component parallel to the surface ("horizontal") is

$$\begin{aligned}
\Delta F_{\text{horiz}} &= \Delta \mathbf{F} \cdot \hat{\boldsymbol{\theta}} \\
&= \frac{GMmR}{r^3} \left(2\cos\theta \, \hat{\mathbf{x}} - \sin\theta \, \hat{\mathbf{y}} \right) \cdot \left(-\sin\theta \, \hat{\mathbf{x}} + \cos\theta \, \hat{\mathbf{y}} \right) \\
&= -3\frac{GMmR}{r^3} \sin\theta \cos\theta \qquad (5.3)
\end{aligned}$$

Given the sign convention that $\hat{\theta}$ points "north," the horizontal force ΔF_{horiz} always points toward the equator. Here are a few additional notes:

- The maximum "pull up" (at the equator) is twice the maximum "push down" (at the poles).
- The horizontal component of the tidal force is largest at midlatitudes.
- Relative to the surface (i.e., in terms of vertical and horizontal components), the tidal force is the same on the near and far sides of the planet.
- The maximum strength of the tidal force occurs along the line between the objects and is given by

$$\Delta F_{\text{max}} = \frac{2GMmR}{r^3}$$

5.2 Effects of Tidal Forces

Since the strength scales as $\Delta F \propto R/r^3$, the tidal force is important for *large* objects that are *near* the source of gravity. There are variety of systems in which tidal forces play an interesting role.

5.2.1 Earth/Moon

Like Newton, we first consider the Earth. We mentioned tidal forces from the Moon, but in principle there could be tidal forces from the Sun as well. Which are more important on Earth? The maximum tidal force from each is

$$\Delta F_{\text{Sun}} = \frac{2GM_{\odot}mR}{r_{\text{Sun}}^3} \quad \text{and} \quad \Delta F_{\text{Moon}} = \frac{2GM_{\text{Moon}}mR}{r_{\text{Moon}}^3}$$

The ratio is

$$\frac{\Delta F_{\text{Moon}}}{\Delta F_{\text{Sun}}} = \frac{M_{\text{Moon}}}{M_{\odot}}\left(\frac{r_{\text{Sun}}}{r_{\text{Moon}}}\right)^3 = \frac{7.35 \times 10^{22}\,\text{kg}}{1.99 \times 10^{30}\,\text{kg}}\left(\frac{1.50 \times 10^{11}\,\text{m}}{3.84 \times 10^8\,\text{m}}\right)^3 = 2.2$$

Although the Moon is much less massive than the Sun, it is so much closer that it exerts the stronger tidal force. Nevertheless, the Sun's effect is not negligible. It modulates the height of tides induced by the Moon, sometimes creating high tides that are higher than average (known as "spring tides") or low tides that are lower than average ("neap tides"; see Problem 5.1).

How does the tidal force from the Moon compare with Earth's own gravity? Let's consider the maximum vertical component of the tidal force:

$$\frac{\Delta F_{\text{Moon}}}{F_{\text{Earth}}} = \frac{2GM_{\text{Moon}}mR_{\text{Earth}}/r_{\text{Moon}}^3}{GM_{\text{Earth}}m/R_{\text{Earth}}^2} = \frac{2M_{\text{Moon}}}{M_{\text{Earth}}}\left(\frac{R_{\text{Earth}}}{r_{\text{Moon}}}\right)^3$$

$$= \frac{2 \times 7.35 \times 10^{22}\,\text{kg}}{5.97 \times 10^{24}\,\text{kg}}\left(\frac{6.38 \times 10^6\,\text{m}}{3.84 \times 10^8\,\text{m}}\right)^3 = 1.1 \times 10^{-7}$$

Another way to think about this is that the maximum tidal force from the Moon would create an acceleration of just $1.1 \times 10^{-6}\,\text{m\,s}^{-2}$ in the vertical direction. Thus, the vertical component of the tidal force would be very difficult to detect against the backdrop of the Earth's gravity.

The horizontal component of the tidal force is a different story, though. Earth's own gravity has a tangential component only to the extent that Earth is not a perfect sphere, which is a small effect. The horizontal tidal force therefore has little opposition beyond the rigidity of material on Earth's surface. It acts on water in the oceans (which, after all, is fluid rather than rigid) to create tidal "bulges" on the near and far sides of Earth (relative to the Moon). Analyzing ocean tides in detail requires advanced material, such as fluid dynamics on a rotating surface, but we can still understand several notable features.

As Earth rotates through the tidal bulges, we see two high tides and two low tides each day. Shore dwellers know the cycle of tides actually lasts longer than 1 day (almost 25 h) because a point on Earth's surface must complete a little more than one full rotation to "catch up" with the Moon moving in its own orbit. Also, friction between rock and water slows Earth's rotation; the length of the day is increasing at a rate of few milliseconds per century. While this effect is small, it can be measured using historical records of eclipse timing [1] as well as geological records of sedimentation that is influenced by tides [2].

As Earth's spin slows, its rotational angular momentum decreases; to keep total angular momentum conserved, the Moon drifts farther away at a rate of about 4 cm per year. We can measure this by timing how long it takes laser pulses to travel out to the Moon and back to Earth, reflecting off mirrors left on the Moon by Apollo astronauts [3]. The changes to Earth's rotation and the Moon's orbit will cease only when Earth's rotation period equals the Moon's orbital period, i.e., when the Earth and Moon are in **synchronous rotation**. At that point we would say Earth is **tidally locked** with the Moon (although we are unlikely to get there because it would take longer than the lifetime of the Sun).

So far we have considered the Moon's effect on Earth, but we can invert the picture and consider the tidal force on the Moon created by gravity from Earth. How do the forces compare? As we have seen, the maximum tidal force on Earth from the Moon is

$$\Delta F_{\text{on Earth}} = \frac{2GM_{\text{Moon}}mR_{\text{Earth}}}{r_{\text{Moon}}^3}$$

while the maximum tidal force on the Moon from Earth is

$$\Delta F_{\text{on Moon}} = \frac{2GM_{\text{Earth}}mR_{\text{Moon}}}{r_{\text{Moon}}^3} \tag{5.4}$$

The ratio is

$$\frac{\Delta F_{\text{on Moon}}}{\Delta F_{\text{on Earth}}} = \frac{M_{\text{Earth}}R_{\text{Moon}}}{M_{\text{Moon}}R_{\text{Earth}}} = \frac{(5.97 \times 10^{24}\,\text{kg}) \times (1.74 \times 10^6\,\text{m})}{(7.35 \times 10^{22}\,\text{kg}) \times (6.38 \times 10^6\,\text{m})} = 22.2$$

The tidal force on the Moon is strong enough to raise tidal bulges in the rock itself. The rotational deceleration has been so strong that the Moon is already tidally locked with Earth, which explains why we always see the same face of the Moon.

5.2.2 Jupiter's Moon Io

Another system that displays fascinating tidal phenomena is Jupiter and its moon Io. Let's examine the tidal force from Jupiter on Io,

$$\Delta F_{\text{on Io}} = \frac{2GM_{\text{Jup}}mR_{\text{Io}}}{r_{\text{Io}}^3}$$

and use the tidal force from Earth on our Moon (Eq. 5.4) as a reference point. Here are the numbers we need to make the comparison:

	M_{planet} (kg)	R_{moon} (km)	r (km)
Moon	5.97×10^{24}	1,737	3.84×10^5
Io	1.90×10^{27}	1,821	4.22×10^5

Io and the Moon are fairly similar in terms of their size and distance from the planet, but of course Io's planet is much more massive than the Moon's planet. That causes the ratio of tidal forces to be

$$\frac{\Delta F_{\text{on Io}}}{\Delta F_{\text{on Moon}}} = \frac{M_{\text{Jup}}}{M_{\text{Earth}}}\frac{R_{\text{Io}}}{R_{\text{Moon}}}\left(\frac{r_{\text{Moon}}}{r_{\text{Io}}}\right)^3 = 250$$

Tidal effects are *much* stronger on Io than on the Moon. They have caused Io to be tidally locked with Jupiter.

Io's orbit is slightly eccentric, with $e = 0.0041$. This may not seem like a lot, but it means Io's distance varies by 0.8 % between pericenter and apocenter, which translates into a 2.4 % change in the strength of the tidal force. This may not seem like a lot either, but 2.4 % of a strong tidal force is significant. Plus, the variation happens over the course of Io's orbital period, which is just 1.8 days. In essence, Io has been flexing every few days for more than 4 billion years, creating a significant amount of internal heat from friction[1] [4]. The cumulative heating has been strong enough to create volcanoes that have been observed by several spacecraft visiting Jupiter and its moons [5, 6]. The volcanoes on Io are perhaps the most striking manifestation of the amount of energy associated with tidal forces.

5.2.3 Extrasolar Planets

Tidal forces can be relevant for planets as well—especially hot Jupiters, since they are large and close to their stars. One interesting case is the planet that transits HD 209458 (see Sect. 4.3.2). Careful observations have shown that gas is escaping from the planet [7]. Heat from the star allows some of the gas molecules to exceed the planet's escape velocity (see Chap. 12 for more discussion), but the tidal force from the star contributes by helping to counteract the planet's gravity.

Many hot Jupiters are expected to be tidally locked to their stars. While direct evidence is difficult to obtain, there is indirect evidence for tidal locking from studies that examine how heat from a star is distributed across a planet by atmospheric circulation [8,9]. It is even conceivable that a star could be tidally locked to a planet. This may be the case for Tau Boötis: the rotation period of the star (measured from flux variability) is consistent with the orbital period of its planet [10, 11].

5.3 Tidal Disruption

When the tidal force from a planet pulls "up" on a moon's surface, it acts against the moon's self gravity. If the tidal force gets strong enough, it could actually tear the moon apart. To estimate when this occurs, let's adopt a simple criterion: if the tidal force "up" at the equator exceeds the gravitational force "down," the surface will be ripped off. (We remark on a more realistic criterion below.)

Consider a moon with mass M_m and radius R_m, which is orbiting a planet with mass M_p and radius R_p. Suppose the moon is a distance r from the planet. As we have seen, the tidal force up at the equator of the moon is

$$F_{\text{tidal}} = \frac{2GM_p m R_m}{r^3}$$

while the gravitational force down is

$$F_{\text{grav}} = \frac{GM_m m}{R_m^2}$$

---

[1] Think of repeatedly bending a paper clip back and forth.

According to the simple criterion we are using, tidal disruption will occur when $F_{grav} \lesssim F_{tidal}$, or

$$\frac{GM_m m}{R_m^2} \lesssim \frac{2GM_p m R_m}{r^3}$$

$$r^3 \lesssim 2\frac{M_p}{M_m} R_m^3$$

$$\lesssim 2\frac{\rho_p}{\rho_m} R_p^3$$

where we switch from mass to mean density using $M = (4/3)\pi R^3 \rho$. We find that the moon will be torn apart if it gets closer than

$$r \lesssim \left(2\frac{\rho_p}{\rho_m}\right)^{1/3} R_p \tag{5.5}$$

Notice that the threshold depends on the density of the moon, but not its size.

The preceding analysis applies to a moon that is rigid enough to maintain its shape as it is peeled away layer by layer. Edouard Roche considered a more general scenario in which the moon gets distorted even before it is disrupted. Tidal distortion stretches the moon in the radial direction (relative to the planet), which enhances the difference between the surface and center of the moon and causes disruption to occur at a somewhat larger distance from the planet. Roche found that a loosely bound moon would be disrupted when

$$r \lesssim 2.4 \left(\frac{\rho_p}{\rho_m}\right)^{1/3} R_p \tag{5.6}$$

This condition is now called the **Roche limit**. The most conspicuous consequence of tidal disruption is Saturn's rings. You can explore this idea and some other interesting scenarios in the problems below.

Problems

5.1. Spring tides occur when the Sun is oriented in a way that reinforces the Moon's tidal force, while neap tides occur when the Sun partially cancels the Moon's effect. Sketch the arrangements of the Earth, Moon, and Sun that lead to spring and neap tides.

5.2. Saturn has mass 5.7×10^{26} kg and radius 60,300 km.

(a) Find an image of Saturn and estimate the radius of the outer edge of the rings, in units of Saturn's radius.
(b) Compute Saturn's average mass density.

(c) What is the minimum density that a moon of Saturn orbiting at the outer edge of the rings must have to resist tidal disruption?

(d) It is thought that Saturn's rings are composed of bodies made of water ice. Is this consistent with your answer from part (c)?

5.3. Neptune has mass 1.02×10^{26} kg and radius 24,764 km. Its moon Triton has mass 2.14×10^{22} kg, radius 1,353 km, and orbital period 5.88 day. Triton's orbit is "backwards" (retrograde) relative to Neptune's spin, so tidal forces are causing the orbit to shrink. Simulations predict that Triton will cross Neptune's Roche limit in a few billion years [12].

(a) Where is the Roche limit for the Neptune/Triton system?

(b) Assuming Triton will reach the Roche limit in 2 billion years, approximately how fast is its orbit shrinking?

5.4. You may have heard that a person falling feet-first into a black hole would be stretched out by the tidal force, in a process affectionately called "spaghettification." But would the effect actually be dramatic? Let's consider:

(a) Use scaling relations to determine whether the tidal force at the event horizon gets stronger or weaker as the black hole mass increases.

(b) It seems reasonable to say that we would "feel" the stretching only if the tidal acceleration exceeds the familiar acceleration due to gravity on Earth. Find the black hole mass that would produce such a tidal acceleration at the event horizon.

(c) Use your results from (a) and (b) to say whether we would be spaghettified by the black hole at the center of the Milky Way.

(d) What about by the black hole in the binary system M33-X7 ($M \approx 16\,M_\odot$)?

5.5. In 1994 the comet Shoemaker-Levy 9 collided with Jupiter. The comet was actually a set of fragments that hit Jupiter one after the other, producing a series of explosions that visibly scarred the planet. Why fragments? It is believed that the comet had been tidally disrupted during a previous close pass by Jupiter (probably in 1992). How close must the comet have come to Jupiter?

5.6. Suppose an asteroid is headed straight for Earth. People have talked about using a rocket or bomb to divert the asteroid. You need not only to prevent a collision, but also to avoid having the asteroid be tidally disrupted. (Creating a bunch of asteroid rubble around Earth would be no good!) If you reach the asteroid when it is 1 AU from Earth, how much "sideways" velocity would you need to impart? How about if you reach it when it is 384,000 km from Earth (the same distance as the Moon)?

You may assume the Earth and asteroid form an isolated system; in other words, you can neglect the effects of the Moon, the Sun, and everything else in the Solar System. You may assume the asteroid started from rest infinitely far from Earth.

Hint: think about the trajectory the asteroid will follow once you have diverted it, and about energy and angular momentum.

References

1. F.R. Stephenson, Astron. Geophys. **44**(2), 2.22 (2003)
2. G.E. Williams, Rev. Geophys. **38**, 37 (2000)
3. J.O. Dickey et al., Science **265**(5171), 482 (1994)
4. S.J. Peale, P. Cassen, R.T. Reynolds, Science **203**, 892 (1979)
5. L.A. Morabito, S.P. Synnott, P.N. Kupferman, S.A. Collins, Science **204**, 972 (1979)
6. D.A. Williams, L.P. Keszthelyi, D.A. Crown, J.A. Yff, W.L. Jaeger, P.M. Schenk, P.E. Geissler, T.L. Becker, Icarus **214**, 91 (2011)
7. A. Vidal-Madjar, A. Lecavelier des Etangs, J.M. Désert, G.E. Ballester, R. Ferlet, G. Hébrard, M. Mayor, Nature **422**, 143 (2003)
8. H.A. Knutson, D. Charbonneau, L.E. Allen, J.J. Fortney, E. Agol, N.B. Cowan, A.P. Showman, C.S. Cooper, S.T. Megeath, Nature **447**, 183 (2007)
9. S. Faigler, L. Tal-Or, T. Mazeh, D.W. Latham, L.A. Buchhave, Astrophys. J. **771**, 26 (2013)
10. S.L. Baliunas, G.W. Henry, R.A. Donahue, F.C. Fekel, W.H. Soon, Astrophys. J. Lett. **474**, L119 (1997)
11. R.P. Butler, G.W. Marcy, E. Williams, H. Hauser, P. Shirts, Astrophys. J. Lett. **474**, L115 (1997)
12. C.F. Chyba, D.G. Jankowski, P.D. Nicholson, Astron. Astrophys. **219**, L23 (1989)

Chapter 6
Gravitational Three-Body Problem

After solving the one- and two-body problems, generalizing to the three-body problem should be easy, right? No! In fact, it was the gravitational three-body problem that led Henri Poincaré to discover dynamical "chaos" [1]. Some systems are so sensitive to initial conditions that a tiny shift today can dramatically change the long-term behavior. The Solar System is actually an example: despite being well-approximated by the two-body problem, planetary motion is formally chaotic because of gravitational interactions among planets [2]. We cannot solve the three-body problem in general, but we can gain valuable insights from two cases that are simplified but still relevant for systems ranging from satellites near Earth to planets around distant stars.

6.1 Two "Stars" and One "Planet"

First consider a three-body problem in which two of the objects are much more massive than the third. Let's use the language of a "planet" (mass m) moving around two "stars" (masses M_1 and M_2), although we will examine a variety of systems. We assume $m \ll M_1, M_2$ so the planet does not affect the stars' motion. Let's further assume the stars have circular orbits, and the planet moves in their orbital plane. This is clearly a restricted version of the three-body problem, but it is one that has some interesting applications.

6.1.1 Theory: Lagrange Points

To analyze this problem, it is convenient to work in a reference frame that rotates at the same angular frequency as the stars so we can keep the stars fixed and focus on the planet. We have to be careful, though, because Newton's laws in their usual

C. Keeton, *Principles of Astrophysics: Using Gravity and Stellar Physics to Explore the Cosmos*, Undergraduate Lecture Notes in Physics, DOI 10.1007/978-1-4614-9236-8_6,
© Springer Science+Business Media New York 2014

form hold only in an inertial (i.e., non-rotating) reference frame. In Sect. A.3 we find that acceleration measured in a rotating reference frame (a_{rot}) is related to acceleration measured in a fixed reference frame (a_{fixed}) via

$$\mathbf{a}_{fixed} = \mathbf{a}_{rot} + \boldsymbol{\Omega} \times (\boldsymbol{\Omega} \times \mathbf{r}) + 2\boldsymbol{\Omega} \times \mathbf{v}_{rot} + \frac{d\boldsymbol{\Omega}}{dt} \times \mathbf{r}$$

where $\boldsymbol{\Omega}$ is a vector that points along the rotation axis and has an amplitude equal to the rotational frequency $\omega = 2\pi/P$. Newton's second law relates the true force to the acceleration in the fixed frame: $\mathbf{F}_{true} = m\,\mathbf{a}_{fixed}$. We can retain the form of this law in the rotating frame if we define an effective force such that $\mathbf{F}_{eff} = m\,\mathbf{a}_{rot}$. Clearly we need

$$\mathbf{F}_{eff} = \mathbf{F}_{true} - m\,\boldsymbol{\Omega} \times (\boldsymbol{\Omega} \times \mathbf{r}) - 2m\,\boldsymbol{\Omega} \times \mathbf{v}_{rot} - m\,\frac{d\boldsymbol{\Omega}}{dt} \times \mathbf{r} \qquad (6.1)$$

The second term is known as the **centrifugal force**, and it is what you feel "pulling" you outward on a merry-go-round. The third term is known as the **Coriolis force**, and it is important for systems like airplanes and weather moving around the rotating Earth. The fourth term is known as the **Euler force**, and it applies only if the rotation rate is not uniform; it vanishes for problems like ours in which $\boldsymbol{\Omega}$ is constant.

It is important to remember that these are not real forces; they are just consequences of working in a rotating reference frame, and are sometimes called "fictitious forces." Nevertheless, they do need to be taken into account when working in a rotating frame.[1]

With the planet's motion restricted to the orbital plane of the stars, $\mathbf{r}$ is perpendicular to $\boldsymbol{\Omega}$ and the centrifugal force simplifies: $\boldsymbol{\Omega} \times (\boldsymbol{\Omega} \times \mathbf{r}) = -\omega^2\,\mathbf{r}$. If we neglect the Coriolis force (because it depends on the speed of the planet and is generally small for the systems we consider), then the effective force is

$$\mathbf{F}_{eff} = \mathbf{F}_{true} + m\,\omega^2\,\mathbf{r}$$

The associated potential energy is

$$U_{eff} = -\int \mathbf{F}_{eff} \cdot d\mathbf{r} = -\int \mathbf{F}_{true} \cdot d\mathbf{r} - \int m\,\omega^2\,\mathbf{r} \cdot d\mathbf{r} = U_{true} - \frac{1}{2}m\,\omega^2 r^2$$

(In principle there is a constant of integration, but it only affects the unobservable zeropoint of the potential.) We know U_{true} for the two stars (see Eq. 2.12), so we can write down the effective potential,

---

[1] And they certainly don't feel fictitious when you make a sharp, fast turn in a car!

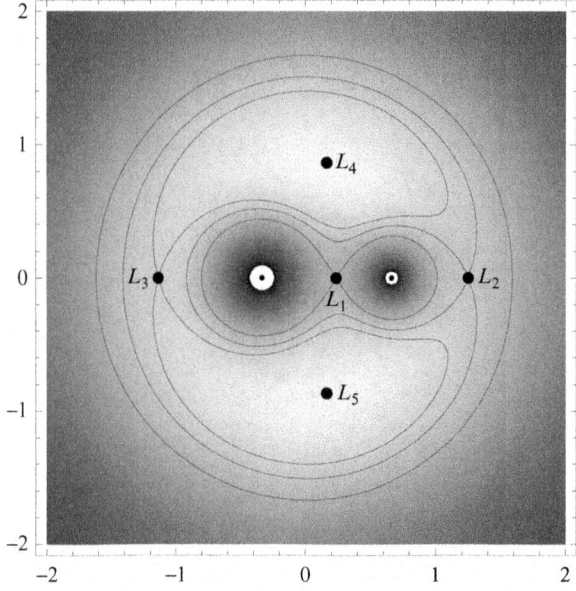

Fig. 6.1 Contour plot of the effective potential for a restricted three-body problem in which the primary objects have a 2:1 mass ratio. The Lagrange points are labeled; L_1, L_2, and L_3 are saddle points, while L_4 and L_5 are local maxima. Contours are chosen to pass through L_1–L_3. The *two small dots* mark the primary masses; they are surrounded by white regions only because the *grayscale* does not capture the divergence $\Phi_{\text{eff}} \rightarrow -\infty$ near M_1 and M_2

$$\Phi_{\text{eff}} = \frac{U_{\text{eff}}}{m} = -\frac{GM_1}{|\mathbf{r} - \mathbf{r}_1|} - \frac{GM_2}{|\mathbf{r} - \mathbf{r}_2|} - \frac{1}{2}\omega^2 r^2 \qquad (6.2)$$

This function is plotted in Fig. 6.1 for an illustrative example. Since Φ_{eff} is a function in two dimensions, it can have three types of critical points where the derivatives vanish: local minima, local maxima, and saddle points. In the restricted three-body problem, the effective potential has three saddle points, which all lie on the line joining the two stars, and two local maxima, which make equilateral triangles with the two stars (regardless of the stars' masses; see Problem 6.3). These are collectively known as **Lagrange points** after Joseph-Louis Lagrange, and they are labeled as follows:

- L_1: between the two stars
- L_2: "behind" the less massive star
- L_3: "behind" the more massive star
- L_4/L_5: leading/trailing by 60°

Formally, the saddle points L_1–L_3 are unstable equilibria: a particle at rest can stay put, but any little nudge will cause it to roll away. It is possible, though, to find small orbits around L_1, L_2, or L_3 [3]. Despite being local maxima, L_4 and L_5 turn out to

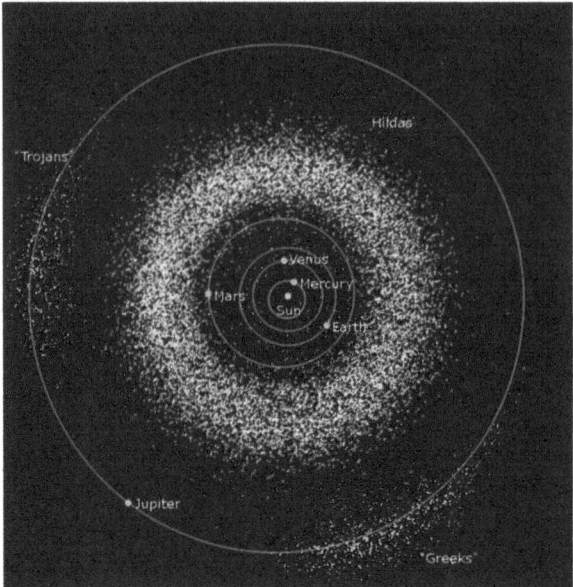

Fig. 6.2 Locations of known asteroids in the inner Solar System. Main belt asteroids are shown in *white*, while Trojan asteroids associated with Jupiter are colored *green*. The Trojans are divided into the "Trojan" and "Greek" camps (Credit: Wikimedia Commons)

be stable equilibria if the mass ratio is $M_1/M_2 > 24.96$. The analysis of stability involves the Coriolis force, which goes beyond the level of detail we are considering here [4].

6.1.2   Applications

Lagrange points are important for natural and artificial objects in our own Solar System, and for certain types of binary star systems as well.

Sometimes we want to place a satellite away from Earth but in a location where it will not drift off. The Lagrange points for the Sun/Earth system are natural choices. Several satellites observing the Sun, including the Solar and Heliospheric Observatory, are at L_1. Several telescopes, including the Wilkinson Microwave Anisotropy Probe and the Planck spacecraft (both observing the Cosmic Microwave Background radiation; see Sect. 20.1), are at L_2.

The L_4 and L_5 Lagrange points of the Sun/Jupiter system host a few thousand asteroids collectively known as "Trojan" asteroids (see Fig. 6.2). These objects are not actually fixed at L_4 or L_5; they move in sizable regions but are trapped in stable orbits around the Lagrange points (again, see [4] for more about stability). There are also some Trojan asteroids associated with Neptune, Mars, and even Earth [5].

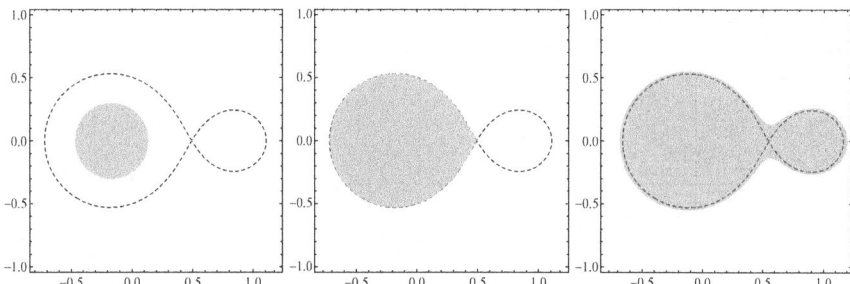

Fig. 6.3 If one component of a binary (*gray*) expands and fills its Roche lobe (*middle*), mass can flow out and envelop the companion (*right*). Here the *dashed line* shows the effective potential contour that passes through the Lagrange point L_1

The Lagrange point L_1 plays a prominent role in binary systems with two stars close together. If one star puffs up (for example, when it becomes a red giant; see Sect. 16.3), then matter near the surface might actually feel stronger gravity from the companion than from its own star. In that case mass can begin to flow from one star to the other. The equipotential contour running through L_1 marks the transition zone, which we call the **Roche lobe**. This scenario, which is pictured in Fig. 6.3, can have several consequences:

- *Accretion.* To conserve angular momentum, the transferred matter often settles into a disk around the second star and then slowly spirals in.
- *Energy release.* When matter drops from L_1 onto the accretion disk or star, potential energy is converted into kinetic energy, which is further converted into heat and light; we can observe X-rays from accretion onto neutron stars and black holes.
- *Nova.* If the second star is a white dwarf, hydrogen can accumulate and heat up to the point that nuclear fusion occurs on the surface; this can make the system much brighter for a few weeks or months, in a phenomenon we call a nova.
- *Supernova.* If enough mass accumulates on a white dwarf, it can carry the white dwarf over the "Chandrasekhar limit" of about $1.4\,M_\odot$ (see Sect. 17.2) and cause the white dwarf to explode as a Type Ia supernova; these objects have become important probes of the expanding universe (see Sect. 18.2).

6.2   One "Planet" and Two "Moons"

Now consider a different limit in which one object far outweighs the other two ($M_1 \gg m_2, m_3$). This limit can describe a variety of systems, but we will initially use the language of a "planet" with two "moons."

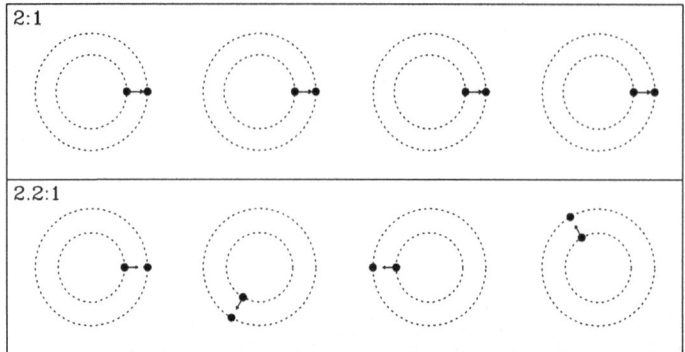

Fig. 6.4 The *top row* shows four successive points of closest approach for two bodies in a 2:1 orbital resonance. At each snapshot in time, the *dots* show the positions of the two bodies, and the *arrow* indicates the direction of the force exerted by the outer body on the inner body. The *bottom row* show a case that is not in resonance, with a frequency ratio of 2.2:1

6.2.1 Theory: Resonances

The planet dominates the gravitational field and keeps the moons moving in Keplerian orbits, but whenever the moons approach one another they exchange a small gravitational "kick." In general, the kicks occur at different locations in the orbits, so they have different directions and tend to average out over time (see the bottom row of Fig. 6.4). Suppose, however, that the inner moon completes exactly two orbits while the outer moon completes one:

$$P_2 = \frac{1}{2}\, P_3 \quad \Leftrightarrow \quad \omega_2 = 2\omega_3$$

In this case the kicks happen at the same place in the orbit and in the same direction (see the top row of Fig. 6.4) so they tend to add up over time. Any *integer* combination of orbits can likewise let the gravitational kicks combine coherently. If the inner moon goes around m times[2] while the outer moon goes around n times, then

$$m P_2 = n P_3 \quad \Leftrightarrow \quad \frac{\omega_2}{\omega_3} = \frac{m}{n}$$

and we call this an **m:n resonance** (for example, a 2:1 resonance, 3:2 resonance, etc.). In any single orbit the gravity between the moons is weak compared with the gravity from the planet, but the accumulated perturbations can have some interesting consequences.

---

[2]Here we briefly use m as an integer, not a mass.

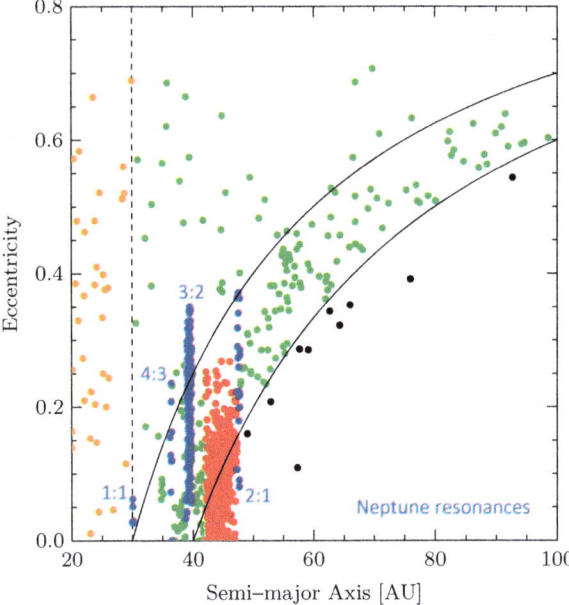

Fig. 6.5 Orbital properties of objects in the outer Solar System. Different colors indicate different classes of objects; we focus on the *blue points*, which are in orbital resonances with Neptune. The *dashed line* marks the semimajor axis of Neptune; the *solid curves* indicate orbits whose perihelion is interior to the semimajor axis of Neptune (*upper curve*) or Pluto (*lower curve*) (Credit: David Jewitt, UCLA)

6.2.2   Applications

One effect of orbital resonances is to lock groups of objects into related orbits. We see this in the Jupiter system; here are the orbital periods and frequencies for three of the moons that Galileo discovered:

	P (day)	ω (day^{-1})
Io	1.769	3.552
Europa	3.551	1.769
Ganymede	7.155	0.878

These moons are in a joint 4:2:1 resonance. The mutual gravitational interaction causes Io's orbit to be more elongated than it would have been otherwise, which couples with the tidal force from Jupiter (Sect. 5.2.2) to make Io the most geologically active body in the Solar System. We also see resonances in the outer Solar System. Figure 6.5 shows the distribution of semimajor axes and eccentricies for known "trans-Neptunian objects." There are notable groupings of objects in 1:1, 4:3, 3:2, and 2:1 resonances with Neptune. (Pluto is part of the group in the 3:2 resonance.)

We might ask how objects come to be in resonance. With the trans-Neptunian objects, an intriguing possibility is that Neptune migrated outward during the planet formation process, causing the location of the resonance to travel outward as well. If the moving resonance captured an object like Pluto, the gravitational interaction would have caused Pluto to migrate such that it remained trapped in resonance [6]. In this way Neptune's migration may have swept a number of objects into resonance. Ongoing research is examining whether a similar process happened among Jupiter's moons to create the resonance between Io, Europa, and Ganymede [7].

In a complementary action, resonances can also clear gaps in extended structures. One example is the dark band called the **Cassini division** in Saturn's rings. Objects cannot stay in this region because they would be in resonance with one of Saturn's moons (see Problem 6.5); the gravitational kicks would elongate the orbit, move the apocenter into the outer ring, and cause these objects to collide with other ring constituents [8]. Another example involves asteroids that lie in the "main belt" between the orbits of Mars and Jupiter. While the distribution of positions in space looks fairly continuous (Fig. 6.2), the distribution of semimajor axes has conspicuous dips at certain values (notably 2.5 and 3.3 AU; see Fig. 6.6).[3] These **Kirkwood gaps** are associated with Jupiter resonances (especially 3:1, 3:2, and 7:3). Even the outer edge of the asteroid belt seems to have been sculpted by a 2:1 resonance with Jupiter.

Problems

6.1. A staple of science fiction is the idea that you could spin a spaceship or space station so that the centrifugal force simulates gravity. How fast would a spaceship with a radius of 10 m have to spin to mimic the gravity on the surface of Earth? How about a space station with a radius of 100 m?

6.2. There are a few known three-body solutions beyond the restricted three-body problem and resonances. Lagrange found a solution with the three bodies forming an equilateral triangle. For this problem, assume the masses are the same.

(a) If the initial velocities are zero, what will happen to the system? Estimate how long it takes the system to reach its final state.
(b) Find the rotational velocity required to balance the gravitational attraction and keep the masses moving along a circular obit.

---

[3]You might ask why the Kirkwood gaps are not apparent in a snapshot of positions in space. Since asteroid orbits can be moderately elliptical, an asteroid with a given semimajor axis can be found at a range of radii. The gaps in a plot of semimajor axis get smeared out in a plot of position.

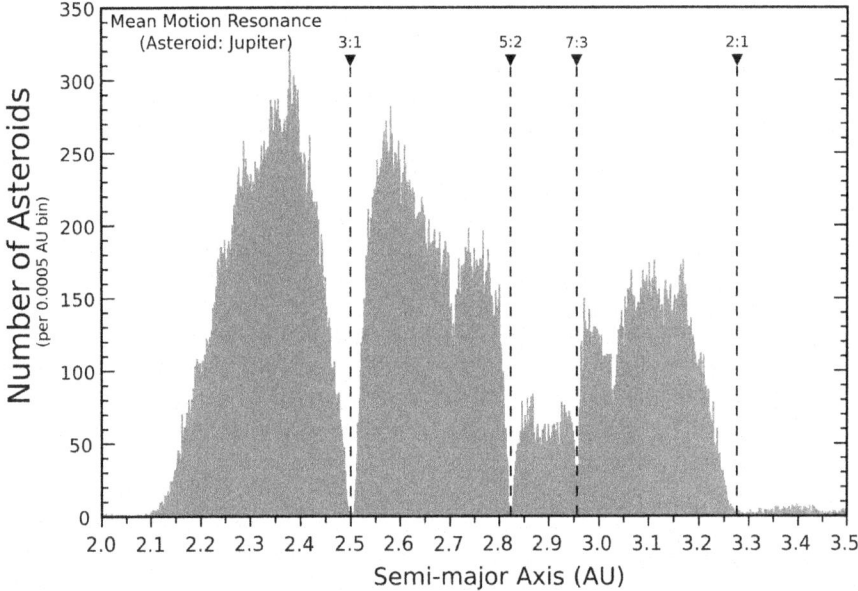

Fig. 6.6 Distribution of semimajor axes for main-belt asteroids. The "Kirkwood gaps" in the distribution coincide with orbital resonances with Jupiter (*dashed lines*) (Credit: NASA/JPL-Caltech)

(c) Is the circular rotating configuration stable? Give a qualitative description of what happens if the velocity is slightly larger or smaller than the "critical" velocity from part (b), or if one of the masses moves slightly inward or outward.

6.3. Consider the restricted three-body problem from Sect. 6.1. Let's show that $\mathbf{F}_{\mathrm{eff}} = 0$ at the Lagrange point L_4. Recall that L_4 makes an equilateral triangle with the two masses.

(a) What is the net gravitational force on a particle of mass m at L_4? Work in Cartesian coordinates, and express your answer in terms of M_1, M_2, and a.
(b) Convert the result from part (a) into polar coordinates centered on the M_1/M_2 center of mass. You should find that the force is radial.
(c) Compute the centrifugal force at L_4 in terms of M_1, M_2, and a. Hint: use Kepler's laws to find ω.
(d) Show that the effective force vanishes at L_4.

6.4. Here is a way to find the locations of closest approach for two orbiting bodies. Consider a planet going around a star in a circular orbit. Its phase angle increases steadily with time, going from $0°$ to $360°$ in one period, then jumping back down to $0°$ and repeating. If the orbital period is 1 yr, the plot looks like this:

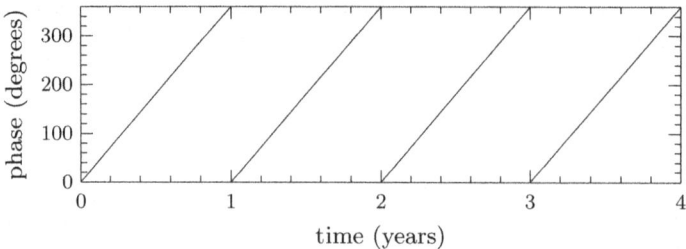

(a) Overplot the phase angle for a planet in a 2:1 resonance with the first. Show that the closest approaches always occur at the same phase (top row of Fig. 6.4).
(b) Now consider a frequency ratio of 2.2:1. Show that the closest approaches do *not* occur at the same phase (bottom row of Fig. 6.4).
(c) What does a 3:1 resonance look like? Show both the phase plot and the closest approach configurations.

6.5. The Cassini division is approximately 118,000 km from the center of Saturn. Below are the orbital radii of some of Saturn's major moons. Which one is responsible for the Cassini division? How do you know? (Hint: you do not explicitly need the orbital periods.)

Mimas	185,000 km
Enceladus	238,000 km
Tethys	295,000 km
Dione	377,000 km
Rhea	527,000 km
Titan	1,222,000 km
Iapetus	3,560,000 km

6.6. How common are resonances between planets in extrasolar planetary systems? Use exoplanet data available online to look for resonances.

References

1. P. Galison, *Einstein's Clocks, Poincaré's Maps: Empires of Time* (W.W. Norton, New York, 2003)
2. J. Laskar, Icarus **196**, 1 (2008)
3. R.W. Farquhar, A.A. Kamel, Celest. Mech. **7**, 458 (1973)
4. N. Cornish, WMAP Education and Outreach, http://wmap.gsfc.nasa.gov/mission/observatory_12.html (1998)
5. M. Connors, P. Wiegert, C. Veillet, Nature **475**, 481 (2011)
6. R. Malhotra, Nature **365**, 819 (1993)
7. D.P. Hamilton, in *AAS/Division of Dynamical Astronomy Meeting #42*, Pasadena, 2011, p. 101 Austin, TX. See: http://dda.aas.org/meetings/2011/
8. P. Goldreich, S.D. Tremaine, Icarus **34**, 240 (1978)

Chapter 7
Extended Mass Distributions: Spiral Galaxies

There is much to say about galaxies,[1] but with our current theme we focus on motion and mass. The stars in a galaxy are always moving, but the sheer number of them means the galaxy's overall mass distribution hardly changes with time.[2] To a good approximation we can take the mass distribution to be static, which makes the gravitational force static and effectively puts us back in the realm of the one-body problem. The difference now is that the mass distribution is spatially extended, which affects the gravitational force and therefore the motion.

7.1 Galaxy Properties

Before we analyze the action, let's set the stage by reviewing the general properties of galaxies. Observed galaxies generally fall into three categories (see Fig. 7.1 for examples):

- **Spiral galaxies** contain stars, gas, and dust that is mostly confined to a thin, rotating disk, although some of the mass may lie in a central bulge. Spiral arms run through the disk, and a straight "bar" may or may not be present in the middle.
- **Elliptical galaxies** contain mostly stars (little gas or dust) in a smooth, feature-less, ellipsoidal distribution of light.
- **Irregular galaxies** include everything that is not spiral or elliptical.

Edwin Hubble introduced an organizational scheme known as the "tuning fork" diagram, which is shown in Fig. 7.2. Elliptical galaxies are placed on the left and classified by their degree of flattening. Spiral galaxies are divided into barred and

---

[1] See the book by Sparke and Gallagher [1] for a more thorough discussion of galaxies.

[2] Unless the galaxy is undergoing some dramatic event such as a collision. We will examine interactions between galaxies in Sect. 8.3.

C. Keeton, *Principles of Astrophysics: Using Gravity and Stellar Physics to Explore the Cosmos*, Undergraduate Lecture Notes in Physics, DOI 10.1007/978-1-4614-9236-8_7, © Springer Science+Business Media New York 2014

Fig. 7.1 (*Left*) Spiral galaxy M101 (Credit: NASA, ESA, K. Kuntz (JHU), F. Bresolin (University of Hawaii), J. Trauger (Jet Propulsion Lab), J. Mould (NOAO), Y.-H. Chu (University of Illinois, Urbana), and STScI). (*Right*) Elliptical galaxy NGC 4458 (Credit: NASA, ESA, and E. Peng (Peking University, Beijing))

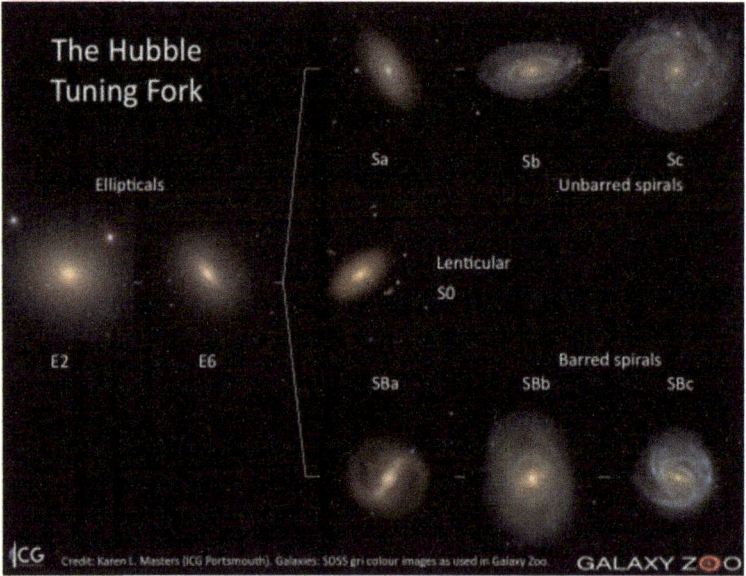

Fig. 7.2 A modern version of Edwin Hubble's "tuning fork" diagram of galaxy types. Elliptical galaxies (*left*) are classified by shape. Spiral galaxies are divided into barred (*bottom*) and unbarred (*top*) families. Lenticular galaxies lie at the transition (Figure created by Karen Masters using astronomical images from the Sloan Digital Sky Survey. Reproduced by permission)

unbarred families and further classified by the size of the bulge relative to the disk and the degree to which spiral arms are wound tightly or loosely. (Some galaxies labeled "lenticular" have intermediate structures with disks but no apparent spiral arms.) For historical reasons, galaxies toward the left are referred to as "early-type" and galaxies toward the right as "late-type," but please be aware that the names are not meant to have any temporal connotations (see [2]).

7.1.1 Luminosity Profiles

When quantifying galaxy properties, the most salient distinction is between **disks** and **spheroids**. Disks are just what you think. Spheroids are roundish (spherical or ellipsoidal) distributions of stars like those found in elliptical galaxies and the bulges of spiral galaxies.

Face-on disks seem to be quite symmetric, apart from the spiral arms. The measured brightness profile is well described by the **exponential disk** model,

$$I(R) = I_0 \, e^{-R/h_R} \tag{7.1}$$

The quantity $I(R)$ is called **surface brightness**; it has dimensions of luminosity per unit area and is often measured in $L_\odot \, pc^{-2}$. Also, I_0 is the surface brightness at the center of the disk, and h_R is the disk **scale length**. If we want to speak about the **surface mass density** (mass per unit area, often in $M_\odot \, pc^{-2}$), we write

$$\Sigma(R) = \Sigma_0 \, e^{-R/h_R} \tag{7.2}$$

We usually assume the light and mass distributions have the same scale length so $I(R)$ and $\Sigma(R)$ are proportional to one another. However, the value of the proportionality constant—called the **mass-to-light ratio**—is not well known. Even though we understand the relation between luminosity and mass for individual stars, at least during the main stage of their lives (see Sect. 16.2), we have limited information about the mix of stars that make up a given galaxy.

The exponential disk model has two notable limitations. First, it explicitly omits spiral arms. While spiral structure stands out in the light distribution, it is less dramatic in the mass distribution. We will ignore spiral arms initially but consider them in Sect. 7.4.4. Second, the model does not account for the thickness of the disk. In edge-on disks, the vertical extent is much smaller than the horizontal size, so we often approximate disks as being infinitesimally thin. We study disk thickness in Sect. 7.4.2.

Spheroids have some depth along the line of sight, but all we can measure is the projected surface brightness distribution. Spheroids typically follow what is called the **de Vaucouleurs $R^{1/4}$ law** after Gérard de Vaucouleurs,

$$I(R) = I_0 \, e^{-7.67(R/R_e)^{1/4}} \tag{7.3}$$

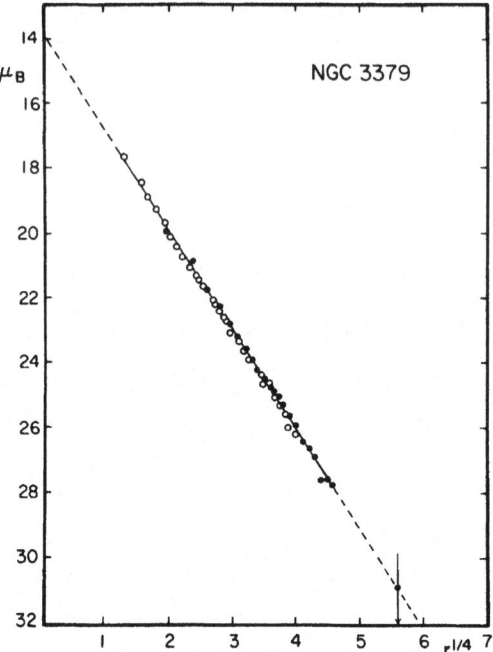

Fig. 7.3 Surface brightness profile of the galaxy NGC 3379. The vertical axis is $\mu_B =$ $-2.5\log_{10} I +$const where I is the surface brightness and B indicates that the measurements were taken through a filter that transmits *blue light*. Because of the minus sign, brighter regions have smaller values of μ_B. For the de Vaucouleurs model we expect $\mu_B = -8.3268\,(R/R_e)^{1/4}+$const, which is shown as the *dashed line* (Credit: de Vaucouleurs and Capaccioli [3]. Reproduced by permission of the AAS)

This is an empirical fit to the data, and it is written with a factor of 7.67 in the exponent so the **effective radius** R_e also winds up being the **half-light radius**, or the radius that encloses half of the light. The de Vaucouleurs profile is shown in Fig. 7.3. As written, Eq. (7.3) describes a galaxy that looks circular, but it is can be generalized to handle galaxies that look elliptical by replacing R with the elliptical radius $(x^2 + y^2/q^2)^{1/2}$ where $q = b/a$ is the axis ratio of the ellipse.

7.1.2 Concepts of Motion

Distinguishing between disks and spheroids also makes sense in terms of motion. In a disk, the stars move on orbits that are nearly circular and coplanar, so the disk rotates coherently. We can plot a **rotation curve** showing orbital speed as a function

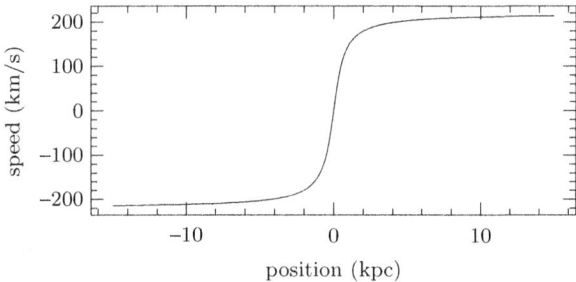

Fig. 7.4 Hypothetical rotation curve for an edge-on spiral galaxy with $v_c = 220 \, \mathrm{km \, s^{-1}}$, based on the model discussed in Sect. 7.3.2. The horizontal axis is position relative to the center of the galaxy. Here the left side of the galaxy is rotating toward us and the right side is rotating away

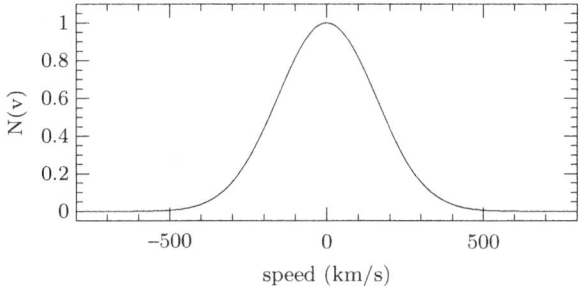

Fig. 7.5 Hypothetical velocity distribution for a galaxy that has a velocity dispersion of $\sigma = 155 \, \mathrm{km \, s^{-1}}$. The vertical axis is the number of stars with a given velocity relative to the center of the galaxy, scaled to a peak of 1

of position in the galaxy (see Fig. 7.4). As we discussed in Sect. 4.2.1, if a disk is inclined by angle i then what we measure with the Doppler effect is $v_{\mathrm{obs}} = v_{\mathrm{int}} \sin i$, where v_{int} is the intrinsic speed. We can estimate the inclination because a circular disk will appear in projection as an ellipse whose axis ratio is $\cos i$. Thus, it is usually feasible to correct for inclination and recover the intrinsic rotation curve of a disk galaxy.

In a spheroid, the star orbits have random orientations, so stars in any small region of the galaxy are moving every which way. Since there is no coherent rotation to measure, we plot the distribution of velocities instead (see Fig. 7.5). The distribution is usually close to Gaussian: if v is the Doppler speed relative to the center of the galaxy, the number of stars as a function of v is approximately $N(v) \propto \exp(-v^2/2\sigma^2)$. The standard deviation of the distribution, σ, is typically referred to the **velocity dispersion** in galaxy dynamics. When analyzing spheroids, we must keep in mind that the measured distribution includes all stars throughout the thickness of the galaxy.

7.2 Equations of Motion

For the rest of this chapter we focus on spiral galaxies (we study elliptical galaxies in Chap. 8). The disk defines a preferred plane, and the main component of motion is tangential in that plane, but there are small components of radial motion within the disk and vertical motion perpendicular to the disk. A flat disk has axial symmetry, but we begin with the case of spherical symmetry to connect with our previous work. As we will see, there is good evidence that galaxies are embedded in "dark matter halos" that are fairly round, so spherical models do have some relevance for spiral galaxies.

7.2.1 Spherical Symmetry

In Sect. 3.1 we studied the equation of motion for a point mass. We now consider a case that retains spherical symmetry but allows an arbitrary radial dependence, so the gravitational potential Φ depends on r but not on θ or ϕ. The analysis follows Sect. 3.1 except that the acceleration is replaced by

$$\mathbf{a} = -\nabla \Phi = -\frac{d\Phi}{dr} \hat{\mathbf{r}}$$

so the three components of the equation of motion are

$$\frac{d^2 r}{dt^2} - r \left(\frac{d\theta}{dt} \right)^2 - r \sin^2 \theta \left(\frac{d\phi}{dt} \right)^2 = -\frac{d\Phi}{dr} \tag{7.4a}$$

$$r \frac{d^2 \theta}{dt^2} + 2 \frac{dr}{dt} \frac{d\theta}{dt} - r \sin \theta \cos \theta \left(\frac{d\phi}{dt} \right)^2 = 0 \tag{7.4b}$$

$$r \sin \theta \frac{d^2 \phi}{dt^2} + 2 \sin \theta \frac{dr}{dt} \frac{d\phi}{dt} + 2r \cos \theta \frac{d\theta}{dt} \frac{d\phi}{dt} = 0 \tag{7.4c}$$

As before the motion is confined to a plane that we can take to be the equatorial plane, and angular momentum is conserved. The radial equation (7.4a) is then

$$\frac{d^2 r}{dt^2} - r \left(\frac{d\phi}{dt} \right)^2 = -\frac{d\Phi}{dr}$$

If $M(r)$ is the mass enclosed within r, the generalization of Eq. (2.13) for the gravitational potential is[3]

$$\Phi(r) = -G \int \frac{M(r)}{r^2} dr$$

---

[3] We can write this as an indefinite integral because Φ is only defined up to an arbitrary constant.

so the radial equation of motion can be written as

$$\frac{d^2 r}{dt^2} - r \left(\frac{d\phi}{dt} \right)^2 = -\frac{GM(r)}{r^2} \tag{7.5}$$

This is the generalization of Eq. (3.7) to an extended, spherical mass distribution.

7.2.2 Axial Symmetry

In cylindrical coordinates (R, ϕ, z), the acceleration vector can be expressed as (see Sect. A.2)

$$\mathbf{a} = \left[\frac{d^2 R}{dt^2} - R \left(\frac{d\phi}{dt} \right)^2 \right] \hat{\mathbf{R}} + \frac{1}{R} \frac{d}{dt} \left(R^2 \frac{d\phi}{dt} \right) \hat{\boldsymbol{\phi}} + \frac{d^2 z}{dt^2} \hat{\mathbf{z}}$$

Suppose the mass distribution and gravitational potential are symmetric about the z-axis, which means they are independent of the azimuthal angle ϕ. This is not strictly true in the presence of spiral arms, but it is a reasonable approximation that captures the key physics. In this model, the gravitational potential can only depend on R and z:

$$\Phi = \Phi(R, z)$$

The three vector components of Newton's second law are

$$\frac{d^2 R}{dt^2} - R \left(\frac{d\phi}{dt} \right)^2 = -\frac{\partial \Phi}{\partial R} \tag{7.6a}$$

$$\frac{1}{R} \frac{d}{dt} \left(R^2 \frac{d\phi}{dt} \right) = 0 \tag{7.6b}$$

$$\frac{d^2 z}{dt^2} = -\frac{\partial \Phi}{\partial z} \tag{7.6c}$$

We will examine each of these equations below.

7.3 Rotational Dynamics

Since the main component of spiral motion is ordered rotation, let's begin our analysis there. Suppose for the time being that all stars move on perfect circular orbits. How does the mass determine the motion, and what can we learn by observing that motion?

7.3.1 Predictions

If the mass distribution is spherically symmetric, we can analyze the motion using
Eq. (7.5). For pure circular motion, the radius is constant so $d^2r/dt^2 = 0$ and we
can solve the equation to find the angular speed

$$\frac{d\phi}{dt} = \left[\frac{GM(r)}{r^3}\right]^{1/2}$$

The corresponding physical speed is

$$v(r) = r\frac{d\phi}{dt} = \left[\frac{GM(r)}{r}\right]^{1/2} \tag{7.7}$$

where we write $v(r)$ to emphasize that speed may vary with radius. (This result can
also be derived by setting the centripetal acceleration for a circular orbit, $a = v^2/r$,
equal to the acceleration due to gravity, $a = GM(r)/r^2$.) Equation (7.7) is useful
if we know the mass distribution and want to compute the corresponding rotation
curve. If instead we measure the rotation curve, we can invert the relation to find the
mass:

$$M(r) = \frac{r\,v(r)^2}{G} \tag{7.8}$$

This is the motion/mass principle applied to rotating spherical objects. Note that
outside an object with a finite extent, $M(r)$ becomes constant and Eq. (7.7) recovers
the Keplerian rotation curve $v \propto r^{-1/2}$.

The analysis of a disk is more involved. In the idealized case of an infinitesimally
thin disk, the density is zero everywhere except in the $z = 0$ plane. The approach
is to solve the Laplace equation $\nabla^2\Phi = 0$ for $z \neq 0$ and then apply appropriate
boundary conditions at $z = 0$. See Sect. 2.6 of *Galactic Dynamics* by Binney and
Tremaine [4] for the complete analysis. We are most interested in motion within the
disk, so we quote the general expression for the gravitational potential in the $z = 0$
plane,

$$\Phi(R,0) = -2\pi G \int_0^\infty dk\, J_0(kR) \int_0^\infty dR'\, R'\, J_0(kR')\, \Sigma(R')$$

where $\Sigma(R')$ is the surface mass density in the disk, and J_0 is a Bessel function.
For an exponential disk with $\Sigma(R') = \Sigma_0 \exp(-R'/h_R)$, the integrals can be
evaluated to yield

$$\Phi(R,0) = -\pi G \Sigma_0 R \left[I_0\left(\frac{R}{2h_R}\right) K_1\left(\frac{R}{2h_R}\right) - I_1\left(\frac{R}{2h_R}\right) K_0\left(\frac{R}{2h_R}\right) \right]$$

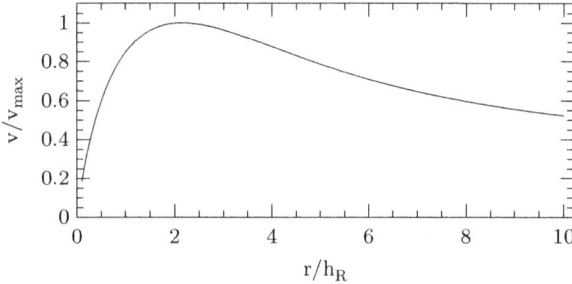

Fig. 7.6 Predicted rotation curve for an exponential disk, plotted in scaled units

where I_0, K_0, I_1, and K_1 are modified Bessel functions. For pure circular motion, the equation of motion (7.6a) lets us compute the circular speed to find

$$v(R)^2 = \pi G \Sigma_0 \frac{R^2}{h_R} \left[I_0 \left(\frac{R}{2h_R} \right) K_0 \left(\frac{R}{2h_R} \right) - I_1 \left(\frac{R}{2h_R} \right) K_1 \left(\frac{R}{2h_R} \right) \right] \quad (7.9)$$

This rotation curve is plotted in Fig. 7.6. The important qualitative features are that the curve peaks at

$$r_{max} = 2.15\,h_R \quad \text{and} \quad v_{max} = 1.56(G\Sigma_0 h_R)^{1/2}$$

and then declines with radius. Since the disk mass is finite, the rotation curve approaches the Keplerian form at large radius.

7.3.2 Observations and Interpretation

Real rotation curves may be more complicated than Fig. 7.6 because disks need not be perfectly exponential, and stellar bulge or gaseous components can also affect the motion. Even so, as a general rule rotation curves should decrease in the outer part of disks if spiral galaxies contain only the stars and gas we see. It therefore came as a surprise in the 1970s when Vera Rubin and others began to discover that observed rotation curves do *not* match predictions. Today we see that some rotation curve fall but not as much as expected, others rise all the way to the largest radii at which they are measured, and many remain approximately constant over a wide range of radii (see Fig. 7.7). The shapes have been seen so many times that the term **flat rotation curves** has entered the lexicon of astronomy.

What is going on? If the observed rotation speed is higher than expected, then the gravitational force must be stronger than expected, so there must be more mass

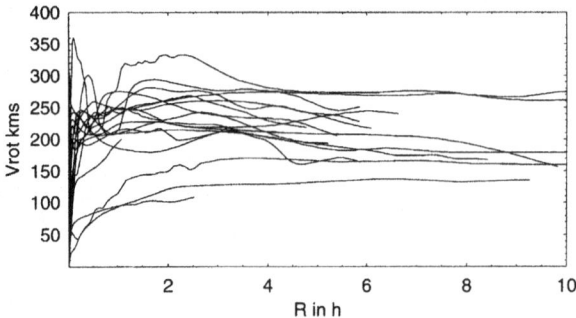

Fig. 7.7 Observed rotation curves for a sample of galaxies. The horizontal axis is plotted in units of the disk scale length (Credit: Sofue et al. [5]. Reproduced by permission of the AAS)

than expected. Whatever this mass is, we seem not to detect any light from it. That, in a nutshell, is the original argument for **dark matter**.[4]

Let's look for the simplest scenario that could give rise to rotation curves similar to what is observed. As a toy model, let's suppose the rotation curve is perfectly flat at all radii: $v(r) = v_c$ where v_c is the constant circular velocity. Let's also suppose the mass distribution is spherical. This is obviously wrong—the stellar distribution is manifestly not spherical—so why should we make the assumption?

- It is simple, and simple can often be good for capturing key ideas without getting bogged down in details.
- We do see some stars (individually and in globular clusters) in a round stellar "halo" around the disk.
- We also see satellite galaxies whose motions imply a roundish halo.
- Gravitational lensing provides evidence for roundish halos (see Chap. 9).
- Galaxy formation models suggest that disks should be embedded in dark matter halos that are fairly round.

With the spherical assumption, Eq. (7.8) gives the enclosed mass as

$$M(r) = \frac{r\,v_c^2}{G}$$

The corresponding density is

$$\rho(r) = \frac{1}{4\pi r^2}\frac{\mathrm{d}M(r)}{\mathrm{d}r} = \frac{v_c^2}{4\pi G}\frac{1}{r^2}$$

---

[4]Evidence for "missing mass" appeared as early as the 1930s, from an analysis of motions in the Coma cluster of galaxies by Fritz Zwicky [6] and an analysis of vertical motions of stars in the Milky Way by Jan Oort [7]. Those analyses were hindered, especially by poor knowledge of mass-to-light ratios, but notice that they too were based on the motion→mass principle.

This model is known as an **isothermal sphere** because any gas in the system will reach an equilibrium with the same ("iso") temperature ("therm") everywhere.[5] While this is admittedly a toy model, it is a very useful one that we will see several times in the next few chapters. The simplicity does raise a few concerns:

- In the model, ρ diverges as $r \to 0$; this is not devastating, but it is inconvenient.
- In the model, v remains constant all the way down to the origin, whereas in real galaxies the rotation speed tends to be small near the center.
- In the inner parts of real galaxies, there is probably more stellar matter than dark matter (more on this in a moment).

One way to address these concerns is to modify the density profile slightly and write

$$\rho(r) = \frac{v_c^2}{4\pi G} \frac{1}{a^2 + r^2} \tag{7.10}$$

where a is referred to as the **core radius**, because when $r \ll a$ the model has a central "core" where the density is approximately constant. When $r \gg a$ the model reduces to $\rho \propto r^{-2}$. This model is referred to as a **softened isothermal sphere**, although the word "softened" is sometimes dropped. Let's derive the rotation curve for a softened isothermal sphere. First, the enclosed mass is

$$
\begin{aligned}
M(r) &= 4\pi \int_0^r (r')^2 \rho(r') \, dr' = \frac{v_c^2}{G} \int_0^r \frac{(r')^2}{a^2 + (r')^2} \, dr' \\
&= \frac{v_c^2 a}{G} \int_0^{r/a} \frac{x^2}{1 + x^2} \, dx = \frac{v_c^2 a}{G} \int_0^{r/a} \left(1 - \frac{1}{1 + x^2}\right) dx \\
&= \frac{v_c^2}{G} \left(r - a \tan^{-1} \frac{r}{a}\right)
\end{aligned}
$$

where we change variables using $x = r'/a$ to make the integral dimensionless. The rotation speed is then

$$v(r) = \left[\frac{GM(r)}{r}\right]^{1/2} = v_c \left(1 - \frac{a}{r} \tan^{-1} \frac{r}{a}\right)^{1/2} \tag{7.11}$$

It is useful to understand the limiting behavior. If $r \gg a$ then a/r approaches zero while $\tan^{-1}(r/a)$ approaches $\pi/2$, so the second term in parentheses vanishes. This means $v(r)$ approaches a constant at large radii, so the rotation curve is asymptotically flat. At small radii $r \ll a$, we can use a Taylor series expansion: $\tan^{-1}(x) \approx x - \frac{x^3}{3} + \frac{x^5}{5} - \ldots$ This gives $v \propto r$ at small radii, which seems to match observed rotation curves.

---

[5] We will study gas in a gravitational potential in Sect. 12.2.

We have been focusing on the physical speed of the stars, but let's briefly consider the angular speed $\omega = v/r$. At small radii, where the rotation curve rises linearly,

$$v(r) = \frac{v_c}{\sqrt{3}} \frac{r}{a} \quad \Rightarrow \quad \omega(r) = \frac{v_c}{\sqrt{3}\,a} = \text{constant}$$

In other words, stars at different radii all take the same amount of time to go around. This is known as **solid body rotation** because it describes the rotation of an object (such as a compact disk) in which all the atoms are connected to one another. By contrast, in the flat part of the rotation curve,

$$v(r) = v_c \quad \Rightarrow \quad \omega(r) = \frac{v_c}{r}$$

which is not constant. This corresponds to **differential rotation**, and it is generic for spiral galaxies in the sense that it occurs even if the rotation curve is not perfectly flat. Differential rotation will be crucial when we study spiral structure in Sect. 7.4.4.

7.3.3 Cold Dark Matter

While the spherical model was instructive, it omitted known parts of the galaxy: the stellar disk and bulge, and perhaps gas as well. If we want to study dark matter in any detail, we need to build models that account for all the components of a galaxy, and in order to do that we need to consider how multiple components combine. By the principle of superposition, densities and masses just add:

$$M_{tot} = M_{disk} + M_{bulge} + M_{gas} + M_{halo}$$

We have seen that expressions for mass involve v^2, so the sum of masses translates into

$$v_{tot}^2 = v_{disk}^2 + v_{bulge}^2 + v_{gas}^2 + v_{halo}^2 \tag{7.12}$$

To quantify the disk, bulge, and gas components, we can take the observed distributions and apply a mass-to-light ratio to obtain model mass distributions. If the mass-to-light ratio is not well known (see Sect. 7.1.1), it can be treated as a free parameter when fitting models to data.

To quantify the dark matter component, people have taken two basic approaches. One is to look for the simplest model that can reproduce the data. The softened isothermal sphere fits the bill. By increasing the core radius, we can reduce the density of dark matter at small radii and let stars dominate the mass there. Then we can adjust the v_c parameter for the halo component to keep the circular velocity high at large radii (where the contributions from stars and gas are falling off). Whether or not this model has a deep physical motivation, it seems to be successful in fitting the data. This is the type of model shown in the left panel of Fig. 7.8.

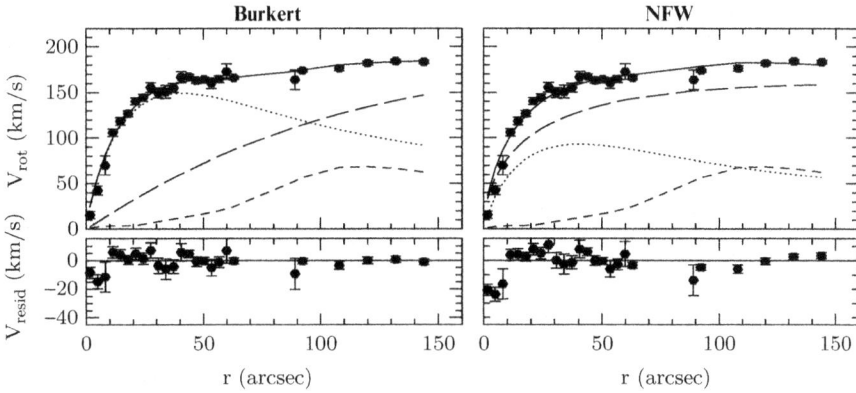

Fig. 7.8 Points show the measured rotation curve for the galaxy ESO 287-G13, while the *solid curve* shows a fit that includes contributions from the stellar disk (*dotted*), the gaseous disk (*short dash*), and the dark matter halo (*long dash*). In the *left panel*, the halo is treated with a Burkert model, which is similar to the isothermal model discussed in the text. In the *right panel*, the halo is treated with an NFW model. The *small bottom panels* show the residuals, or the differences between data and model. Note that 1 kpc corresponds to 5.8″ (Credit: Gentile et al. [8], reproduced by permission of Oxford University Press on behalf of the Royal Astronomical Society)

The second approach is to try to *predict* the properties of dark matter halos and compare those predictions with observations. What do we know about dark matter?

- It must be non-relativistic; otherwise it would move too fast to collect around galaxies.
- As a starting point, we assume that dark matter feels gravity but is not affected by any other forces.
- From studies of "nucleosynthesis" in the early universe (see Sect. 20.2), we know that most of the dark matter cannot be composed of protons, neutrons, and electrons. It must be something exotic—probably some other kind of fundamental particle.

These are the tenets of the **Cold Dark Matter (CDM)**[6] paradigm, which has become the foundation for modern cosmology. In Chap. 11 we will see that this model is remarkably successful at describing the global structure of the universe.[7]

Since the 1980s, people have used computer simulations to study how galaxies form in a universe dominated by cold dark matter. They find that simulated dark matter halos can be described by a density profile of the form

$$\rho = \frac{A}{r(r_s + r)^2} \tag{7.13}$$

---

[6]"Cold" refers to the fact that the particles are slow compared with the speed of light. As we will see in Chap. 12, the temperature of a gas is related to the typical speed of its constituent particles.

[7]With one important modification: dark energy.

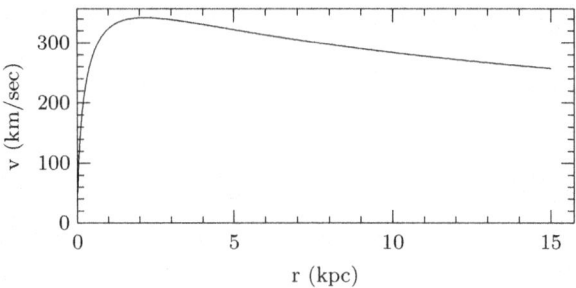

Fig. 7.9 Rotation curve for a spherical NFW model with $A = 10^{10}\,M_\odot$ and $r_s = 1$ kpc

where r_s is a "scale radius" and A is a constant that has dimensions of mass (although it should not be interpreted as the total mass of the halo). This is called the **Navarro-Frenk-White (NFW)** profile after the scientists who first made the prediction [9, 10]. An important feature of the NFW profile is that the density diverges as $\rho \propto r^{-1}$ at small radius. This is referred to as a *cusp*, in contrast with the core in the softened isothermal model.

To test the prediction, we need to compute the NFW rotation curve. First, the enclosed mass is

$$M(r) = 4\pi \int_0^r (r')^2\,\rho(r')\,dr' = 4\pi A \int_0^r \frac{r'}{(r_s + r')^2}\,dr'$$

$$= 4\pi A \int_{r_s}^{r_s+r} \frac{w - r_s}{w^2}\,dw = 4\pi A \int_{r_s}^{r_s+r} \left(\frac{1}{w} - \frac{r_s}{w^2}\right)\,dw$$

$$= 4\pi A \left[\ln\left(1 + \frac{r}{r_s}\right) - \frac{r}{r_s + r}\right]$$

In the third step we change variables using $r' = w - r_s$. The rotation speed is then

$$v(r) = \left[\frac{G\,M(r)}{r}\right]^{1/2} = \left(4\pi G A \left[\frac{1}{r}\ln\left(1 + \frac{r}{r_s}\right) - \frac{1}{r_s + r}\right]\right)^{1/2} \qquad (7.14)$$

This rotation curve is shown in Fig. 7.9. The presence of the central cusp causes the rotation curve to increase more quickly at small radius ($v \propto r^{1/2}$ when $r \ll r_s$) than it does for the isothermal model with a flat core. The dependence $\rho \propto r^{-3}$ for $r \gg r_s$ is steeper than the isothermal model, so the rotation curve declines slowly at large radius.

You might think this would lead to a nice application of the scientific method: we have both a prediction and data to test it. The situation is murky, though.

For ESO 287-G13 (Fig. 7.8), an isothermal model is formally better than an NFW model, although the main differences are at small radii where the motion may be complicated. For other galaxies, NFW fits seem to be favored. The challenge here is dealing with devils that lurk in the details for both observations and interpretation. With the measurements, we must worry about systematic effects such as the placement of the slit used to measure the spectrum, and blurring from the atmosphere. With the analysis, we usually assume the dark matter halo is spherical, the disk is thin, and the orbits are perfectly circular; while those assumptions may seem reasonable, they might not be strictly true, and relaxing them could affect the conclusions. The last point to recall is that there are uncertainties in the rotation curve data themselves, the inclination, and the mass-to-light ratio. All together, these effects can permit a range of successful models, and it is difficult to say for certain whether rotation curves "prefer" cusps or cores.

To bypass some of the details, we could just ask how much dark matter is found in the central regions of galaxies. There *seems* to be less dark matter than CDM models would predict. However, it is not clear if that represents a fundamental problem with the CDM paradigm. It may just indicate that there are aspects of galaxy formation— which is complicated, after all—that are not fully understood. For our purposes, what is important is to follow the physical reasoning that astronomers use to find evidence for dark matter in galaxies and deduce its abundance and distribution.

7.3.4 Is Dark Matter Real?

Throughout the preceding analysis we relied on Newton's laws of gravity and motion to connect rotation curves with the underlying mass distribution. When we saw a discrepancy, we imagined that we have the right laws of physics but the wrong ideas about how mass is distributed. That approach seems reasonable because Newton's laws (and Einstein's generalizations of them; see Chap. 10) have been well tested. However, most of the tests have taken place on Earth and in the Solar System, where the accelerations are much larger than the accelerations of stars in galaxies:

Situation	Acceleration ($m\,s^{-2}$)
Surface of Earth	9.8
Moon orbiting Earth	0.003
Earth orbiting Sun	0.006
Sun orbiting Galaxy	2×10^{-10}

In the 1980s, Mordehai Milgrom asked: What if Newton's laws break down at low accelerations? After all, we already know they fail at high speeds (for relativity) or short distances (for quantum mechanics). Milgrom proposed to modify Newton's second law when the acceleration is smaller than some value a_0:

$$
F = \begin{cases} m\,a & (a \gg a_0) \\[2mm] m\,\dfrac{a^2}{a_0} & (a \ll a_0) \end{cases} \tag{7.15}
$$

This idea is known as **Modified Newtonian Dynamics (MOND)** because what changes is not the force of gravity but rather a particle's response to the force. In a series of papers, Milgrom argued that applying MOND below $a_0 \sim 10^{-10}\,\mathrm{m\,s^{-2}}$ could explain galaxy rotation curves as well as an observed correlation between the rotation speeds and luminosities of spiral galaxies [11–13]. You can explore these ideas in Problem 7.5.

Another possibility is to modify Newton's law of gravity so the force is something other than $F = GMm/r^2$. We know the usual force law works very well on scales ranging from labs on Earth to the Solar System, so the idea would be to have a force law that is equivalent to $F = GMm/r^2$ at "small" radii but has a different form at radii larger than some value r_0.

Most astronomers prefer the idea of dark matter to that of modified dynamics or gravity. While MOND can successfully explain galaxy dynamics, it faces more trouble with galaxy clusters (most famously, the Bullet Cluster; see Sect. 9.4) and the universe as a whole. Even MOND requires some amount of dark matter to explain these systems. Supporters of MOND suggest the additional mass could be provided by massive neutrinos, but it remains to be seen whether this hypothesis works out in detail (e.g., [14–16]). In my view, strong evidence supports the conventional theory of dark matter. Still, there is value in exploring alternatives because scientific disputes like dark matter versus MOND are ultimately settled by developing different models and testing them with observations.

7.4 Beyond Rotation

To this point we have focused on tangential motion, which is the predominant form of motion in spiral galaxy disks. Stars can, however, have small components of motion in the radial and vertical directions. We can analyze the additional motion using Eq. (7.6).

7.4.1 Tangential Motion

The tangential component of the equation of motion is

$$
\frac{1}{R}\frac{d}{dt}\left(R^2\frac{d\phi}{dt}\right) = 0
$$

This is conservation of angular momentum—but only of the component of angular momentum that corresponds to motion around the z-axis,

$$\ell_z = R^2 \frac{d\phi}{dt} = R v_\phi \qquad (7.16)$$

This conservation law follows from axisymmetry (similar to the way in which conservation of the full angular momentum vector follows from spherical symmetry in Sect. 2.2).

7.4.2 Vertical Motion

The vertical component of the equation of motion is

$$\frac{d^2 z}{dt^2} = -\frac{\partial \Phi}{\partial z} \qquad (7.17)$$

We cannot solve this equation in general without knowing the gravitational potential $\Phi(R, z)$. However, we can learn a lot if we consider *small* motions. Since disks are thin, the stars never get very far from the midplane, so we might consider z to be small and make a Taylor series expansion of the potential:

$$\Phi(R, z) \approx \Phi_0(R) + \left.\frac{\partial \Phi}{\partial z}\right|_0 z + \frac{1}{2} \left.\frac{\partial^2 \Phi}{\partial z^2}\right|_0 z^2 + \ldots \qquad (7.18)$$

If we take the "middle" of the disk (indicated by the subscript 0) to be the place where $\partial \Phi / \partial z = 0$, the second term vanishes and Eqs. (7.17) and (7.18) combine to give

$$\frac{d^2 z}{dt^2} = -\nu^2 z \quad \text{where} \quad \nu^2 \equiv \left.\frac{\partial^2 \Phi}{\partial z^2}\right|_0 \qquad (7.19)$$

This is an equation for simple harmonic motion (The angular frequency ν may depend on R but it is independent of z.) Physically, any star above the disk will be pulled down. The star will pass through the disk, come out the other side, and then be pulled back up. The star will keep going back and forth, oscillating in the vertical direction with a period of $P = 2\pi / \nu$.

Our Sun is presently about 25 pc out of the midplane of the Milky Way and moving away at a speed of about 7 km s^{-1} [17, 18].

Example: Uniform Disk

Consider a simple model in which the disk density is uniform. (This can be viewed as an approximation that is valid in a small region of a more realistic disk model.) Let's explicitly derive the equation of motion and check that the preceding analysis makes sense. We start with the following formal analysis:

$$\mathbf{a} = -\nabla\Phi$$

$$\nabla \cdot \mathbf{a} = -\nabla^2\Phi = -4\pi G\rho$$

$$\int (\nabla \cdot \mathbf{a})\, \mathrm{d}V = -4\pi G \int \rho\, \mathrm{d}V$$

$$\oint \mathbf{a} \cdot \mathrm{d}\mathbf{A} = -4\pi G M_{\mathrm{in}} \tag{7.20}$$

In the second line, we use the Poisson equation, $\nabla^2\Phi = 4\pi G\rho$. From the third to the fourth line, we use Gauss's divergence theorem to rewrite the volume integral on the left as a surface integral. (You may have seen a similar analysis in electromagnetism.) The fourth line tells us the surface integral of the acceleration vector is given by the mass enclosed by that surface.

Let's take the surface of integration to be a small box that extends from $-z$ to $+z$ and has cross sectional area S (the shape of S is arbitrary). For vertical acceleration, the integral on the left-hand side of Eq. (7.20) has aS for the top and another aS for the bottom, giving a total of $2aS$. The mass inside the box is the density, ρ, times the volume, $2zS$. Putting the pieces together, we have

$$2aS = -4\pi G \times 2zS\rho \quad \Rightarrow \quad a = -4\pi G\rho z$$

This is the vertical acceleration at height z in a uniform density disk. Since the vertical acceleration is $a = \mathrm{d}^2z/\mathrm{d}t^2$, the key equation is

$$\frac{\mathrm{d}^2z}{\mathrm{d}t^2} = -4\pi G\rho z$$

This is the equation for a simple harmonic oscillator, as expected. The vertical oscillation frequency is $\nu = (4\pi G\rho)^{1/2}$.

Application: Disk Thickness

Real spiral galaxy disks have finite thicknesses. Empirically, the vertical distribution is often characterized as an exponential function,

$$\rho(z) \propto e^{-|z|/h_z}$$

where h_z is the **scale height**. At any given time, some stars are moving up and others down, so there is a distribution of vertical velocities and it is natural to characterize the distribution using the vertical velocity dispersion σ_z. When quantified in this way, the Milky Way seems to have two disk components (see [17, 19] and references therein). The "thin" disk has $h_z \approx 300\,\mathrm{pc}$ and $\sigma_z \approx 18\,\mathrm{km\,s^{-1}}$, and it tends to contain younger stars. The "thick" disk has $h_z \approx 900\,\mathrm{pc}$ and $\sigma_z \approx 35\,\mathrm{km\,s^{-1}}$, and it tends to contain older stars. The disk scale radius is $h_R \approx 3.5\,\mathrm{kpc}$ so even the "thick" disk is still thin in comparison with its radial extent. Other spiral galaxies show similar structures [20].

Disk thickness can be created by a variety of mechanisms. When a star encounters an object such as another star, a gas cloud, or a spiral arm, the gravitational interaction can give the star a "kick" in the vertical direction. Also, if a star migrates out from the center of the galaxy, any vertical motion can be amplified. Finally, an external event such as a small galaxy falling into the Milky Way can generate vertical motion. There is a lot of interest in using the vertical structure of galaxy disks to understand the processes that have driven their evolution over billions of years (see [21] and references therein).

7.4.3 Radial Motion

Finally, we come to the radial component of the equation of motion (7.6a). Using Eq. (7.16) to rewrite $d\phi/dt$ in terms of the constant ℓ_z, we obtain

$$\frac{d^2 R}{dt^2} = -\frac{\partial \Phi}{\partial R} + \frac{\ell_z^2}{R^3} = -\frac{\partial \Phi_{\mathrm{eff}}}{\partial R} \tag{7.21}$$

In the last step we introduce the **effective potential**

$$\Phi_{\mathrm{eff}}(R) \equiv \Phi(R) + \frac{\ell_z^2}{2R^2} \tag{7.22}$$

As with vertical motion, we cannot solve the equation of motion in general without knowing the potential, but we can make progress by considering small deviations from a constant radius. A circular orbit has $d^2 R/dt^2 = 0$, so by Eq. (7.21) the derivative $\partial \Phi_{\mathrm{eff}}/\partial R$ must be zero at the radius of the circular orbit. Let's call this radius R_0, and then write the radius more generally as

$$R = R_0 + \Delta R$$

where we expect ΔR to be small. Then we can make a Taylor series expansion:

$$\Phi_{\mathrm{eff}}(R) \approx \Phi_{\mathrm{eff}}(R_0) + \left.\frac{\partial \Phi_{\mathrm{eff}}}{\partial R}\right|_0 \Delta R + \frac{1}{2}\left.\frac{\partial^2 \Phi_{\mathrm{eff}}}{\partial R^2}\right|_0 \Delta R^2 + \dots \tag{7.23}$$

The first term disappears when we take the derivative. The second term vanishes because we just noted that $\partial \Phi_{\text{eff}}/\partial R = 0$ at R_0. Thus the only meaningful term is the third, and using it in Eq. (7.21) yields

$$\frac{d^2(\Delta R)}{dt^2} = -\kappa^2 \, \Delta R \tag{7.24}$$

This is again an equation for simple harmonic motion with angular frequency

$$\kappa^2 \equiv \left. \frac{\partial^2 \Phi_{\text{eff}}}{\partial R^2} \right|_0 \tag{7.25}$$

Stars can oscillate in and out (in addition to up and down), all while orbiting the center of the galaxy. The radial oscillations can actually be viewed as a small circle superimposed on the main circular orbit—in other words, as an epicycle. The idea originally introduced by ancient Greeks to explain the retrograde motion of planets (see Sect. 2.1) has reemerged, albeit in a different form! Because of this connection, κ is called the **epicycle frequency**.

Example: Point Mass

Consider motion around a point mass. Obviously this is not a good model for a galaxy, but it serves as an instructive example. A point mass has spherical symmetry, but we can think of that as a type of axial symmetry as well. The gravitational potential is

$$\Phi(r) = -\frac{GM}{r}$$

so the effective potential is

$$\Phi_{\text{eff}}(r) = -\frac{GM}{r} + \frac{\ell^2}{2r^2}$$

and the epicycle frequency is

$$\kappa = \left(\frac{\partial^2 \Phi_{\text{eff}}}{\partial r^2} \right)^{1/2} = \left[\frac{\partial}{\partial r} \left(\frac{GM}{r^2} - \frac{\ell^2}{r^3} \right) \right]^{1/2} = \left(-\frac{2GM}{r^3} + \frac{3\ell^2}{r^4} \right)^{1/2}$$

We know a circular orbit at radius r has velocity $v_\phi = (GM/r)^{1/2}$ and hence $\ell = r v_\phi = (GMr)^{1/2}$. Plugging this in yields

$$\kappa = \left(\frac{GM}{r^3} \right)^{1/2} = \omega$$

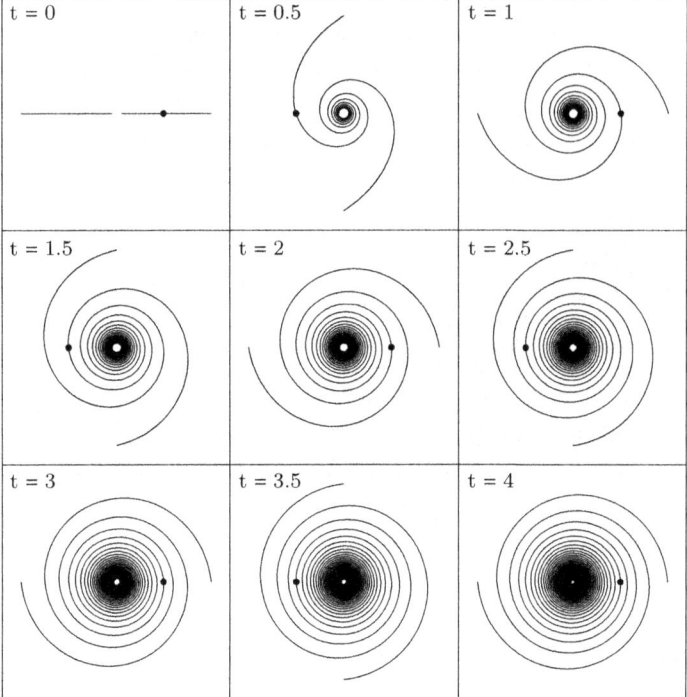

Fig. 7.10 Illustration of the winding problem. The *dot* denotes the Sun, and the time is in units of the time it takes the Sun to orbit the Milky Way (which is addressed in Problem 7.1)

where $\omega = v_\phi/r$ is the angular frequency. In other words, around a point mass the epicycle frequency exactly equals the angular frequency. That, in turn, allows the orbit to be perfectly closed (the object returns exactly to its starting point). We already knew this from our analysis of Kepler's laws and the one-body problem, but now we see it in a different context.

7.4.4   Application to Spiral Arms

We finally have the tools to examine the defining feature of spiral galaxies, namely spiral arms. We noted in Sect. 7.3.2 that spiral galaxy disks have differential rotation, meaning the orbital period varies with radius. As a result, if we paint a stripe on a galaxy, it will wrap up and look like a spiral, as shown in Fig. 7.10. This is good, right? Well, yes and no.

The "yes" applies to certain kinds of spiral galaxies called **flocculent spirals**, like NGC 4414 shown in Fig. 7.11. Flocculent means fluffy; this term is applied to galaxies with little wisps of spiral structure, rather than grand spiral arms. Here the

Fig. 7.11 Hubble Space Telescope image of the flocculent spiral galaxy NGC 4414 (Credit: NASA and The Hubble Heritage Team (STScI/AURA))

idea is that if you have a little cloud of gas that forms some stars, differential rotation can stretch the cloud out into a wispy structure like what is seen in these galaxies.

The "no" applies to **grand design spirals**, or galaxies where the spiral arms run through the whole disk, like the one shown in Fig. 7.1. The problem is that differential rotation causes spirals to wind up way too fast to survive for billions of years. This is known as the **winding problem**, and it means the simplest imaginable explanation of spiral arms cannot be correct.

How can we proceed? Imagine creating an arrangement of stars labeled by $j = 1, \ldots, N$. From Sect. 7.4.3 we know each star will execute radial oscillations given by

$$R_j(t) = a + b \cos(\kappa t + \alpha_j)$$

where a is the radius of the reference circle, b is the amplitude of the radial oscillations, κ is the epicycle frequency, and α_j is the initial phase of the oscillations. While the star is doing this, it is also moving around the galaxy with angle

$$\phi_j(t) = \phi_{j0} + \omega t$$

where ω is the angular speed and ϕ_{j0} is the starting angle for star j. The star's x and y positions as a function of time are then

$$x_j(t) = R_j(t) \cos \phi_j(t) \qquad\qquad y_j(t) = R_j(t) \sin \phi_j(t)$$

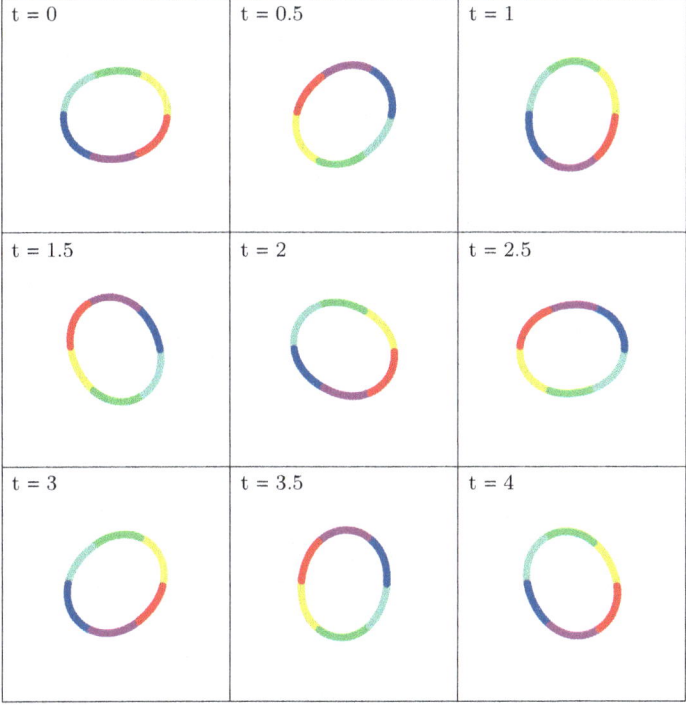

Fig. 7.12 Evolution of a collection of stars following epicyclic orbits. The *pattern* rotates at a different speed than the stars themselves. The *color coding* illustrates that the stars move *through* the pattern

To get stars evenly distributed around the galaxy, we set

$$\phi_{j0} = \frac{2\pi j}{N}$$

To get a nice oval-shaped pattern of stars, let's relate α_j to ϕ_{j0} by setting

$$\alpha_j = 2\phi_{j0}$$

When we do all this, the arrangement at $t = 0$ looks like the first panel in Fig. 7.12. This example has $a = 1$ and $b = 0.1$.

What happens at later times? Each star follows its epicyclic orbit, oscillating in radius as it orbits the galaxy. But the *pattern* appears to rotate more slowly, as shown in the remaining panels of Fig. 7.12. It is crucial to understand that *the motion of the pattern is different from the motion of the individual stars.*

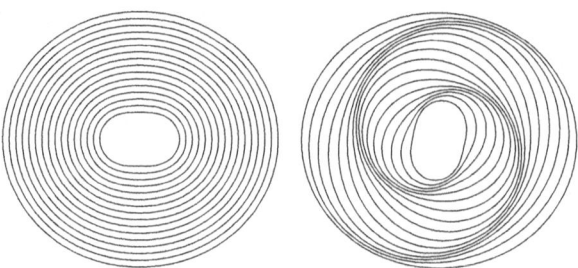

Fig. 7.13 How to set up nested ovals to create a spiral pattern (see [22])

How can we think about the pattern motion? The position of star j at time t is given by

$$R_j(t) = a + b\cos(\kappa t + 2\phi_{j0}) \tag{7.26a}$$

$$\phi_j(t) = \phi_{j0} + \omega t \tag{7.26b}$$

The star that is farthest from the center of the galaxy (greatest R) is the one that has

$$\kappa t + 2\phi_{j0} = 0 \quad \Rightarrow \quad \phi_{j0} = -\frac{\kappa t}{2}$$

Plugging this into Eq. (7.26b), we find that the angular position of this farthest star is

$$\phi_j(t) = \left(\omega - \frac{\kappa}{2}\right)t$$

In other words, the long axis of the oval pattern rotates with an angular frequency given by the **pattern speed**

$$\Omega_p = \omega - \frac{\kappa}{2} \tag{7.27}$$

The example in Fig. 7.12 has $\omega = 6.28$ and $\kappa = 10.05$, yielding $\Omega_p = 1.26$. The time it takes for the pattern to rotate once is $P = 2\pi/\Omega_p = 5.0$. Thus, the *pattern* of stars rotates five times more slowly than any *individual* star. The color coding in the figure is designed to show this. Notice, for example, that at $t = 0.5$ each star has moved halfway around the galaxy, but the oval pattern has rotated by only 36°.

How does this help with spiral structure? At $t = 0$ we can set up nested ovals to create a spiral pattern, as shown in Fig. 7.13. When we let this evolve, as shown in Fig. 7.14, the spiral winds up much less quickly than before. Working with *patterns* that can occur thanks to epicyclic orbits, we can mitigate the winding problem.

There is still more that can be said about the dynamics of spirals. To this point we have imagined that the stars move in a smooth, constant gravitational field, but in

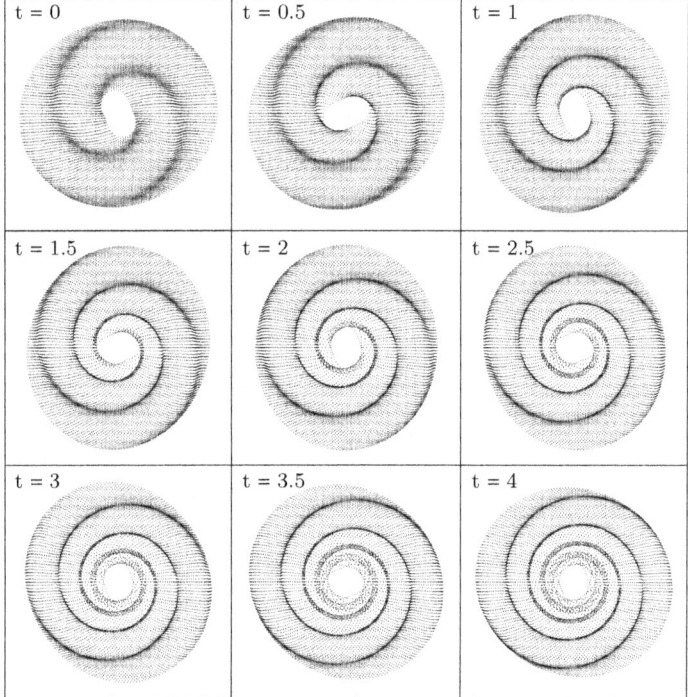

Fig. 7.14 If stars move on epicyclic orbits, they can form spiral patterns that rotate more slowly than the stars themselves, which helps mitigate the winding problem

fact they feel an additional force from the local overdensity of matter in a spiral arm. Lin and Shu [23,24] developed this notion into a hypothesis that spiral arms are self-sustaining **density waves** propagating through galaxy disks. The physical picture is often likened to a traffic jam, with stars as cars that catch the jam from behind, slow down as they move through it, and then escape out the front and keep going. The Lin-Shu hypothesis has stimulated extensive work on the theory of density waves in disks (see [24]), but whether it truly describes real spirals is still unclear. In general, the hypothesis can explain why there is more star formation in spiral arms than elsewhere in the disk: the buildup and compression of gas in the arms can kick-start the formation process (see Chap. 19). In detail, though, the predictions may not be consistent with new observations of the spatial distribution of features associated with star formation [25]. Regardless of how the story turns out, the key point for us is that spiral arms are *patterns* (rather than fixed groups of stars) whose behavior seem to be connected to the epicyclic motion we have studied here.

Problems

7.1. Consider the exponential disk model in Eq. (7.1).

(a) Show that the total brightness of the exponential disk is $I_{\text{total}} = 2\pi I_0 h_R^2$. Hint: change variables to $x = R/h_R$, and use integration by parts.
(b) What fraction of the total light is contained within one disk scale length ($R \leq h_R$)? Within three disk scale lengths ($R \leq 3h_R$)?

7.2. The Milky Way has a rotation curve that is approximately flat with a circular speed of about $220 \, \text{km s}^{-1}$. The Sun is about 8 kpc from the center of our Galaxy. In this problem you may assume the mass distribution of the Milky Way is spherical.

(a) How much mass is enclosed by the orbit of the Sun (in $M_\odot$)?
(b) Assuming an appropriate mass distribution, what is the density of mass in the vicinity of the Sun (in $M_\odot \, \text{pc}^{-3}$)?
(c) How long does it take the Sun to make one orbit of our Galaxy?

7.3. Recall the rotation curve data and models for the galaxy ESO 287-G13 shown in Fig. 7.8. Answer the following questions for both types of models.

(a) What is the mass of dark matter within $50''$?
(b) What fraction of the total mass within $50''$ is dark matter? (Here you may assume all the mass is in a spherical distribution.)

7.4. Here is the rotation curve for the galaxy UGC 5166, along with a model that includes a spheroidal bulge component (data from [26], figure courtesy Kristine Spekkens).

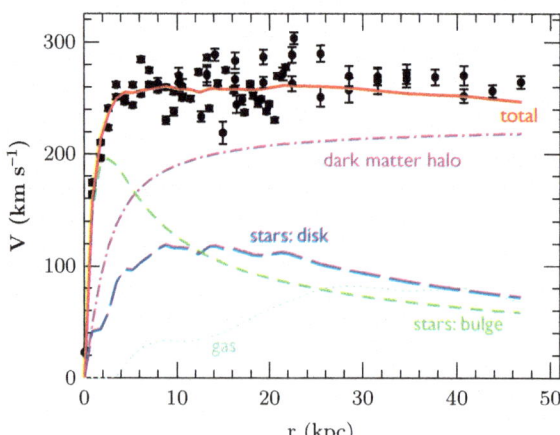

(a) What is the total mass of the bulge? Hint: check several values of the radius to make sure you've gotten all the enclosed bulge mass.

(b) A popular model for the mass distribution in a spheroid is the Hernquist model,

$$\rho(r) = \frac{M_{total}\, a}{2\pi r(a+r)^3}$$

where M_{total} is the total mass and a is a constant with dimensions of length. Derive the rotation curve for this model.

(c) Fit the Hernquist model to the bulge component in UGC 5166 and plot your derived rotation curve (with r in kpc and v in $km\,s^{-1}$). Hint: use the total bulge mass from part (a), and use trial and error to find a reasonable value for a.

7.5. In Modified Newtonian Dynamics, Newton's second law is replaced by $F = ma^2/a_0$ for accelerations smaller than some value a_0 (see Eq. 7.15).

(a) In this scheme, what is the rotation curve around a point mass?
(b) Assuming that all spiral galaxies have the same ratio of mass to light, what is the scaling relation between circular velocity and luminosity in MOND?

7.6. The vertical motion of stars in a spiral galaxy depends on the gravity exerted by the disk, so it allows us to "weigh" the disk.

(a) Use dimensional analysis to derive an estimate of the mass density ρ of a spiral galaxy disk, in terms of its scale height h_z, its vertical velocity dispersion σ_z, and a relevant physical constant.
(b) Use the disk parameters given in Sect. 7.4.2 to estimate the mass density of the Milky Way's disk, in $M_\odot\,pc^{-3}$. Do the thin and thick disks give a consistent results to the level of precision we might expect from dimensional analysis?

7.7. Recall from Sect. 7.4.2 that the vertical of motion for a uniform density disk corresponds to simple harmonic motion with angular frequency $\nu = (4\pi G\rho)^{1/2}$. The motion can therefore be written as

$$z(t) = A\sin(\nu t) + B\cos(\nu t)$$

where A and B are constants.[8]

(a) The mass density near the Sun is about $\rho = 0.1\, M_\odot\,pc^{-3}$ [28]. The Sun is about 25 pc above the midplane of the disk and rising at $7\,km\,s^{-1}$. Find the constants A and B, then plot the Sun's vertical motion. Label the axes and be quantitative.
(b) Some people have suggested that the Sun's motion through spiral arms and/or the Galactic disk may affect climate and even mass extinctions on Earth [29,30]. (Passing through higher-density regions increases the chance of encounters with other stars or gas clouds that could send comets toward the inner Solar System.) When did the Sun last cross the midplane of the disk?

7.8. How does the epicycle frequency compare with the angular frequency for an isothermal sphere? The gravitational potential is $\Phi(r) = v_c^2 \ln(r) + $ constant. Using the isothermal model for the Milky Way, what is the epicycle period for the Sun?

---

[8]This question is inspired in part by a problem in the book by Carroll and Ostlie [27].

References

1. L.S. Sparke, J.S. Gallagher, III, *Galaxies in the Universe: An Introduction* (Cambridge University Press, Cambridge, 2007)
2. I.K. Baldry, Astron. Geophys. **49**(5), 050000 (2008)
3. G. de Vaucouleurs, M. Capaccioli, Astrophys. J. Suppl. Ser. **40**, 699 (1979)
4. J. Binney, S. Tremaine, *Galactic Dynamics: Second Edition* (Princeton University Press, Princeton, 2008)
5. Y. Sofue, Y. Tutui, M. Honma, A. Tomita, T. Takamiya, J. Koda, Y. Takeda, Astrophys. J. **523**, 136 (1999)
6. F. Zwicky, Astrophys. J. **86**, 217 (1937)
7. J.H. Oort, Bull. Astron. Inst. Neth. **6**, 249 (1932)
8. G. Gentile, P. Salucci, U. Klein, D. Vergani, P. Kalberla, Mon. Not. R. Astron. Soc. **351**, 903 (2004)
9. J.F. Navarro, C.S. Frenk, S.D.M. White, Astrophys. J. **462**, 563 (1996)
10. J.F. Navarro, C.S. Frenk, S.D.M. White, Astrophys. J. **490**, 493 (1997)
11. M. Milgrom, Astrophys. J. **270**, 365 (1983)
12. M. Milgrom, Astrophys. J. **270**, 371 (1983)
13. M. Milgrom, Astrophys. J. **270**, 384 (1983)
14. R.H. Sanders, Mon. Not. R. Astron. Soc. **342**, 901 (2003)
15. R.H. Sanders, Mon. Not. R. Astron. Soc. **380**, 331 (2007)
16. P. Natarajan, H. Zhao, Mon. Not. R. Astron. Soc. **389**, 250 (2008)
17. M. Jurić et al., Astrophys. J. **673**, 864 (2008)
18. R. Schönrich, J. Binney, W. Dehnen, Mon. Not. R. Astron. Soc. **403**, 1829 (2010)
19. Y. Karataş, R.J. Klement, New Astron. **17**, 22 (2012)
20. P. Yoachim, J.J. Dalcanton, Astron. J. **131**, 226 (2006)
21. P.C. van der Kruit, K.C. Freeman, Annu. Rev. Astron. Astrophys. **49**, 301 (2011)
22. A.J. Kalnajs, Proc. Astron. Soc. Aust. **2**, 174 (1973)
23. C.C. Lin, F.H. Shu, Astrophys. J. **140**, 646 (1964)
24. C.C. Lin, F.H. Shu, Proc. Natl. Acad. Sci. **55**, 229 (1966)
25. K. Foyle, H.W. Rix, C.L. Dobbs, A.K. Leroy, F. Walter, Astrophys. J. **735**, 101 (2011)
26. K. Spekkens, R. Giovanelli, Astronom. J. **132**, 1426 (2006)
27. B.W. Carroll, D.A. Ostlie, *An Introduction to Modern Astrophysics*, 2nd edn. (Addison-Wesley, San Francisco, 2007)
28. J. Holmberg, C. Flynn, Mon. Not. R. Astron. Soc. **313**, 209 (2000)
29. M.R. Rampino, Celest. Mech. Dyn. Astron. **69**, 49 (1997)
30. M.D. Filipović, J. Horner, E.J. Crawford, N.F.H. Tothill, G.L. White, Serbian Astronom. J. (2013)

Chapter 8
N-Body Problem: Elliptical Galaxies

In Chap. 7 we treated galaxies as if each star orbits in a fixed gravitational field. In fact, the gravity comes from other stars and particles of dark matter, which themselves are moving. How can we analyze a collection of particles that move under the influence of each other's gravity? While we can write down the exact equations of motion, we should not expect to solve them by hand. Where we can make progress is understanding *statistical* properties of the motion. The statistical approach is applicable to elliptical galaxies and a range of other systems.

8.1 Gravitational N-Body Problem

We begin by developing a formal framework for analyzing the collective motion of N particles interacting via gravity.

8.1.1 Equations of Motion

The force on any one object in the system is a vector sum of contributions from all the other objects. When we write the force vectors, it is useful to avoid unit vectors by writing $\hat{\mathbf{r}} = \mathbf{r}/|\mathbf{r}|$. In the one-body problem, the force can then be written as

$$\mathbf{F} = -\frac{GMm}{|\mathbf{r}|^2}\hat{\mathbf{r}} = -\frac{GMm}{|\mathbf{r}|^3}\mathbf{r}$$

In the N-body problem, the force on star α from star β is

$$\mathbf{f}_{\alpha\beta} = -\frac{Gm_\alpha m_\beta}{|\mathbf{r}_\alpha - \mathbf{r}_\beta|^3}(\mathbf{r}_\alpha - \mathbf{r}_\beta)$$

C. Keeton, *Principles of Astrophysics: Using Gravity and Stellar Physics to Explore the Cosmos*, Undergraduate Lecture Notes in Physics, DOI 10.1007/978-1-4614-9236-8_8, © Springer Science+Business Media New York 2014

Notice that if we reverse the indices we have

$$\mathbf{f}_{\alpha\beta} = -\mathbf{f}_{\beta\alpha} \tag{8.1}$$

which is Newton's third law of motion. The total force on star α is the sum of forces from all the other stars,

$$\mathbf{F}_{\alpha} = \sum_{\beta;\ \beta\neq\alpha} \mathbf{f}_{\alpha\beta} \tag{8.2}$$

The sum notation means that β runs over all values from 1 to N, except for the case when β matches α. Thus, the equation of motion for star α is

$$\frac{d^2\mathbf{r}_{\alpha}}{dt^2} = -\sum_{\beta;\ \beta\neq\alpha} \frac{Gm_{\beta}}{|\mathbf{r}_{\alpha}-\mathbf{r}_{\beta}|^3} (\mathbf{r}_{\alpha}-\mathbf{r}_{\beta}) \tag{8.3}$$

With α running from 1 to N, this constitutes the complete set of equations of motion for the system. It is a system of N coupled second-order differential equations.

8.1.2 Conservation of Energy

As we have seen before, analyzing energy can be a good way to understand a system. The kinetic and potential energies for the N-body problem are

$$K = \frac{1}{2} \sum_{\alpha} m_{\alpha}|\mathbf{v}_{\alpha}|^2 \quad \text{and} \quad U = -\frac{1}{2} \sum_{\alpha,\beta;\ \alpha\neq\beta} \frac{Gm_{\alpha}m_{\beta}}{|\mathbf{r}_{\alpha}-\mathbf{r}_{\beta}|} \tag{8.4}$$

In the expression for potential energy, the sum runs over all values of α and β except when they match each other. The factor of $1/2$ enters because the sum in this form counts each pair of stars twice. When we compute a sum like this, we can exchange the indices α and β everywhere and still get the same result. In other words for the potential energy we can write

$$U = -\frac{1}{2} \sum_{\alpha,\beta;\ \alpha\neq\beta} \frac{Gm_{\alpha}m_{\beta}}{|\mathbf{r}_{\alpha}-\mathbf{r}_{\beta}|} = -\frac{1}{2} \sum_{\alpha,\beta;\ \alpha\neq\beta} \frac{Gm_{\beta}m_{\alpha}}{|\mathbf{r}_{\beta}-\mathbf{r}_{\alpha}|}$$

We will use this index switching trick a few times below.

In Sect. 2.2 we saw conservation of energy as a general principle, but now we ask whether and how it holds in a complicated N-body system. To test energy conservation, we clearly need to compute the time derivative of the total energy. Evaluating dK/dt and dU/dt from Eq. (8.4) gives:

$$\frac{dE}{dt} = \sum_\alpha m_\alpha \mathbf{v}_\alpha \cdot \frac{d\mathbf{v}_\alpha}{dt} + \frac{1}{2} \sum_{\alpha,\beta;\ \alpha\neq\beta} \frac{Gm_\alpha m_\beta}{|\mathbf{r}_\alpha - \mathbf{r}_\beta|^3} (\mathbf{r}_\alpha - \mathbf{r}_\beta) \cdot (\mathbf{v}_\alpha - \mathbf{v}_\beta)$$

$$= \sum_\alpha \mathbf{F}_\alpha \cdot \mathbf{v}_\alpha - \frac{1}{2} \sum_{\alpha,\beta;\ \alpha\neq\beta} \mathbf{f}_{\alpha\beta} \cdot (\mathbf{v}_\alpha - \mathbf{v}_\beta) \tag{8.5}$$

Let's see what we can do with the second term:

$$\frac{1}{2} \sum_{\alpha,\beta;\ \alpha\neq\beta} \mathbf{f}_{\alpha\beta} \cdot (\mathbf{v}_\alpha - \mathbf{v}_\beta) = \frac{1}{2} \sum_{\alpha,\beta;\ \alpha\neq\beta} \mathbf{f}_{\alpha\beta} \cdot \mathbf{v}_\alpha - \frac{1}{2} \sum_{\alpha,\beta;\ \alpha\neq\beta} \mathbf{f}_{\alpha\beta} \cdot \mathbf{v}_\beta$$

$$= \frac{1}{2} \sum_{\alpha,\beta;\ \alpha\neq\beta} \mathbf{f}_{\alpha\beta} \cdot \mathbf{v}_\alpha - \frac{1}{2} \sum_{\alpha,\beta;\ \alpha\neq\beta} \mathbf{f}_{\beta\alpha} \cdot \mathbf{v}_\alpha$$

$$= \frac{1}{2} \sum_{\alpha,\beta;\ \alpha\neq\beta} \mathbf{f}_{\alpha\beta} \cdot \mathbf{v}_\alpha + \frac{1}{2} \sum_{\alpha,\beta;\ \alpha\neq\beta} \mathbf{f}_{\alpha\beta} \cdot \mathbf{v}_\alpha$$

$$= \sum_{\alpha,\beta;\ \alpha\neq\beta} \mathbf{f}_{\alpha\beta} \cdot \mathbf{v}_\alpha$$

In the second step we exchange the indices α and β in the second term. In the third step we use Eq. (8.1). In the last step, we notice that the two terms are identical and combine them.

Now going back to Eq. (8.5), we can write

$$\frac{dE}{dt} = \sum_\alpha \mathbf{F}_\alpha \cdot \mathbf{v}_\alpha - \sum_{\alpha,\beta;\ \alpha\neq\beta} \mathbf{f}_{\alpha\beta} \cdot \mathbf{v}_\alpha$$

$$= \sum_\alpha \left(\mathbf{F}_\alpha - \sum_{\beta;\ \alpha\neq\beta} \mathbf{f}_{\alpha\beta} \right) \cdot \mathbf{v}_\alpha$$

The quantity in parentheses vanishes by Eq. (8.2), leaving

$$\frac{dE}{dt} = 0$$

Thus energy is in fact conserved. While the final result is not surprising, the analysis itself is enlightening as we learn to handle complex systems. Seeing the analysis also helps us remember that what is conserved is the *total* energy of all stars. Stars can exchange energy, both between kinetic and potential and among different stars, so conservation of energy does not apply to individual objects. But it does apply to the system as a whole.

8.1.3 Virial Theorem

Let's now examine the quantity

$$Q = \sum_\alpha m_\alpha \, \mathbf{r}_\alpha \cdot \mathbf{v}_\alpha$$

which we call the **virial**.[1] Its time derivative is

$$\frac{dQ}{dt} = \sum_\alpha m_\alpha \, \mathbf{v}_\alpha \cdot \mathbf{v}_\alpha + \sum_\alpha m_\alpha \, \mathbf{r}_\alpha \cdot \frac{d\mathbf{v}_\alpha}{dt} = 2K + \sum_\alpha \mathbf{F}_\alpha \cdot \mathbf{r}_\alpha \qquad (8.6)$$

What can we do with the second term?

$$\sum_\alpha \mathbf{F}_\alpha \cdot \mathbf{r}_\alpha = - \sum_{\alpha,\beta;\ \alpha \neq \beta} \frac{G m_\alpha m_\beta}{|\mathbf{r}_\alpha - \mathbf{r}_\beta|^3} (\mathbf{r}_\alpha - \mathbf{r}_\beta) \cdot \mathbf{r}_\alpha$$

$$= -\frac{1}{2} \sum_{\alpha,\beta;\ \alpha \neq \beta} \frac{G m_\alpha m_\beta}{|\mathbf{r}_\alpha - \mathbf{r}_\beta|^3} (\mathbf{r}_\alpha - \mathbf{r}_\beta) \cdot \mathbf{r}_\alpha$$

$$\quad - \frac{1}{2} \sum_{\alpha,\beta;\ \alpha \neq \beta} \frac{G m_\beta m_\alpha}{|\mathbf{r}_\beta - \mathbf{r}_\alpha|^3} (\mathbf{r}_\beta - \mathbf{r}_\alpha) \cdot \mathbf{r}_\beta$$

$$= -\frac{1}{2} \sum_{\alpha,\beta;\ \alpha \neq \beta} \frac{G m_\alpha m_\beta}{|\mathbf{r}_\alpha - \mathbf{r}_\beta|^3} (\mathbf{r}_\alpha - \mathbf{r}_\beta) \cdot (\mathbf{r}_\alpha - \mathbf{r}_\beta)$$

$$= -\frac{1}{2} \sum_{\alpha,\beta;\ \alpha \neq \beta} \frac{G m_\alpha m_\beta}{|\mathbf{r}_\alpha - \mathbf{r}_\beta|}$$

$$= U \qquad (8.7)$$

In the second step we split the sum into two identical terms and then exchange the indices α and β in the second term. In the third and fourth lines we combine the two terms and simplify. In the last step we realize that what we have is the total gravitational potential energy. So when we go back to Eq. 8.6 we have

$$2K + U = \frac{dQ}{dt} \qquad (8.8)$$

Now consider averaging over time τ. If $f(t)$ is some function of time, we define the time average to be

$$\langle f \rangle \equiv \frac{1}{\tau} \int_0^\tau f(t) \, dt \qquad (8.9)$$

---

[1] You can think of the quantity Q as the time derivative of something like a moment of inertia (see Sect. A.4). If we put $I = \sum_\alpha \frac{1}{2} m_\alpha |\mathbf{r}_\alpha|^2$ then $Q = dI/dt$.

When we take the time average of Eq. (8.8) we obtain:

$$2\langle K\rangle + \langle U\rangle = \frac{1}{\tau}\int_0^\tau \frac{dQ}{dt}\,dt = \frac{1}{\tau}\Big[Q(\tau) - Q(0)\Big] = 0 \qquad \text{for } \tau \to \infty$$

At the last step we assume that Q remains finite, so the whole quantity vanishes if we average over a long enough time interval (because of the τ in the denominator). We do not actually have to know anything about Q except that it remains finite, as it should for a well-behaved system.

The bottom line from this analysis is:

$$2\langle K\rangle + \langle U\rangle = 0 \tag{8.10}$$

This is known as the **virial theorem**. It looks a little bit like conservation of energy, but it is quite different. Conservation of energy is instantaneous, whereas the virial theorem describes the *average* properties of a system. Also, the virial theorem has that funny factor of two.

One way in which the virial theorem is like conservation of energy is that it is *exact* (at least for time-averaged quantities). It is not a result of dimensional analysis, estimation, or approximation. While we will sometimes employ it for estimation, the virial theorem is really much deeper.

8.1.4 A Simple Application: N = 2

The derivation above holds for any $N \geq 2$, so the virial theorem ought to apply to the familiar two-body problem. For simplicity, first consider a two-body problem with a large central mass M, and a small mass $m \ll M$ in a circular orbit with radius R; this is effectively a one-body problem. We have seen that the orbital speed is $v = \sqrt{GM/R}$, so the kinetic and potential energies are

$$K = \frac{1}{2}mv^2 = \frac{GMm}{2R} \quad \text{and} \quad U = -\frac{GMm}{R}$$

Because the orbit is circular, these quantities are constant in time, so $\langle K\rangle = K$ and $\langle U\rangle = U$. Clearly the virial theorem is satisfied: $2\langle K\rangle + \langle U\rangle = 0$.

Now let's relax the conditions and allow for elliptical orbits with two bodies of arbitrary masses, m_1 and m_2. We showed in Chap. 4 that this problem can be expressed as an equivalent one-body problem with a reduced mass μ orbiting a fixed total mass M. With elliptical orbits, the speed and separation, and hence the kinetic and potential energies, vary with time but we can compute their average as follows. From Sect. 4.1.3 we know the potential energy is

$$U = -\frac{GM\mu}{r}$$

where the orbit follows the ellipse

$$r(\phi) = \frac{\ell^2}{GM} \frac{1}{1 + e\cos\phi} \tag{8.11}$$

To compute the average, we just need to integrate over one full orbital period; after that, averaging over more orbits will not change the results. Therefore we can take $\tau = P$ in Eq. (8.9):

$$
\begin{aligned}
\langle U \rangle &= -\frac{1}{P}\int_0^P \frac{GM\mu}{r(\phi(t))}\,dt \\
&= -\frac{GM\mu}{P}\int_0^{2\pi} \frac{1}{r(\phi)}\frac{dt}{d\phi}\,d\phi \\
&= -\frac{GM\mu}{P\ell}\int_0^{2\pi} r(\phi)\,d\phi \\
&= -\frac{\mu\ell}{P}\int_0^{2\pi} \frac{d\phi}{1 + e\cos\phi} \\
&= -\frac{GM\mu(1-e^2)^{1/2}}{2\pi a} \times \frac{2\pi}{(1-e^2)^{1/2}} \\
&= -\frac{GM\mu}{a}
\end{aligned}
$$

In the second step we change integration variables from t to ϕ, and in the third step we use $d\phi/dt = \ell/r^2$. In the fourth step we use Eq. (8.11), and finally we carry out the integration[2] and also use our expressions for ℓ and P from Eqs. (3.11) and (3.14).

For the kinetic energy, we can use the components of the velocity vector from Sect. 4.1.4 to write

$$K = \frac{\mu}{2}\left(v_r^2 + v_\phi^2\right) = \frac{GM\mu}{2a(1-e^2)}\left(1 + e^2 + 2e\cos\phi\right)$$

Then the average is

$$\langle K \rangle = \frac{GM\mu}{2a(1-e^2)P}\int_0^{2\pi}\left(1 + e^2 + 2e\cos\phi\right)\frac{dt}{d\phi}\,d\phi$$

---

[2]With help from Mathematica [1].

$$= \frac{GM\mu(1 - e^2)^{1/2}}{4\pi a} \int_0^{2\pi} \frac{1 + e^2 + 2e\cos\phi}{(1 + e\cos\phi)^2} \, d\phi$$

$$= \frac{GM\mu(1 - e^2)^{1/2}}{4\pi a} \times \frac{2\pi}{(1 - e^2)^{1/2}}$$

$$= \frac{GM\mu}{2a}$$

In the first step we change integration variables as before, in the second step we again use $d\phi/dt = \ell/r^2$ and substitute for ℓ and P, and in the third step we carry out the integration. The bottom line is:

$$\langle K \rangle = \frac{GM\mu}{2a} \quad \text{and} \quad \langle U \rangle = -\frac{GM\mu}{a} \quad \Rightarrow \quad 2\langle K \rangle + \langle U \rangle = 0$$

The virial theorem is satisfied for the general two-body problem (as it must be).

8.2 Elliptical Galaxies

The $N = 2$ example was a case in which we already knew the complete motion. The power of the virial theorem becomes more apparent when $N \gg 2$ and a complete, exact description of the motion is not available. Let's see how it leads to a motion → mass principle for elliptical galaxies. We will study a spherical system with radius R and total mass M, but similar concepts apply to ellipsoidal systems (with some geometric complications that are not essential here).

8.2.1 Potential Energy

When we average over time, the individual stars blur out and we get what looks like a smooth mass distribution. If the system is in equilibrium, we can compute the potential energy for the corresponding smooth case and take that to be $\langle U \rangle$.

Let's imagine building the distribution of stars by assembling spherical shells like the layers of an onion. Suppose we have a sphere of radius r and enclosed mass $M(r)$, and we add to it a shell of thickness dr and density $\rho(r)$. The mass of this new layer is

$$dm = 4\pi r^2 \rho(r) \, dr$$

If we bring the shell in from infinity, its gravitational potential energy is

$$dU(r) = -\frac{GM(r) \, dm}{r} = -4\pi \, GM(r) \, \rho(r) \, r \, dr$$

The total potential energy is found by summing all the shells,

$$U = -4\pi G \int_0^R M(r)\,\rho(r)\,r\,dr \qquad (8.12)$$

For an infinite distribution, we extend the integral to $R \to \infty$ as long as the density falls off faster than $\rho \propto r^{-2}$ (otherwise the integral diverges). The potential energy depends on how the mass is distributed, so let's examine two simple examples.

Example: Uniform Density Sphere with Radius R

Here the density is

$$\rho(r) = \rho \quad \text{(constant)}$$

so the enclosed mass is

$$M(r) = \frac{4}{3}\pi\rho r^3 \quad \Rightarrow \quad \text{total mass } M = \frac{4}{3}\pi\rho R^3$$

The potential energy integral is then:

$$U = -4\pi G \int_0^R \frac{4}{3}\pi\rho r^3\,\rho\,r\,dr$$

$$= -\frac{16}{3}\pi^2 G\rho^2 \times \frac{1}{5}R^5$$

$$= -\frac{16}{15}\pi^2 G R^5 \left(\frac{3M}{4\pi R^3}\right)^2$$

$$= -\frac{3}{5}\frac{GM^2}{R} \qquad (8.13)$$

Example: Finite Isothermal Sphere with Radius R

Here the density is

$$\rho(r) = \frac{v^2}{4\pi G r^2} \qquad (r < R)$$

so the enclosed mass is

$$M(r) = \frac{r v^2}{G} \quad \Rightarrow \quad \text{total mass } M = \frac{R v^2}{G}$$

The potential energy integral is then:

$$
\begin{aligned}
U &= -4\pi G \int_0^R \frac{rv^2}{G} \frac{v^2}{4\pi Gr^2} \, r \, dr \\
&= -\frac{v^4 R}{G} \\
&= -\left(\frac{GM}{R}\right)^2 \frac{R}{G} \\
&= -\frac{GM^2}{R}
\end{aligned}
\tag{8.14}
$$

From these two examples, and from dimensional analysis, we deduce that the potential energy for a sphere of mass M and radius R has the general form

$$
\langle U \rangle = -\eta \frac{GM^2}{R}
\tag{8.15}
$$

where η is a dimensionless factor of order unity that depends on the density profile. We have seen that $\eta = 1$ for an isothermal sphere and $\eta = 3/5$ for a uniform sphere; other distributions can lead to other values for η.

8.2.2 Kinetic Energy

In a reference frame centered on the galaxy, the total kinetic energy is

$$
\langle K \rangle = \frac{1}{2} \sum_\alpha m_\alpha \left\langle v_{\alpha x}^2 + v_{\alpha y}^2 + v_{\alpha z}^2 \right\rangle
\tag{8.16}
$$

What we can measure with the Doppler effect is the dispersion in the component of velocity along the line of sight (which we take to be the z-axis). The dispersion among all stars is

$$
\sigma^2 = \frac{1}{N} \sum_\alpha v_{\alpha z}^2
\tag{8.17}
$$

Stars may not contribute equally to our measurements, though. Brighter stars will contribute more of the light we receive, and fainter stars less. Therefore it may be better to think of a luminosity-weighted dispersion,

$$
\sigma^2 = \frac{\sum_\alpha L_\alpha v_{\alpha z}^2}{\sum_\alpha L_\alpha}
\tag{8.18}
$$

where L_α is the luminosity of star α. If the stars are identical, this reduces to Eq. (8.17). As defined, σ is instantaneous. We do not actually see σ change (on human time scales) because we are summing over so many stars, so we take σ^2 and $\langle \sigma^2 \rangle$ to be equivalent.

In order to relate $\langle K \rangle$ to what we can measure, we need to make two assumptions. First, we need to assume something about how the components of velocity in the x and y directions relate to what we measure in the z direction. The simplest possibility is that the orbits are arranged so the motion is *isotropic* and $\langle v_{\alpha x}^2 \rangle = \langle v_{\alpha y}^2 \rangle = \langle v_{\alpha z}^2 \rangle$. Second, we need to assume something about the distribution of star masses and luminosities. The simplest possibility is that the stars are identical. If we assume identical stars undergoing isotropic motion, we can combine Eqs. (8.16) and (8.17) to write

$$\langle K \rangle \;=\; \frac{3}{2} m \sum_\alpha \langle v_{\alpha z}^2 \rangle \;=\; \frac{3}{2} m N \sigma^2 \;=\; \frac{3}{2} M \sigma^2$$

Relaxing the two assumptions, we write the general case as

$$\langle K \rangle = \frac{3}{2} \beta M \sigma^2 \tag{8.19}$$

where β is a dimensionless factor that depends on the arrangement of orbits and population of stars.

8.2.3 Mass Estimate

Plugging Eqs. (8.15) and (8.19) into the virial theorem yields

$$3\beta M \sigma^2 - \eta \frac{GM^2}{R} = 0$$

Rearranging, we obtain the **virial mass estimate**

$$M = \frac{3\beta}{\eta} \frac{R\sigma^2}{G} \tag{8.20}$$

The factor $R\sigma^2/G$ is what we would derive from dimensional analysis. Now we identify the dimensionless factors (β and η) that encode the internal properties of the system. For an isothermal sphere composed of identical stars with isotropic motion, $\beta = \eta = 1$. For other scenarios, β and η can take on different values.

Equation (8.20) gives the motion/mass connection for elliptical galaxies. Once we measure the size (R) and motion (σ), we can use the formula to estimate the mass. When astronomers discuss elliptical galaxies, they usually quote the velocity dispersion because it is directly measurable, and—as we now know—it is a good indicator of the mass.

Here are some typical numbers for elliptical galaxies:

$$\sigma = 200\,\mathrm{km\,s^{-1}}$$

$$R = 10\,\mathrm{kpc}$$

$$M = 3 \times \frac{(10 \times 3.09 \times 10^{19}\,\mathrm{m}) \times (200 \times 10^3\,\mathrm{m\,s^{-1}})^2}{6.67 \times 10^{-11}\,\mathrm{m^3\,kg^{-1}\,s^{-2}}}$$

$$= 6 \times 10^{41}\,\mathrm{kg}$$

$$= 3 \times 10^{11}\,M_\odot$$

(This mass estimate is derived with $\beta = \eta = 1$.)

8.3   Galaxy Interactions

Now let's shift attention from a single, isolated galaxy to a pair of galaxies that interact with one another. This is a natural step for two reasons. First, elliptical galaxies are thought to be formed by collisions between spiral galaxies. Second, studying galaxy interactions illustrates how the virial theorem can teach us something interesting and perhaps unexpected.

8.3.1   Fly-By

First consider an interaction in which two galaxies fly past one another but do not collide.[3] The encounter has two effects: the gravitational attraction gives each galaxy a global impulse in the direction of the other galaxy, and the tidal force squeezes each galaxy. What happens to the *internal* properties of each galaxy?

Suppose the encounter is fast enough that the stars within a given galaxy do not move very much during the event. Then the potential energy is the same just before and just after the encounter. The kinetic energy changes, though, because each star gets a little velocity "kick." Part of this is collective motion (the whole galaxy moves), but part of it is internal motion (from tidal squeezing). The key implication is that the encounter increases the internal kinetic energy.[4]

---

[3]This analysis follows Binney and Tremaine [2].

[4]Note that the internal energy of each galaxy is not conserved during the encounter. The total energy of the system is conserved, though; the "new" internal energy comes at the expense of the translational kinetic energy of the two galaxies moving past each other.

Before the encounter, a galaxy is in equilibrium with

$$\text{state \#1} \qquad \text{kinetic energy } \langle K \rangle = K_i$$
$$\text{potential energy } \langle U \rangle = U_i$$
$$\text{equilibrium } \Rightarrow 2K_i + U_i = 0 \qquad (8.21)$$

Immediately after the encounter, the galaxy has

$$\text{state \#2} \qquad \text{kinetic energy } K_i + \Delta K$$
$$\text{potential energy } U_i$$

The system no longer satisfies the virial theorem, so it is not in equilibrium. It must redistribute the energy to achieve a new equilibrium. It will settle into a final state with

$$\text{state \#3} \qquad \text{kinetic energy } \langle K \rangle = K_f$$
$$\text{potential energy } \langle U \rangle = U_f$$
$$\text{equilibrium } \Rightarrow 2K_f + U_f = 0 \qquad (8.22)$$

We can combine the virial theorem with conservation of energy[5] to obtain two equations that allow us to find the final state:

$$\text{virial theorem} : 2K_f + U_f = 0$$
$$\text{conservation of energy} : K_f + U_f = K_i + \Delta K + U_i$$

Subtract these two equations:

$$K_f = -K_i - \Delta K - U_i$$
$$= -K_i - \Delta K - (-2K_i)$$
$$= K_i - \Delta K$$

In the second step we use Eq. (8.21) to replace $U_i = -2K_i$. Once we know K_f we can use Eq. (8.22) to find U_f:

$$U_f = U_i + 2\Delta K$$

---

[5]Once the encounter is complete, energy must again be conserved. Thus, energy is conserved between states #2 and #3 even if it is not conserved between states #1 and #2.

Long after the encounter, the galaxy settles into a state with *lower* kinetic energy and *higher* potential energy. Since $U = -\eta GM/R^2$, the system must increase R and/or redistribute mass (thereby changing η) in order to increase the potential energy. Either way, we realize that the galaxy has been "puffed up" by the encounter. This is a conclusion that is interesting and not at all obvious, and it comes from very general reasoning. Such is the power of the virial theorem.

8.3.2 Collision

What happens when two galaxies meet each other head-on? Galaxies are mostly empty space, so they pass right through one another; it is very unlikely that one star will actually hit another (see Problem 8.4). But stars can pass close enough to change each other's motions. The stellar orbits are dramatically disrupted, creating a final state that can be very different from either of the initial galaxies.

There are so many stars, and the interactions are so complicated, that it is difficult to make much progress analytically. Theoretical studies of galaxy collisions therefore rely on numerical simulations [3,4]. Computers are good at the large-scale computations needed to track the motions of stars throughout a collision. The benefit of the numerical approach is that we can make movies and see all the stages of the merger event. The drawback is that the details depend on many factors, including the galaxies' masses, velocities, rotation rates, and orientations. Each simulation takes a lot of computer time, so the idea is to do a plausible range of examples and then try to extract some general conclusions. For example, one common aspect is the formation of long, coherent structures called "tidal tails" that last for a few hundred million years before settling back into the final galaxy.

Observational studies investigate systems that appear to be collisions in progress (see Fig. 8.1). While we cannot follow a single event from start to finish, we can find observed systems that look like various stages of the simulated mergers, and vice versa. By identifying such matches, we can validate the simulations and also make educated guesses about what happened in the past and what will occur in the future for each system. The general lessons are that galaxies can collide, the collision process drastically changes the motions of stars in the galaxies, and the end product is a system full of stars with seemingly random motions that resembles conventional elliptical galaxies (e.g., [5]).

Fig. 8.1 Hubble Space Telescope images of 12 ongoing galaxy collisions. The *top* and *bottom* panels in the *third column* show particularly clear examples of tidal tails (Credit: NASA, ESA, the Hubble Heritage (STScI/AURA)-ESA/Hubble Collaboration, and A. Evans)

8.4 Other N-Body Problems

Gravitational N-body problems also occur in many other astrophysical settings, ranging from star clusters within galaxies, to the formation and evolution of galaxies themselves, to clusters and superclusters of galaxies, and finally to the large-scale distribution of matter in the universe. Computer simulations, coupled with analytic tools like the virial theorem, allow us to understand astrophysical processes over a huge range of scales. You can explore a few examples in the problems below.

Problems

8.1. You should be able to answer these questions using the virial theorem, without doing any detailed calculations.

(a) For a system in virial equilibrium, is the total energy positive, negative, or zero?
(b) Suppose a system is in equilibrium. In order for the system to shrink in size, how must the total energy change?

8.2. The Plummer model for a spherical star cluster has density

$$\rho(r) = \frac{3M}{4\pi} \frac{a^2}{(r^2 + a^2)^{5/2}}$$

where M is the total mass, and a is a core radius.

(a) Compute the enclosed mass profile $M(r)$. Hint: by changing variables, you can express the integral in a form that can be evaluated using Sect. A.7.
(b) Now compute the total potential energy in terms of G, M, and a. Hint: again change variables and use Sect. A.7.
(c) If the mass distribution is in equilibrium, what is the total kinetic energy? What is the total energy? Give your answers in terms of G, M, and a.
(d) The globular cluster ω Centauri can be described by a Plummer model with a total mass $M = 5 \times 10^6 \, M_\odot$ and core radius $a = 4.5 \, \text{pc}$. Assuming identical stars in isotropic orbits, find the cluster's radial velocity dispersion σ in km s^{-1}.

8.3. Some time in the future, the Milky Way and Andromeda galaxies will collide. While we need computer simulations to study the process in detail, we can get an idea of what the end product will be like.

(a) For an isothermal sphere with a radius R and circular velocity v, express the potential and total energies in terms of M and v.
(b) Consider two identical finite isothermal spheres, each with initial mass M_i and initial circular velocity v_i, that are at rest a distance d apart. What is the total energy of this system? Hint: consider the total energy for each object in isolation, and then the potential energy between the two.
(c) Suppose the two spheres fall toward each other and merge, and after some time they equilibrate into a single isothermal sphere. Use conservation of mass and energy along with the virial theorem to derive the following quantities, and explain your results in words:

- The final mass M_f (in terms of the initial mass M_i)
- The final circular velocity v_f (in terms of v_i, R_i, and d)
- The final radius R_f (in terms of R_i and d)

(d) Apply your results to a system like the Milky Way and Andromeda, approximating the galaxies as isothermal spheres with circular velocities of $250 \, \text{km s}^{-1}$ and radii of $150 \, \text{kpc}$, which fall from rest at an initial separation of $780 \, \text{kpc}$. What are the mass (in $M_\odot$), radius (in kpc), and circular velocity (in km s^{-1}) of the final galaxy? (This is not a perfect model of the Milky Way/Andromeda system, because the two galaxies are not identical and they are already heading toward each other. Nevertheless, it gives a reasonable idea of how things will go.)

8.4. I mentioned that when two galaxies come together they do not physically hit each other, but gravitational interactions change the star motions. Let's make some estimates to understand what would happen if an interloper passed through the Milky Way's disk in the vicinity of the Sun.[6] The mass density in stars near the Sun is about $0.05\,M_\odot\,\mathrm{pc}^{-3}$ [7]. For simple estimates you may assume the disk has a uniform density and is 1 kpc thick, and all stars are like the Sun.

(a) As seen from above, what fraction of the area of the disk is covered by stars? This can be interpreted as the probability that a star passing through the disk would hit a disk star.

(b) Even if the interloper does not hit a disk star, it may pass close enough to change the disk star's motion. Let's suppose this happens if the gravity from the interloper ever exceeds the gravity from the galaxy. Show that this occurs if the interloper comes within a distance

$$d \sim \frac{(G m R_{\mathrm{gal}})^{1/2}}{v_c}$$

of the disk star, where m is the mass of the disk star, R_{gal} is the distance of the disk star from the center of the galaxy, and v_c is its circular rotation speed.

(c) Now estimate the probability that the interloper perturbs a disk star while it passes through the disk.

The key result here is that bona fide collisions between stars are rare, but interactions that change stars' motions are common.

8.5. The Coma cluster of galaxies has a velocity dispersion of about $1{,}000\,\mathrm{km\,s}^{-1}$ and a radius of about 3 Mpc. Estimate its total mass assuming an isothermal sphere with isotropic orbits of identical galaxies. Would your estimate increase, decrease, or stay the same if you used a constant density model?

References

1. Wolfram Research, Inc., *Mathematica*, 8th edn. (Wolfram Research, Champaign, 2010)
2. J. Binney, S. Tremaine, *Galactic Dynamics*, 2nd edn. (Princeton University Press, Princeton, 2008)
3. J.E. Barnes, L. Hernquist, Annu. Rev. Astron. Astrophys. **30**, 705 (1992)
4. J.E. Barnes, J.E. Hibbard, Astronom. J. **137**, 3071 (2009)
5. R.S. Remus, A. Burkert, K. Dolag, P.H. Johansson, T. Naab, L. Oser, J. Thomas, Astrophys. J. **766**, 71 (2013)
6. B.W. Carroll, D.A. Ostlie, *An Introduction to Modern Astrophysics*, 2nd edn. (Addison-Wesley, San Francisco, 2007)
7. J. Holmberg, C. Flynn, Mon. Not. R. Astron. Soc. **313**, 209 (2000)

[6]This problem is an extension of Problem 26.1 in the book by Carroll and Ostlie [6].

Chapter 9
Bending of Light by Gravity

To this point we have examined how massive objects move under the influence of gravity. Einstein taught us that light's motion is affected by gravity as well. Despite being relativistic, gravitational light bending can be studied with a quasi-Newtonian framework to obtain a new way to probe mass in the universe.

9.1 Principles of Gravitational Lensing

The gravitational deflection of light can be treated as a variant of the Newtonian one-body problem. A full relativistic analysis gives a deflection angle that is twice as large (see Sect. 10.6.5), but for most astrophysical purposes we can insert the factor of 2 by hand and proceed in the Newtonian framework. In this section we identify observable effect of light bending including distortion, magnification, and multiple imaging.

9.1.1 Gravitational Deflection

Consider a particle of mass m passing near a massive body $M \gg m$. The particle's trajectory is curved, but asymptotically (i.e., far from M) it is a straight line. We can quantify the bending in terms of the angle $\hat{\alpha}$ between the asymptotic segments, as shown in Fig. 9.1. To compute $\hat{\alpha}$, strictly speaking we need to solve a differential equation characterizing the motion.[1] If the bending is small, however, we can obtain a good approximation much more simply, by computing the change in velocity perpendicular to the original motion.

---

[1] This is related to the analysis in Sect. 3.1, but now applied to an unbound orbit.

C. Keeton, *Principles of Astrophysics: Using Gravity and Stellar Physics to Explore the Cosmos*, Undergraduate Lecture Notes in Physics, DOI 10.1007/978-1-4614-9236-8_9,
© Springer Science+Business Media New York 2014

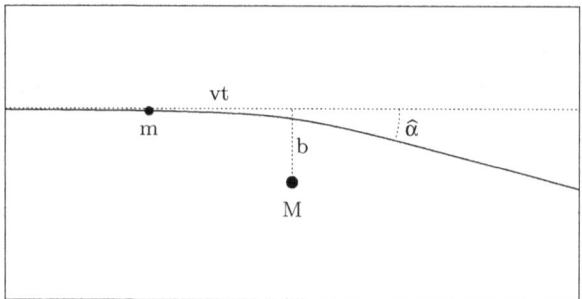

Fig. 9.1 Setup for calculating the deflection when a particle of mass m passes near a massive body M. The particle moves from left to right, and we define $t = 0$ to be the time at the point of closest approach on the original trajectory. The position shown has $t < 0$

Let the particle's speed be v. Consider the point of closest approach on the original trajectory: let the distance of this point from M, known as the **impact parameter**, be b; and let the time at this point be $t = 0$. The component of the force equation perpendicular to the (original) direction of motion is then

$$m \frac{dv_\perp}{dt} = \frac{GMm}{b^2 + v^2 t^2} \frac{b}{(b^2 + v^2 t^2)^{1/2}}$$

The first factor is the strength of the gravitational force, while the second factor gives (by trigonometry) the component in the perpendicular direction. The net change in the component of velocity perpendicular to the original motion can be found by integrating:

$$\Delta v_\perp = \int_{-\infty}^{\infty} \frac{dv_\perp}{dt} \, dt = \int_{-\infty}^{\infty} \frac{GMb}{(b^2 + v^2 t^2)^{3/2}} \, dt$$
$$= \frac{GM}{vb} \int_{-\infty}^{\infty} \frac{dx}{(1 + x^2)^{3/2}} = \frac{2GM}{vb}$$

In the third step we change variables $x = vt/b$ to make the integral dimensionless. The integral can then be evaluated by changing variables again to $x = \tan\theta$. Using $\Delta v_\perp$, we can write the **deflection angle** as

$$\hat\alpha \approx \tan\hat\alpha = \frac{\Delta v_\perp}{v} = \frac{2GM}{v^2 b}$$

where we use the small-angle approximation. Notice that the deflection angle is independent of the mass of the moving particle. It must apply to arbitrarily low masses, and even to the limit $m \to 0$ as appropriate if we think of light as a photon. Therefore, we expect light to be bent by gravity.[2]

This analysis used Newtonian gravity. The analysis with general relativity (see Sect. 10.6.5) gives a bending angle that is the same for a massive particle, but a factor of 2 larger for a massless particle (like light). Therefore we can say:

$$\hat{\alpha} = \begin{cases} \dfrac{2GM}{v^2 b} & \text{massive, non-relativistic particle} \\[2ex] \dfrac{4GM}{c^2 b} & \text{massless particle} \end{cases} \tag{9.1}$$

It is possible to develop the theory of gravitational lensing in a relativistic framework (e.g., [1]), but for lensing by stars and galaxies it is adequate (and much simpler) to work in the Newtonian framework and insert the factor of 2 for light.[3]

Example: Deflection of Light by the Sun

The nearest object that creates measurable light bending is the Sun. Light from a distant star that passes just outside the surface of the Sun is deflected by the angle

$$\hat{\alpha}_\odot = \frac{4GM_\odot}{c^2 R_\odot} = \frac{4 \times (6.67 \times 10^{-11}\, \text{m}^3\,\text{kg}^{-1}\,\text{s}^{-2}) \times (1.99 \times 10^{30}\,\text{kg})}{(3.0 \times 10^8\,\text{m s}^{-1})^2 \times (6.96 \times 10^8\,\text{m})}$$

$$= 8.5 \times 10^{-6}\,\text{rad} \times \frac{180\,\text{deg}}{\pi\,\text{rad}} \times \frac{3{,}600\,\text{arcsec}}{1\,\text{deg}} = 1.75\,\text{arcsec}$$

Such starlight is normally swamped by light from the Sun, but it becomes visible during a solar eclipse. Frank Dyson and Arthur Eddington led expeditions to measure the positions of stars during an eclipse in 1919 [3]. They found that the positions were shifted (relative to the standard positions when the Sun is not present) by amounts that were consistent with Einstein's predictions (see Fig. 9.2). This measurement and a similar one by Campbell et al. during a 1922 eclipse [4] are considered to be among the classic tests of general relativity (see Sect. 10.4).

---

[2] You might wonder whether it makes sense to take the limit of the gravitational force as $m \to 0$, but in general relativity we learn that energy gravitates.

[3] Gravitational lensing by black holes does require a full relativistic treatment (see [2] for a review).

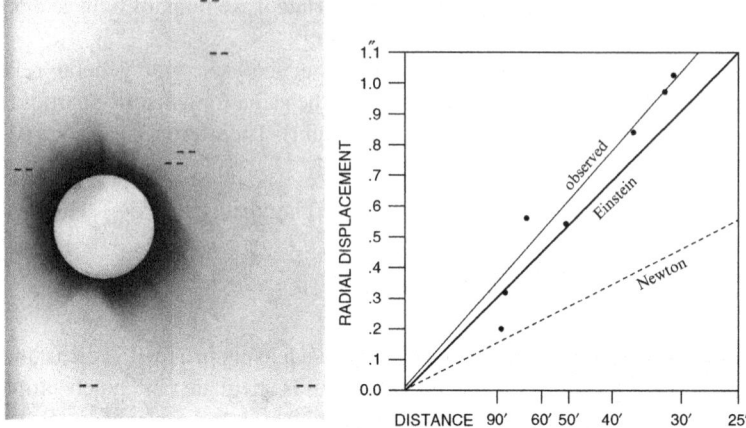

Fig. 9.2 The *left panel* shows a photographic negative from the solar eclipse of 1919. Although they are hard to see, star positions are marked. Comparing the positions in this picture with those measured when the Sun is not in the way yielded the deflections plotted in the *right panel*. (Note that the horizontal axis is inverted so stars closer to the Sun are plotted toward the right) (Credit: Dyson, Eddington and Davidson [3])

9.1.2 Lens Equation

If the impact parameter is small enough, light can go around both sides of the lensing mass and still reach Earth. In such **strong lensing**,[4] we see what appears to be the same light coming from two different directions, so we detect two images of the background source.

To quantify this effect, let D_l and D_s be the distances from the observer to the lens and source, respectively, and D_{ls} be the distance from the lens to the source. Using the small-angle approximation, we can define various distances perpendicular to the line of sight as shown on the left-hand side of Fig. 9.3. If we assume Euclidean geometry, we can write down the relation

$$D_s \beta = D_s \theta - D_{ls} \hat{\alpha}(\theta)$$

where we write $\hat{\alpha}(\theta)$ to remind ourselves that the deflection angle depends on the impact parameter, which in turn depends on θ. Rewriting this very slightly, we have

$$\beta = \theta - \alpha(\theta) \tag{9.2}$$

---

[4]"Strong" is a relative term; the bending angle is still in the small-angle regime.

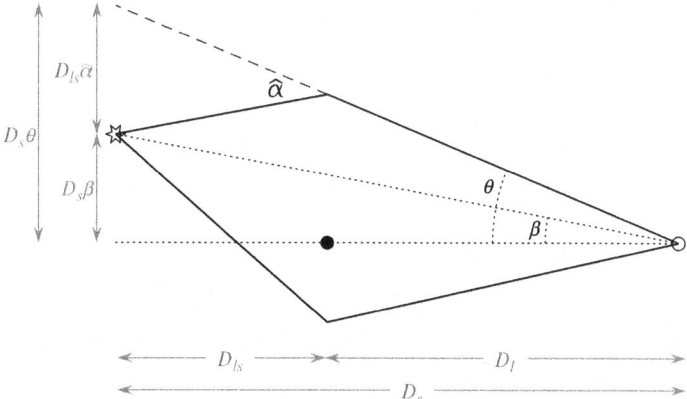

Fig. 9.3 The geometry of strong gravitational lensing. In this example, light can take two paths from the source (on the left) to the observer (on the right). The angle from the lens to the source is β, the angle from the lens to an observed image is θ, and the deflection angle is $\hat{\alpha}$. The distance from the observer to the lens is D_l, from the observer to the source is D_s, and from the lens to the source is D_{ls}. The lengths shown on the left assume the small-angle approximation

where we define the scaled deflection angle

$$\alpha(\theta) = \frac{D_{ls}}{D_s}\, \hat{\alpha}(\theta) \tag{9.3}$$

For lensing by galaxies, we cannot use Euclidean geometry to describe the expanding universe through which the light rays move. However, the bending happens only in close proximity to the galaxy, over a distance that is a small fraction of the total distance traveled. We can therefore view the trajectory as two "straight" lines (as generalized to an expanding universe) that are connected by a sharp bend. This is known as the **thin lens approximation**. It allows us to interpret Eq. (9.2) in a cosmological context provided that we take the distances D_l, D_s, and D_{ls} to be cosmological *angular diameter distances* (see Sect. 11.3.2 for details). The key point for now is that angular diameter distances do not add in a simple way, so $D_{ls} \neq D_s - D_l$ for cosmological lensing

In Fig. 9.3 all the light rays lie in a plane, which is true if the gravitational field is spherically symmetric and the force is purely radial. In general that may not be the case, but we can keep the same form of the lens equation if we interpret all the angles (β, θ, and α) as 2-dimensional vectors in the plane of the sky. In other words, θ has two components (θ_1, θ_2) that measure angles in the east/west and north/south directions, respectively (and similar for β and α). This general form of the **lens equation** serves as the foundation for the theory of gravitational lensing. The vector form of α acts as a 2-d analog of the gravitational force, so in the same way that we define a potential via $\mathbf{F} = -\nabla U$ in 3-d, we can define a **lens potential** in 2-d via

$$\alpha = \nabla \psi \tag{9.4}$$

(Note that we do not include a minus sign when defining the lens potential ψ, because we explicitly incorporate the sign into the lens equation.) Then we can write the lens equation as

$$\boldsymbol{\beta} = \boldsymbol{\theta} - \nabla\psi \tag{9.5}$$

9.1.3 Lensing by a Point Mass

To see some detail, let's consider lensing by a point mass. This is the application of the gravitational one-body problem to light bending. The scaled deflection angle is

$$\alpha = \frac{D_{ls}}{D_s} \frac{4GM}{c^2 b} = \frac{4GM}{c^2} \frac{D_{ls}}{D_l D_s} \frac{1}{\theta} \tag{9.6}$$

where we write the impact parameter as $b = D_l\theta$ in the small-angle approximation. It is convenient to define

$$\theta_E = \left(\frac{4GM}{c^2} \frac{D_{ls}}{D_l D_s} \right)^{1/2} \tag{9.7}$$

We will interpret this quantity momentarily. For now, it lets us write the lens equation as

$$\beta = \theta - \frac{\theta_E^2}{\theta}$$

Rearranging gives

$$\theta^2 - \beta\theta - \theta_E^2 = 0$$

which is a quadratic equation with two solutions,

$$\theta_\pm = \frac{1}{2} \left[\beta \pm \left(\beta^2 + 4\theta_E^2 \right)^{1/2} \right] \tag{9.8}$$

Consider the case $\beta = 0$, so the solutions are $\theta_\pm = \pm\theta_E$. In this case the observer, lens, and source all lie on a line, so we can rotate the system around that line and have perfect symmetry. In other words, there are images that appear *all the way around the lens*, forming a perfect **Einstein ring** image. Since θ_E gives the angular size of the ring, we call it the **angular Einstein radius**.

In the general case $\beta \neq 0$, notice that

$$\theta_+ \geq \theta_E \quad \text{and} \quad -\theta_E \leq \theta_- < 0$$

The $+$ image is always outside the Einstein ring, while the $-$ image is always inside the Einstein ring and on the other side of the lens (as indicated by the minus sign). Now consider:

$$\theta_+ \, \theta_- = \frac{1}{2} \left[\beta + (\beta^2 + 4\theta_E^2)^{1/2} \right] \times \frac{1}{2} \left[\beta - (\beta^2 + 4\theta_E^2)^{1/2} \right]$$

$$= \frac{1}{4} \left[\beta^2 - (\beta^2 + 4\theta_E^2) \right]$$

$$= -\theta_E^2$$

Substituting for θ_E from Eq. (9.7), we can solve for mass:

$$M = \frac{c^2}{4G} \frac{D_l D_s}{D_{ls}} \, |\theta_+ \, \theta_-| \tag{9.9}$$

If we observe two lensed images, and we know the distances involved, we can compute the mass of the lens. This is the motion $\rightarrow$ mass principle for gravitational lensing. What is different now is that we are using the motion of *light* to measure mass.

In Fig. 9.4, the left and middle columns show examples of lensing by a point mass. Each source produces two images, one on the same side of the lens as the source and outside θ_E, the other on the opposite side and inside θ_E. The exception is a source directly behind the lens, which produces a complete Einstein ring. The right column shows an example in which the gravitational field is not spherically symmetric, which we will examine below. In that case lensing can produce four images for certain source positions. Figure 9.5 shows an example of an observed 4-image lens system.

9.1.4 Distortion and Magnification

In Fig. 9.4 we see that lensed images can be stretched, and in Fig. 9.5 we see that images of a single source can have different brightnesses. Thus, the observable effects of lensing include distortion and magnification. To illustrate how these occur, Fig. 9.6 shows the images of a straight arrow source behind a point mass lens. The outer image is created when each piece of the source arrow is pushed radially outward until it lies beyond the Einstein radius. The image subtends the same

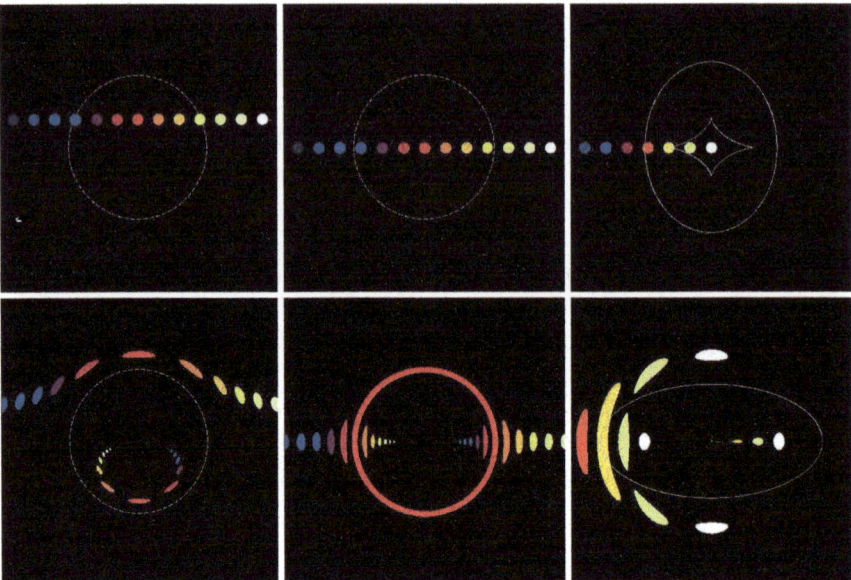

Fig. 9.4 Examples of strong gravitational lensing. The *top row* shows arrays of sources, while the *bottom row* shows the resulting lensed images (colored the same as the sources). In the *left* and *middle columns*, the lens is a point mass; the *dashed circle* indicates the Einstein radius. The difference is whether the sources are offset from or aligned with the middle of the lens. A source directly behind a circular lens produces an Einstein ring (*bottom middle*). In the *right column*, the lens is an ellipsoidal galaxy model; the *dashed curves* indicate the "critical curves" (in the image plane) and "caustics" (in the source plane). A source within the inner caustic produces four images

azimuthal angle as the source,[5] but since it lies at a larger radius it winds up being longer. The inner image is created when each piece of the source is pushed radially "through" the center of the lens. Again the image subtends the same azimuthal angle as the source, but it can lie close to the center and thus be short, or it can lie near (but inside) the Einstein radius and thus be relatively long (as in the example). Notice that the outer image gets distorted but retains the same orientation as the source. By contrast, the inner image gets flipped upside down while keeping the same left/right orientation as the source. There is no way to obtain this image by distorting and rotating the source, so we say the **parity** (or handedness) of the inner image has been reversed.

---

[5]Having the image subtend *exactly* the same azimuthal angle as the source requires a radial deflection and thus is limited to circular lenses. The concept of tangential stretching is general, though.

Fig. 9.5 Hubble Space Telescope image of the four-image gravitational lens SDSS J0924−0219. The *red-orange* object in the middle is the lens galaxy, while the four *blue-white* objects are lensed images of a background quasar (Credit: Keeton et al. [5]. Reproduced by permission of the AAS)

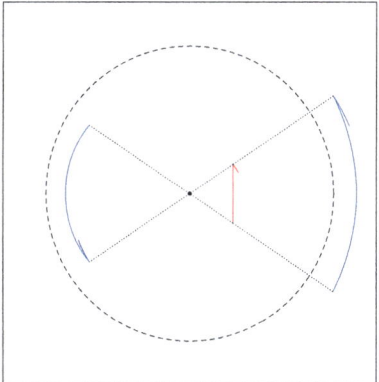

Fig. 9.6 Example of lensing distortion and magnification. The straight *red arrow* shows the source; the *curved blue arrows* show the two lensed images. The outer image has the same parity as the source, but the inner image has the opposite parity. The *dashed circle* indicates the Einstein radius. The *dotted lines* show that, for a spherical lens, each part of the source yields two images on radial lines

To quantify these effects, consider a small displacement $\Delta\boldsymbol{\beta}$ in the source plane. It will map to a small displacement in the image plane given by

$$\Delta\boldsymbol{\theta} = \frac{\partial\boldsymbol{\theta}}{\partial\boldsymbol{\beta}}\,\Delta\boldsymbol{\beta} \qquad (9.10)$$

Since $\Delta\boldsymbol{\theta}$ and $\Delta\boldsymbol{\beta}$ are 2-d vectors, the quantity $\mathbf{A} \equiv \partial\boldsymbol{\theta}/\partial\boldsymbol{\beta}$ is a 2×2 tensor. It specifies how the shape of a (small) source is changed by lensing, so we call it the **amplification tensor**. It is actually easier to compute the inverse using the lens equation:

$$\mathbf{A}^{-1} = \frac{\partial\boldsymbol{\beta}}{\partial\boldsymbol{\theta}} = \begin{bmatrix} 1 - \frac{\partial\alpha_1}{\partial\theta_1} & -\frac{\partial\alpha_1}{\partial\theta_2} \\ -\frac{\partial\alpha_2}{\partial\theta_1} & 1 - \frac{\partial\alpha_2}{\partial\theta_2} \end{bmatrix} = \begin{bmatrix} 1 - \frac{\partial^2\psi}{\partial\theta_1^2} & -\frac{\partial^2\psi}{\partial\theta_1\partial\theta_2} \\ -\frac{\partial^2\psi}{\partial\theta_1\partial\theta_2} & 1 - \frac{\partial^2\psi}{\partial\theta_2^2} \end{bmatrix} \tag{9.11}$$

Here subscripts indicate components of a 2-d vector on the sky (see Sect. 9.1.2). In the last step we use Eq. (9.4) and see explicitly that $\mathbf{A}$ is symmetric.

To characterize the distortion and magnification of a small source, we introduce three quantities $(\kappa, \gamma_+, \gamma_\times)$ defined by

$$\mathbf{A}^{-1} = \begin{bmatrix} 1 - \kappa - \gamma_+ & -\gamma_\times \\ -\gamma_\times & 1 - \kappa + \gamma_+ \end{bmatrix} \tag{9.12}$$

Comparing this with Eq. (9.11) lets us write

$$\kappa = \frac{1}{2}\left(\frac{\partial^2\psi}{\partial\theta_1^2} + \frac{\partial^2\psi}{\partial\theta_2^2}\right) \tag{9.13a}$$

$$\gamma_+ = \frac{1}{2}\left(\frac{\partial^2\psi}{\partial\theta_1^2} - \frac{\partial^2\psi}{\partial\theta_2^2}\right) \tag{9.13b}$$

$$\gamma_\times = \frac{\partial^2\psi}{\partial\theta_1\,\partial\theta_2} \tag{9.13c}$$

In Problem 9.1 you can learn that $\kappa > 0$ makes a source look bigger; it is related to focusing of light, so it is known as **convergence**. By contrast, γ_+ and $\gamma_\times$ cause a source to look distorted, so they are known as **shear**. Strictly speaking, the convergence and shear describe what happens to a source that is small enough for $(\kappa, \gamma_+, \gamma_\times)$ to be constant across the source, but they offer an intuitive sense of what happens to larger sources as well.

Lensing conserves surface brightness (it merely redirects photons, without creating or destroying any), so if a small source has surface brightness I and area dA_{src} and it leads to an image with area dA_{img}, then the ratio of fluxes is

$$\frac{f_{img}}{f_{src}} = \frac{I\,dA_{img}}{I\,dA_{src}} = \frac{dA_{img}}{dA_{src}} = |\det\mathbf{A}|$$

Thus we define

$$\mu \equiv \det\mathbf{A} \tag{9.14}$$

to be the **lensing magnification** such that the ratio of fluxes is $f_{img}/f_{src} = |\mu|$. We could make the absolute value part of the definition of μ, but it is convenient to let μ be a signed quantity because the sign reveals the parity of the image. If the source is large enough that we can resolve the images, then we observe the shapes directly and so we work with $\mathbf{A}$ itself. By contrast, if the source is small and we cannot resolve the images, then we only measure fluxes and so we work with μ. Using Eq. (9.12) we can write the magnification in terms of the convergence and shear:

$$\mu = \left[(1 - \kappa)^2 - \gamma_+^2 - \gamma_\times^2\right]^{-1} \tag{9.15}$$

Circular Symmetry

If the lens has circular symmetry, the potential and deflection are functions of $\theta = (\theta_1^2 + \theta_2^2)^{1/2}$. Then working out the derivatives and using some trigonometry (see Problem 9.2) gives

$$\mu = \left(1 - \frac{\alpha}{\theta}\right)^{-1} \left(1 - \frac{d\alpha}{d\theta}\right)^{-1} \tag{9.16}$$

Recall that the Einstein radius satisfies $\theta_E - \alpha(\theta_E) = 0$, so at the Einstein radius the first factor vanishes and hence the magnification diverges. For an image near but not precisely at θ_E, the magnification will be finite but it can be large. This is reflected in the size and shape of images near the Einstein radius in Fig. 9.4. In multiply-imaged quasars it is not uncommon for the brightest images to have magnifications of 10 or 20, and in some cases of microlensing (Sect. 9.2) magnifications of hundreds or even thousands have been recorded [6, 7].

Point Mass

For a point mass, using $\alpha = \theta_E^2/\theta$ in Eq. (9.16) leads to a magnification

$$\mu = \left(1 - \frac{\theta_E^2}{\theta^2}\right)^{-1} \left(1 + \frac{\theta_E^2}{\theta^2}\right)^{-1} = \frac{\theta^4}{\theta^4 - \theta_E^4}$$

Recall that the $+$ image has $\theta_+ \geq \theta_E$, so the denominator is positive, and indeed the entire quantity is larger than 1; this image is always brighter than the source. By contrast, the $-$ image has $|\theta_-| \leq \theta_E$, so the denominator and hence the magnification is negative. The sign reflects the parity reversal. There is no lower bound on $|\mu|$ for the $-$ image, so this image can be bright or faint. For both images, when θ approaches θ_E the magnification gets arbitrarily large.

9.1.5 Time Delay

Looking back at Fig. 9.3, notice that each light ray is longer than it would have been if it went straight from the source to the observer. Also, each light ray experiences a relativistic phenomenon called gravitational time dilation (see Sect. 10.2.3). The two effects cause the light to take longer to reach us along the lensed path than it would have along the direct route (without lensing). The excess light travel time, which is called the lens **time delay**, is

$$\tau = \frac{1 + z_l}{c} \frac{D_l D_s}{D_{ls}} \left[\frac{1}{2} |\boldsymbol{\theta} - \boldsymbol{\beta}|^2 - \psi(\boldsymbol{\theta}) \right] \qquad (9.17)$$

where z_l is the cosmological redshift of the lens (see Sect. 11.3.1). The first term in square brackets quantifies the extra distance the light has to travel, while the second term encodes gravitational time dilation.

Usually we cannot measure the time delay itself, because we cannot know how long it would have taken the light to reach us without lensing, but we can measure the *differential* time delay between two images. Time delays are thus another observable aspect of lensing, although we will not say much more about them here. One conceptual point is that time delays provide a new way to think about where the lens equation comes from. By Fermat's principle, light will "choose" trajectories that correspond to stationary points of the travel time function.[6] The condition $\nabla \tau = 0$ immediately yields $\boldsymbol{\theta} - \boldsymbol{\beta} - \nabla \psi = 0$, which is the lens equation (9.5). In other words, images form at stationary points of the time delay surface.

9.2 Microlensing

In the remainder of this chapter we examine several ways in which gravitational lensing can be used to investigate matter that is difficult or impossible to observe directly. Let's begin in our own Milky Way galaxy. Once galaxy rotation curves gave evidence for dark matter, people begin to wonder what the extra mass is made of. Two competing hypotheses emerged[7]:

- **MACHOs**, or Massive Astrophysical Compact Halo Objects. According to this hypothesis, dark matter is composed of astrophysical objects that are faint but otherwise familiar. Possibilities include: brown dwarf stars, which are balls of

---

[6]You may be familiar with the principle of least time, but local minima are not the only stationary points. As a function of two dimensions, τ can also have local maxima and saddle points.

[7]Don't blame me—I didn't invent the names! For the record, "WIMP" was introduced first, and "MACHO" was chosen deliberately (see [8]).

gas that are too small to support nuclear fusion, so they do not shine (see Problem 16.5); white dwarf stars, which are dim stellar corpses (see Sect. 17.2); planets; or black holes.

- **WIMPs**, or Weakly Interacting Massive Particles. According to this hypothesis, dark matter is a fundamental particle that is unfamiliar to us. There are many hypothetical particles that could have the right properties to act as dark matter, including neutralinos, axions, gravitinos, and much more (see [9]).

If dark matter is made of MACHOs, the Milky Way should be rife with objects the mass of planets or stars that can cause a form of lensing known as microlensing. If dark matter is instead made of WIMPs, it should be spread more diffusely, which would limit microlensing to events produced by stars. Measuring the rate of lensing in our own galaxy can therefore help us distinguish between MACHO and WIMPy dark matter.

9.2.1 Theory

Consider using a star in the Milky Way as the source of light, and either another star or a MACHO in the foreground as the lens. In a typical scenario the source is a star in the bulge of our galaxy, which is about 8 kpc away, and the lens is a star roughly halfway in between. The Einstein radius for a solar mass star is then

$$
\theta_E = \left[\frac{4 \times (6.67 \times 10^{-11} \, \mathrm{m^3 \, kg^{-1} \, s^{-2}}) \times (1.99 \times 10^{30} \, \mathrm{kg})}{(3.0 \times 10^8 \, \mathrm{m \, s^{-1}})^2} \times \frac{1}{8 \times 3.09 \times 10^{19} \, \mathrm{m}} \right]^{1/2}
$$

$$
= 4.9 \times 10^{-9} \, \mathrm{rad} \times \frac{180 \, \mathrm{deg}}{\pi \, \mathrm{rad}} \times \frac{3600 \, \mathrm{arcsec}}{1 \, \mathrm{deg}}
$$

$$
= 0.001 \, \mathrm{arcsec}
$$

Since the Einstein radius is so small, the images are too close together to be resolved (even with the Hubble Space Telescope). As the source and lens move through the galaxy, though, the positions and brightnesses of the images change with time. We can detect microlensing through variations in the apparent brightness of the source star.

Problem 9.4 you can practice solving the lens equation to predict the changes in brightness as the source moves relative to the lens. For now let us focus on the time scale for variability. The natural scale is the time it takes for the source and lens to move (relative to the each other) by the diameter of the Einstein ring. Since θ_E is the angular Einstein radius, the corresponding length is $D_l \theta_E$. The speed that matters is the relative velocity of the lens and source perpendicular to the line of sight, which we write as $v_\perp$. The typical **Einstein crossing time** is therefore

$$t_E = \frac{2 D_l \theta_E}{v_\perp} = \frac{2}{v_\perp} \left(\frac{4GM}{c^2} \frac{D_l D_{ls}}{D_s} \right)^{1/2} \tag{9.18}$$

For the typical values quoted above and $v_\perp = 200 \, \mathrm{km \, s^{-1}}$, the time scale is $t_E = 70$ days. This is quite convenient: short enough that impatient astronomers do not have to wait too long, but long enough that they can make many measurements during the course of an event even if some nights are lost to bad weather.

9.2.2 Observations

The biggest observational challenge is the low probability for any given star to be microlensed (which you can estimate in Problem 9.5). If you watch enough stars over a long enough period of time, however, you are bound to see some events. The prospect of testing the MACHO hypothesis was tantalizing enough to lead several groups to make a concerted effort to look for microlensing. Three of the main teams were the MACHO Project, the Optical Gravitational Lensing Experiment (OGLE), and Expérience pour la Recherche d'Objets Sombres (EROS). To give a sense of scale: the MACHO Project monitored about 17 million stars toward the center of the Milky Way for 3 years and observed 99 events, and also monitored almost 12 million stars in the Large Magellanic Cloud (LMC, a small galaxy orbiting the Milky Way) for almost 6 years and observed 13–17 events (depending on the selection criteria) [10,11]. Looking toward the Galactic Center raised the odds that the team would see at least a few events and thus validate their observational methods, while looking toward the LMC let them look through the Milky Way's halo to search for MACHO dark matter.

Two sample microlensing events are shown in Fig. 9.7. Each star was observed in both red and blue light to distinguish lensing from other effects. Light bending is independent of wavelength, so a microlensing event ought to look the same in both red and blue light.[8] Most of the time the light curve is constant (revealing the star's natural flux). But during a period of a few months the star brightens, reaches a peak, and then fades back to its original flux. The measured data points nicely follow the predicted microlensing light curve. For each event, we can measure the peak magnification, which depends on how close the source star came to the lens, and the duration of the event, which depends on a combination of the mass of the lens, the distances, and the relative velocity (see Eq. 9.18). While this information does not uniquely determine the mass of the lens star, it does at least confirm that we saw microlensing.

As we said, the idea is to see whether the number of microlensing events is comparable to or higher than the number expected from known populations of stars in the Milky Way. The analysis is necessarily detailed; suffice it to say that the

---

[8]By contrast, variable stars tend to change color as they change brightness.

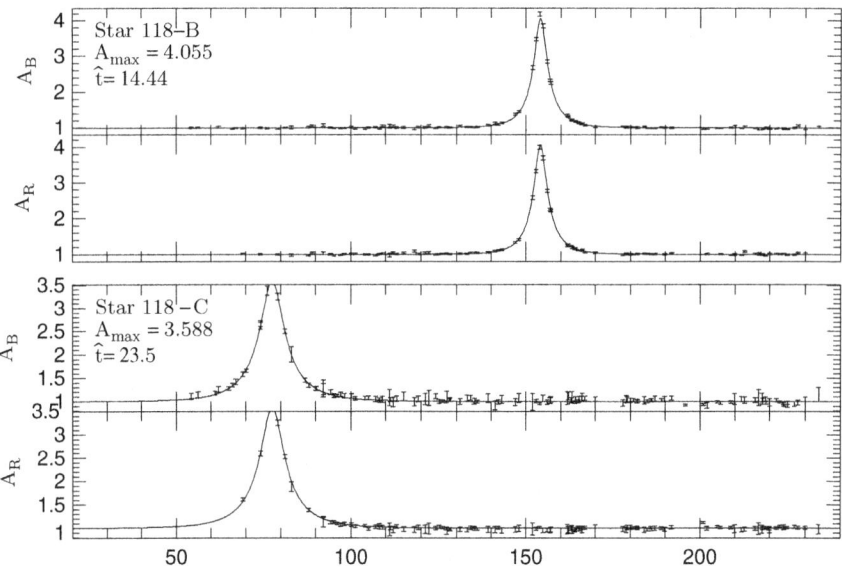

Fig. 9.7 Microlensing light curves for two stars from the MACHO project. Each pair of panels shows the same star in *red* and *blue light*. The horizontal time axis is measured in days. The points with errorbars show the measured brightness, while the curves show microlensing models (Credit: Alcock et al. [12]. Reproduced by permission of the AAS)

microlensing event rate is much lower than expected if all the dark matter were MACHOs [11]. There may be some but not very many MACHOs in the Milky Way. Dark matter, it seems, is mostly WIMPy.

9.2.3 Binary Lenses

Observed light curves do not always match standard predictions as well as the ones in Fig. 9.7. Many of them have features that arise when the lens star has a companion: either another star or a planet. The gravitational field for a binary lens is sufficiently complicated that we cannot predict the light curve analytically. Nevertheless, we can understand some of the distinctive phenomena that occur in binary lensing.

Consider the case of two equal mass stars, and a source star directly behind the center of mass. The resulting image configuration is shown in Fig. 9.8. Image #1 appears right in the middle because the gravity from the star on the right cancels the gravity from the star on the left. Image #2 appears where it does because both stars pull the light to the right; and vice versa for image #3. For image #4, both stars pull down, while the leftward and rightward forces cancel; and vice versa for image #5.

The key concept is that there can be *five* images. This is true not only for a source right between the stars, but also for some other positions. As shown in Fig. 9.9, there

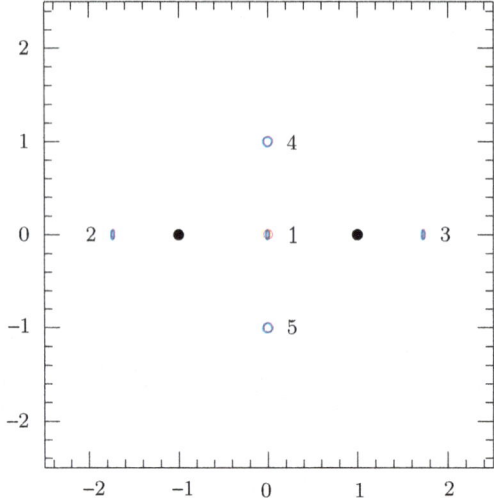

Fig. 9.8 Images produced when a source is directly behind the center of a lens consisting of two equal-mass stars

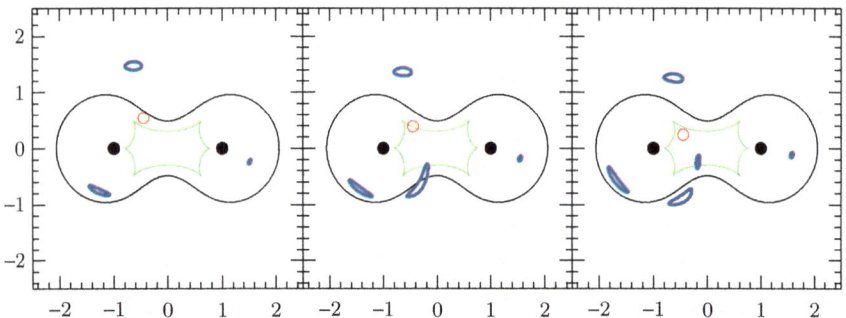

Fig. 9.9 Illustration of binary lensing. In each panel, the *green curve* shows the caustic, the *black curve* shows the corresponding critical curve, the *red circle* shows the source, and the *blue curves* show the lensed images

is a region in the source plane that leads to five images, and another region that leads to three. The boundary between them is called a **caustic curve** in the source plane. Caustics map to **critical curves** in the image plane, which are like the Einstein ring but generalized to scenarios without circular symmetry. Caustics mark where the number of images changes. A source just inside a caustic produces two images near a critical curve that are highly magnified and distorted. (If the source straddles the caustic, the two images merge into one that crosses the critical curve.) Consequently, the lensing magnification can change dramatically from one side of the caustic to the other.

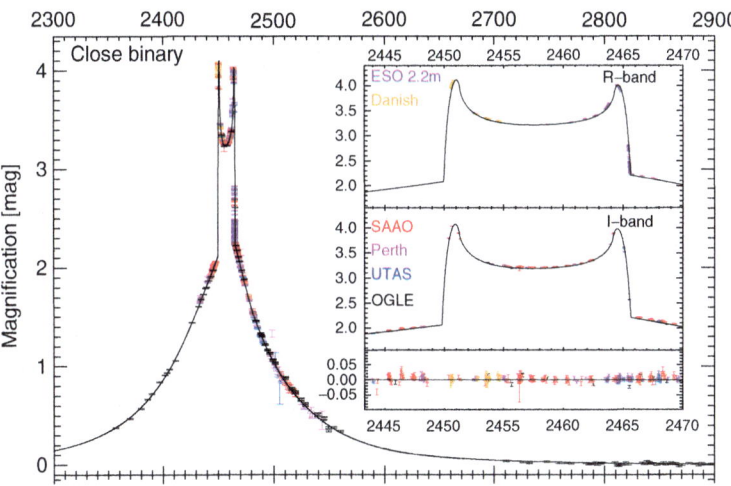

Fig. 9.10 Light curve for a binary microlensing event. The vertical axis is labeled magnification but is actually $2.5 \log |\mu|$, so the highest points correspond to magnifications of around 40. The *colored points* show data from different telescopes. The curve shows a binary lens model. In the inset, the two *upper panels* show results for different filters, while the *bottom panel* shows the residuals between the data and model (Credit: Kubas et al. [13]. Reproduced with permission © ESO)

We can see this as a sharp change in the magnification when a source moves across a caustic during a microlensing event, as shown in Fig. 9.10. The main plot shows the full light curve. The inset shows a close-up of the caustic crossing event; the colored points show the data, while the black curve shows a theoretical prediction. Remarkably, the light curve depends not only on the properties of the lens (the masses and positions of the two stars), but also on the structure of the source. At any given time the part of the source that is right on the caustic is dramatically magnified. As the source moves across the caustic, different portions of its surface are magnified in turn, and the light curve essentially maps the surface of the star (in the direction of motion, at least). In this way microlensing effectively boosts the resolving power of our telescopes to help us study structures that would otherwise be too small to see.

9.2.4 Planets

If we reduce the companion mass to the scale a planet, the caustics shrink to the point that the gravity from the planet just produces a "blip" on the light curve. The time scale for the planetary feature, compared to the full stellar event, is

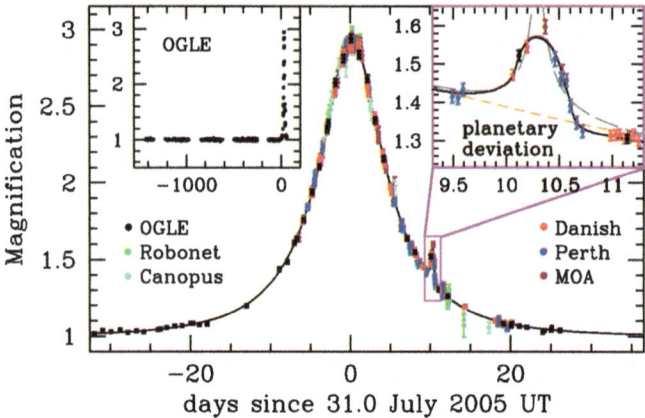

Fig. 9.11 Microlensing light curve revealing a planet estimated to be about 5.5 Earth masses (Reprinted by permission from Macmillan Publishers Ltd: Beaulieu et al. [15], © 2006)

$$\frac{t_{\text{planet}}}{t_{\text{star}}} = \frac{2D_l \theta_{\text{planet}}/v_\perp}{2D_l \theta_{\text{star}}/v_\perp} = \left(\frac{M_{\text{planet}}}{M_{\text{star}}}\right)^{1/2}$$

For the Sun and Jupiter, the mass ratio translates into a duration ratio of $t_{\text{planet}}/t_{\text{star}} = 0.031$. With the typical numbers from Sect. 9.2.1, the planetary event would last about $t_{\text{planet}} \sim 2$ days. For the Sun and Earth, the numbers are $t_{\text{planet}}/t_{\text{star}} = 0.0017$ and $t_{\text{planet}} \sim 3$ h.

Planetary microlensing events are short enough that they require continuous monitoring by telescopes around the world. To do this efficiently, microlensing observers developed a strategy in which the main teams would observe their large samples of stars once a week or so. When they spotted a star in the early stages of a microlensing event, they would broadcast an alert so that other teams could begin using other telescopes to monitor the event very closely. This strategy has paid off with the discovery of more than a dozen microlensing planets so far [14]. Figure 9.11 shows a planetary event discovered in August 2005 after an alert from the OGLE team.

Beyond merely detecting a planet, what did they learn from this event? The most well-constrained quantity is the mass ratio between the planet and star, which basically comes from the ratio of time scales [15]:

$$\frac{M_{\text{planet}}}{M_{\text{star}}} = (7.6 \pm 0.7) \times 10^{-5}$$

To estimate the actual masses of the star and planet, the team had to make a detailed model of the population of stars in the galaxy and figure out which ones are most likely to produce an event like the one seen. This yielded

$$M_{\text{star}} = 0.22^{+0.21}_{-0.11} \, M_\odot \quad \text{and} \quad M_{\text{planet}} = 5.5^{+5.5}_{-2.7} \, M_\oplus$$

The uncertainties are significant: the planet could be only a few times more massive than Earth, or more than 10 times more massive. This is the best that can be done without direct knowledge of the distance to the lens or the relative velocity between the lens and source. Still, it allows the important conclusion that this planet is in the same league as Neptune.

While microlensing has revealed fewer exoplanets than the Doppler and transit techniques (see Sect. 4.3), it serves as a valuable complement. Microlensing involves completely different physical processes and observational methods, so it provides independent confirmation that other stars have planets. Also, microlensing is more sensitive to small planets far from their stars. Finally, microlensing is better able to examine stars that are far from Earth. The main drawback is that we only see a microlensing event once, when the star and planet cross in front of a background source; after the event concludes, it cannot be repeated. Therefore microlensing will probably contribute more to a statistical census of planets rather than to detailed knowledge of individual systems. Nevertheless, microlensing is expected to play an increasingly important role in planet searches in the coming decade.

9.3 Strong Lensing

With stars and planets it is reasonable to use the point mass approximation, but when we turn to galaxies and clusters of galaxies we must consider extended mass distributions.

9.3.1 Extended Mass Distribution

We can still work in the thin lens approximation, so what matters is the projected surface mass density of the galaxy, Σ. A small patch of the lens at position θ' has mass $\Sigma(\theta')\,d\theta'$, so the amount of bending it creates at θ is

$$d\alpha(\theta) = \frac{4G}{c^2} \frac{D_{ls}}{D_l D_s} \Sigma(\theta') \frac{\theta - \theta'}{|\theta - \theta'|^2}\,d\theta'$$

(This is the 2-d vector form of Eq. 9.6.) We can therefore write the total scaled deflection as

$$\alpha(\theta) = \frac{1}{\pi} \int \frac{\Sigma(\theta')}{\Sigma_{\text{crit}}} \frac{\theta - \theta'}{|\theta - \theta'|^2}\,d\theta' \tag{9.19}$$

where we have collected multiplicative factors and defined

$$\Sigma_{\text{crit}} = \frac{c^2}{4\pi G} \frac{D_l D_s}{D_{ls}} \tag{9.20}$$

We call this the **critical surface density** for lensing, for reasons that will become clear shortly. If we take the divergence $\nabla \cdot \boldsymbol{\alpha}$ and recall that the deflection is related to the lens potential by $\boldsymbol{\alpha} = \nabla \psi$, we obtain

$$\nabla^2 \psi = 2 \frac{\Sigma}{\Sigma_{\text{crit}}} \tag{9.21}$$

This has the form of the Poisson equation for the gravitational potential, but in two dimensions. It provides the general framework for lensing by arbitrary 2-d mass distributions. By comparing Eqs. (9.13a) and (9.21), we see that the convergence is the surface mass density scaled by the critical density:

$$\kappa = \frac{\Sigma}{\Sigma_{\text{crit}}} \tag{9.22}$$

9.3.2 Circular Mass Distribution

For a mass distribution with circular symmetry, we can evaluate Eq. (9.19) using an analog of Newton's theorem about gravity from a spherical mass distribution. Recall from Sect. 2.3 that Newton found $F(r) \propto M(r)/r^2$ where $M(r)$ is the mass enclosed within r. The radial dependence is $1/r^2$ because a certain "amount of gravity" is spread over a spherical shell whose area scales as r^2. By analogy, in 2-d the dependence should be $1/R$, or in terms of the angular impact parameter $1/\theta$. Indeed, the scaled deflection from a circular mass distribution is

$$\alpha(\theta) = \frac{4G}{c^2} \frac{D_{ls}}{D_l D_s} \frac{M(\theta)}{\theta} \tag{9.23}$$

Recalling that the Einstein radius is defined by $\alpha(\theta_E) = \theta_E$, we can write

$$M(\theta_E) = \frac{c^2}{4G} \frac{D_l D_s}{D_{ls}} \theta_E^2 \tag{9.24}$$

If we see an Einstein ring, we can infer the mass enclosed by the ring even if we do not know the density profile. If we do not see a complete ring, the principle still holds that the quantity we measure best is the mass within θ_E. This is how we can use gravitational lensing as a tool to weigh distant galaxies and clusters of galaxies.

Consider the average surface mass density enclosed by the Einstein ring (in angular units, e.g., solar masses per square arcsecond):

$$\langle \Sigma \rangle = \frac{M(\theta_E)}{\pi \theta_E^2} = \frac{c^2}{4\pi G} \frac{D_l D_s}{D_{ls}} = \Sigma_{\text{crit}}$$

All Einstein rings enclose an average density that is given by Σ_{crit} from Eq. (9.20). Put another way, an object must have $\Sigma \geq \Sigma_{crit}$ in order to have an Einstein ring at all. This is the sense in which Σ_{crit} is the critical density for lensing.

9.3.3 Singular Isothermal Sphere

A specific example of an extended, circular mass distribution is the Singular Isothermal Sphere (SIS), which we first encountered when studying spiral galaxies (Sect. 7.3.2). With spiral galaxy rotation curves we used the (softened) isothermal model as one part of a multi-component model that also included contributions from a disk and bulge. With lensing we can often get away with even simpler models, because we mostly deal with elliptical galaxies where the stellar distribution is roundish like the dark matter halo, we focus on the *total* mass distribution (light bending depends only on the total amount of matter, not whether it is luminous or dark), and we only need to know the *projected* surface mass density. For all of these reasons, the singular isothermal sphere (and its generalization to an ellipsoid; Sect. 9.3.4) turns out to be a valuable model for lensing. An isothermal sphere with circular velocity v_c has a 3-d density profile[9]

$$\rho = \frac{v_c^2}{4\pi G r^2} = \frac{v_c^2}{4\pi G (R^2 + z^2)}$$

where r is the spherical radius and (R, z) are cylindrical coordinates. The mass enclosed by the angle θ is obtained by integrating over R out to $D_l \theta$ and integrating over all z:

$$M(\theta) = \int_{-\infty}^{\infty} dz \int_0^{D_l \theta} dR\, 2\pi R\, \frac{v_c^2}{4\pi G(R^2 + z^2)} = \frac{\pi}{2} \frac{v_c^2 D_l \theta}{G}$$

The scaled deflection angle is then

$$\alpha = \frac{4G}{c^2} \frac{D_{ls}}{D_l D_s} \frac{\pi}{2} \frac{v_c^2 D_l}{G} = 2\pi \left(\frac{v}{c}\right)^2 \frac{D_{ls}}{D_s}$$

The deflection is *independent of position*. The constant deflection angle is directly related to the constant circular velocity that we encountered when studying spiral galaxy dynamics (see Sect. 7.3). Clearly the Einstein radius is $\theta_E = \alpha$. To be more precise, we should take into account the direction of the deflection:

$$\alpha(\theta) = \begin{cases} +\theta_E & \theta > 0 \\ -\theta_E & \theta < 0 \end{cases}$$

---

[9] The SIS model can also be expressed in terms of the velocity dispersion, which is $\sigma = v_c / \sqrt{2}$.

We can then write the lens equation as:

$$\theta > 0 : \quad \beta = \theta - \theta_E \tag{9.25a}$$

$$\theta < 0 : \quad \beta = \theta + \theta_E \tag{9.25b}$$

Without loss of generality we can take $\beta \geq 0$. Then we can solve Eq. (9.25a) to find one of the images:

$$\theta_+ = \beta + \theta_E$$

We can also solve Eq. (9.25b):

$$\theta_- = \beta - \theta_E \qquad \text{only if } \beta \leq \theta_E$$

(If $\beta > \theta_E$ this equation would imply $\theta_- > 0$, which would violate the condition in Eq. 9.25b.) A singular isothermal sphere, in other words, can produce three types of configurations:

$$\beta = 0 : \text{ Einstein ring at } \theta_E$$

$$0 < \beta \leq \theta_E : \text{ two images at } \theta_\pm = \beta \pm \theta_E$$

$$\beta > \theta_E : \text{ one image at } \theta_+ = \beta + \theta_E$$

Whereas a point mass lens always produces two images, an SIS lens creates two images only for sources in a finite region behind the lens.

9.3.4   Singular Isothermal Ellipsoid

Few galaxies are perfectly spherical, and new lensing phenomena appear when spherical symmetry is broken (see Figs. 9.4 and 9.9), so it worthwhile to consider the case of ellipsoidal symmetry. With circular symmetry the surface mass density Σ is a function of the polar radius $\theta = (\theta_1^2 + \theta_2^2)^{1/2}$. To make the symmetry elliptical instead, we can write Σ in terms of the ellipse coordinate $\xi = (\theta_1^2 + \theta_2^2/q^2)^{1/2}$ where $0 < q \leq 1$ is a dimensionless parameter that measures the ratio of the short and long axes: for $q = 1$ the model is again spherical, but for $q < 1$ it is flattened.

With elliptical symmetry it can be difficult to evaluate the integral in Eq. (9.19). The singular isothermal ellipsoid (SIE) is one case that can be treated analytically, leading to the lens equation [16]

$$\beta_1 = \theta_1 - \frac{\theta_E q}{(1-q^2)^{1/2}} \tan^{-1}\left[\frac{(1-q^2)^{1/2}\theta_1}{(q^2\theta_1^2 + \theta_2^2)^{1/2}}\right] \tag{9.26a}$$

$$\beta_2 = \theta_2 - \frac{\theta_E q}{(1-q^2)^{1/2}} \tanh^{-1}\left[\frac{(1-q^2)^{1/2}\theta_2}{(q^2\theta_1^2 + \theta_2^2)^{1/2}}\right] \tag{9.26b}$$

Consider a source at the origin ($\beta_1 = \beta_2 = 0$). If we put $\theta_1 = 0$ then Eq. (9.26a) is trivially satisfied and Eq. (9.26b) becomes

$$0 = \theta_2 - \frac{\theta_E q}{(1-q^2)^{1/2}} \tanh^{-1}\left[(1-q^2)^{1/2}\,\mathrm{sgn}(\theta_2)\right]$$

which can be solved by

$$\theta_2 = \pm\frac{\theta_E q}{(1-q^2)^{1/2}} \tanh^{-1}(1-q^2)^{1/2}$$

Alternatively, if we go back to the equations and put $\theta_2 = 0$, then Eq. (9.26b) is trivially satisfied and Eq. (9.26a) becomes

$$0 = \theta_1 - \frac{\theta_E q}{(1-q^2)^{1/2}} \tan^{-1}\left[\frac{(1-q^2)^{1/2}}{q}\,\mathrm{sgn}(\theta_1)\right]$$

which can be solved by

$$\theta_1 = \pm\frac{\theta_E q}{(1-q^2)^{1/2}} \tan^{-1}\frac{(1-q^2)^{1/2}}{q}$$

While these expressions are admittedly non-intuitive, the main conceptual point is straightforward: a source at the origin yields four images, with two on the horizontal axis and two on the vertical axis. Figure 9.4 shows that other source positions can also yield four images.

9.3.5 Spherical Galaxy with External Shear

We can capture a lot of the same phenomenology using simpler algebra if we revert to a spherical model but account for the gravitational influence of other galaxies that happen to lie near the main lens galaxy. If the neighboring galaxies lie more than a few Einstein radii away, their effects can be characterized using a tensor of the form given in Eq. (9.12) where κ, γ_+, and $\gamma_\times$ are constant across the main lens galaxy.[10] If we choose coordinates such that $\gamma_\times = 0$, the lens equation has the form

---

[10]The shear is basically a tidal effect analogous to what we studied in Chap. 5.

$$\beta_1 = \theta_1 - \frac{\theta_E \theta_1}{(\theta_1^2 + \theta_2^2)^{1/2}} - (\kappa + \gamma)\theta_1 \qquad (9.27a)$$

$$\beta_2 = \theta_2 - \frac{\theta_E \theta_2}{(\theta_1^2 + \theta_2^2)^{1/2}} - (\kappa - \gamma)\theta_2 \qquad (9.27b)$$

We can now examine how the "external" shear[11] influences the number of images. Typical values of external shear are $\gamma \sim 0.01$–0.1. In what follows we set $\kappa = 0$ for simplicity, because it does not actually affect the image multiplicity.

For a source at the origin, an analysis similar to what we did in Sect. 9.3.4 yields four images:

$$(\theta_1, \theta_2) = \left(0, \pm\frac{\theta_E}{1 - \gamma}\right) \quad \text{and} \quad (\theta_1, \theta_2) = \left(\pm\frac{\theta_E}{1 + \gamma}, 0\right)$$

A source on the horizontal axis can be treated analytically as well. In Problem 9.6 you can find the following results:

- For $0 \le \beta_1 < 2\gamma\theta_E/(1 - \gamma)$ there are four images. Two are on the θ_1-axis and two are off the axis.
- For $2\gamma\theta_E/(1 - \gamma) < \beta_1 < \theta_E$ there are two images, both on the θ_1-axis.

This helps you understand the different configurations seen in Fig. 9.4, as well as the transition between two and four images. There is one additional type of 4-image configuration that is produced by an off-axis source, but it is usually found numerically.

We have considered ellipticity *or* shear, but real lenses may have both. Quantitatively, both ellipticity and shear are often required to fit observed 4-image lenses in detail. Qualitatively, though, the two models we have considered capture the main phenomenology of 4-image lensing.

9.3.6 Science with Galaxy Strong Lensing

Several hundred cases of strong lensing by galaxies have now been observed; in some the source is a quasar or other compact source that is lensed into multiple distinct images, while in others the source is a galaxy that is lensed into a partial or complete Einstein ring. The majority of lens galaxies are ellipticals because such galaxies tend to be more massive, and hence better lenses, than spirals.

So far in this chapter we have assumed a mass distribution and solved for the image positions. When we study observed lenses, we invert the problem: we take

---

[11]"External" because it comes from outside the main lens galaxy (i.e., from the neighbors). Note that we drop the subscript on γ to simplify the notation.

the images as given and try to solve for the mass distribution that produced them. It impossible to uniquely determine the mass distribution, though; there are just too many unknowns. To make progress, we often adopt assumptions that limit the unknowns.[12] For example, if we assume the lens is a point mass or singular isothermal sphere then we only need to solve for the mass or velocity dispersion (respectively). We can make the model more complicated by adding more parameters: for example, moving from an isothermal profile to a general power law adds one parameter; allowing the mass distribution to be elongated adds two (ellipticity and orientation angle); accounting for external shear adds another two (shear strength and direction); and so forth. A lot of the art and science of strong lens modeling lies in choosing assumptions whose restrictions are useful but not oversimplified, incorporating observational and/or theoretical knowledge from other realms of astrophysics.

Strong lens modeling has taught us a number of lessons about galaxy mass distributions (see the review by Treu [18]; you can explore some aspects of lens modeling in Problem 9.7). The most robust quantity we can measure is the mass within the Einstein radius, $M(\theta_E)$. By comparing the mass inferred from lensing with the mass associated with the starlight, we can find evidence that lens galaxies contain dark matter. The next step is to learn how the dark matter is distributed. One approach is to recognize that θ_E varies from one lens to another (it depends not only on the lens mass but also on the distances between the observer, lens, and source); if we assume lenses follow certain scaling relations, we can use the various $M(\theta_E)$ measurements to infer the average mass profile. Another approach is to combine lensing with an analysis of stellar dynamics, which tends to be sensitive to the mass closer to the center of a galaxy (see Sect. 8.2.3). Having mass measurements at small radii from dynamics and somewhat larger radii from lensing provides important information about the mass profile in individual systems. All told, models suggest that lens mass distributions are nearly isothermal, so the dark matter halos are more extended than the visible galaxies.

We noted above that many lens models require both ellipticity and external shear. Constraints on shear let us investigate the distribution of matter in the vicinity of a lens, which is interesting because lens galaxies often lie in gravitationally bound "groups" containing a few dozen galaxies [19]. Lensing therefore helps us study how galaxies form and evolve in environments that play an important role in galaxy evolution.

---

[12]An alternative approach is to make as few assumptions as possible (although assumptions can never be avoided altogether), and then deal with the large range of mass models that are consistent with the observed images [17].

9.4   Weak Lensing

To this point we have considered situations in which the impact parameter is small and light bending is strong enough to create multiple images. At larger impact parameters, lensing still acts but the effects are more subtle. Consider an array of source galaxies as shown in the left panel of Fig. 9.12. Putting a lens in front yields the picture shown on the right. Only sources near the center are multiply imaged, but sources farther out are still distorted. This is the regime known as **weak lensing**.

There is not much we can learn from *individual* sources that are weakly lensed. The observed shape of an image depends not only on the lensing distortion but also on the intrinsic shape of the source, and it is difficult or impossible to distinguish the two effects on a galaxy-by-galaxy basis (see Fig. 9.13). We can make progress, though, by examining *collections* of galaxies. Weak lensing distortion is predominantly tangential (perfectly so in the case of a spherical lens), whereas intrinsic shapes and orientations are random.[13] Therefore if we measure the shapes of galaxies in polar coordinates centered on a lens, the intrinsic shapes should average out while the lensing distortions will not.

One way to study weak lensing is to collect galaxies into annuli centered on the lens, compute the average shape in each annulus, and relate that to the lensing shear. As you can show in Problem 9.2, the shear is related to the density for a circular lens by

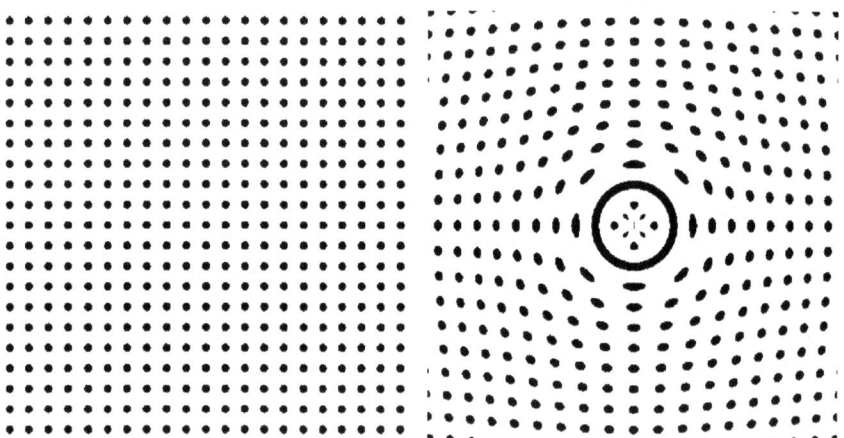

Fig. 9.12   The *left panel* shows an array of source galaxies. The *right panel* shows what we would see if there were a gravitational lens in front. One source produces an Einstein ring, a few are multiply imaged, but most are only slightly distorted ("weak lensing")

---

[13]We hope. Correlations among the intrinsic shapes of galaxies could present a challenge for weak lensing [20, 21], but they are generally expected to be small and there are ways to deal with them in a weak lensing analysis [22].

Fig. 9.13 Similar to Fig. 9.12, but the source galaxies have random positions, shapes, and orientations. The images are no longer perfectly tangential, but the overall pattern of distortion is still apparent

$$\gamma(r) = \frac{\bar{\Sigma}(r) - \Sigma(r)}{\Sigma_{\text{crit}}} \tag{9.28}$$

where $\bar{\Sigma}(r)$ is the average surface mass density within radius r. Measuring the shear profile clearly provides information about the density profile of the lens.

A more sophisticated approach is to observe a large sample of galaxies, collect them in bunches on the sky, and measure the full shear map (at a spatial resolution that is limited by the sample size). If we know both γ_+ and $\gamma_\times$ as a function of position, we can view Eqs. (9.13b) and (9.13c) as a pair of differential equations that can be solved for the lens potential, ψ. We can then use Eq. (9.21) to uncover the underlying mass distribution, Σ. This general analysis cannot be done analytically, but it is well suited to computational methods (e.g., [23]).

The challenge of weak lensing is that its statistical nature provides less detailed information about lens mass distributions, compared with strong lensing. The benefit is that there are many, many more objects in the universe that are weakly lensed than objects that are strongly lensed. As a result, weak lensing has become a widespread and important tool for studying dark matter. This is especially true for clusters of galaxies, which are the most massive bound objects in the universe and thus good targets for weak lensing (see the review by Kneib and Natarajan [24]).

Figure 9.14 shows a famous weak lensing system known as the "bullet cluster," which provides arguably the clearest evidence that dark matter is real. The system contains two clusters of galaxies that passed through each other some 100 million years ago; the cluster on the left is moving to the left, and the one on the right is moving to the right. Each cluster contained hot gas that can be mapped because it

Fig. 9.14 Composite image of the "bullet cluster" system. Superposed on an image of the galaxies is a map of the hot X-ray gas (*colored red*) and the dark matter inferred from weak lensing (*colored blue*) (Credit: X-ray: NASA/CXC/M. Markevitch et al. Optical: NASA/STScI; Magellan/U. Arizona/D. Clowe et al. Lensing Map: NASA/STScI; ESO WFI; Magellan/U.Arizona/D. Clowe et al.)

emits X-rays (colored red in the image). During the "collision" the two gas clouds slammed into one another, but the galaxies and dark matter did not feel gas pressure so they kept on going. As a result, the X-ray gas got separated from the galaxies and dark matter.

How does lensing apply? There are lots and lots of small background galaxies in the field (although they are too small and faint to be apparent in Fig. 9.14). A weak lensing analysis yields the mass distributions indicated in blue in the image [25]. There is a significant offset between the hot gas, which represents the bulk of the normal matter in the clusters, and the source of gravity. This is exactly what we would expect if there is a significant amount of dark matter that exerts gravity but is otherwise inactive. Most astrophysicists conclude that it would be very difficult to explain the weak lensing result in the bullet cluster and similar systems [26] without exotic dark matter (but see [27] for a dissenting view).

Strong and weak lensing are most apparent near massive objects like galaxies and clusters, but gravitational deflection actually affects all light rays in the universe at some level. Inhomogeneities in the large-scale distribution of matter create distortions that are quite small but detectable with a careful statistical analysis of galaxy shapes [28]. This **cosmic shear** is sensitive to the relative abundances of dark matter and dark energy in the universe, so it plays a prominent role in existing and planned probes of cosmology [29]. The analysis methods are more detailed than we want to get into here, but the fundamental principle is just what we have used throughout this chapter: mass creates gravity that bends light, so if we can detect the light bending we can use it to map the matter and weigh the universe.

Problems

9.1. This problem will help you understand the interpretation of κ, γ_+, and $\gamma_\times$ in Eq. (9.12). Let the source be a unit circle and write its boundary as

$$\Delta\beta = (\cos\phi, \sin\phi)$$

where ϕ is an azimuthal angle running from 0 to 2π. Use Eq. (9.10) to find and plot the boundary of the image for the following cases:

(a) $\kappa = 0.2$ and $\gamma_+ = \gamma_\times = 0$
(b) $\gamma_+ = \pm 0.2$ and $\kappa = \gamma_\times = 0$
(c) $\gamma_\times = \pm 0.2$ and $\kappa = \gamma_+ = 0$

9.2. In this problem we consider the lensing properties of a circular mass distribution. In the text we refer to the two components of position on the sky as (θ_1, θ_2), but for the sake of familiarity let's revert to (x, y) and the associated polar coordinates (r, ϕ). With circular symmetry, the lens potential is a function of r only: $\psi(r)$.

(a) Work out the first and second derivatives of the potential with respect to x and y, but expressed in polar coordinates. For example, the chain rule for derivatives gives

$$\frac{\partial\psi}{\partial x} = \frac{\partial r}{\partial x}\psi'(r) + \frac{\partial r}{\partial y}\psi'(r)$$

where $\psi'(r) = d\psi/dr$.
(b) Use Eq. (9.13) to show that the convergence and shear can be written as

$$\kappa = \frac{1}{2}\left(\frac{\psi'}{r} + \psi''\right) \tag{9.29a}$$

$$\gamma_+ = \frac{1}{2}\left(\frac{\psi'}{r} - \psi''\right)\cos 2\phi \tag{9.29b}$$

$$\gamma_- = \frac{1}{2}\left(\frac{\psi'}{r} - \psi''\right)\sin 2\phi \tag{9.29c}$$

where $\psi'' = d^2\psi/dr^2$.
(c) In circular symmetry, the deflection is $\alpha = \psi'$. Use this with Eq. (9.29) to show that the magnification has the form given in Eq. (9.16)
(d) From Eqs. (9.29b) and (9.29c) it is clear that the shear strength is

$$\gamma = \frac{1}{2}\left(\frac{\psi'}{r} - \psi''\right)$$

Now derive Eq. (9.28). You will need to use Eqs. (9.20), (9.22), and (9.23).

9.3. Consider a star orbiting 10 pc from the black hole at the center of the Milky Way (see Sect. 3.2.1). Suppose we view the star's orbit perfectly edge-on.

(a) What is the Einstein radius for this scenario (in arcsec)?
(b) When the star is at a source angle of $\beta = 0.1''$, where are the two gravitationally lensed images?
(c) When the star passes behind the black hole we see a "microlensing" event. How long does it last?
(d) If the star's orbit were larger, how would the answers change? Explain using equations or drawings.

9.4. Let's see how to calculate points on a microlensing light curve. In the figure below, the line denotes the trajectory of a source passing behind a point mass lens. The circle indicates the Einstein radius. All lengths are in units of the Einstein radius. For each of the three marked source positions, find the two images, compute their individual magnifications, and then find the total magnification is $\mu_{tot} = |\mu_+| + |\mu_-|$ (with absolute values because in this problem we do not worry about parities).

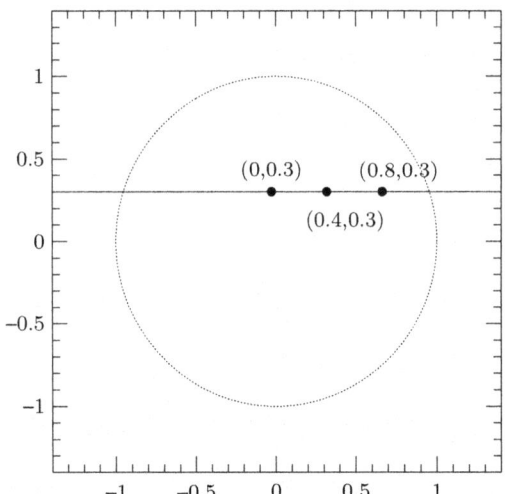

9.5. As discussed in Sect. 9.2, microlensing is used to test the hypothesis that the Milky Way's dark matter is made of MACHOs. In this problem you will estimate the microlensing probability. (This is analogous to Problem 8.4, with the interloper star replaced by a light ray.)

(a) Suppose there is a uniform mass density ρ in MACHOs between us and a source a distance D_s away. Consider a thin slab that is located a distance D_l away and has thickness dD_l. Find the fraction of the area of the slab that is covered by the Einstein rings of MACHOs. This is the probability that the light ray passes within one Einstein radius of one of the MACHOs in the slab, i.e., close enough to be strongly lensed. Hints: θ_E is the angular Einstein radius, but here you need to convert it to a length; the result does not depend on the mass of the MACHOs.

(b) Now sum up all the slabs between us and the source, i.e., integrate over D_l. The resulting quantity is called the "optical depth" for microlensing, often written as τ, and it represents the probability that a light ray passes close enough to a MACHO to be lensed.

(c) To compute a numerical value of τ you need to specify the mass density in MACHOs. To make a simple estimate, assume that dark matter is distributed uniformly between the Sun and the center of the Milky Way, and compute the mean density. Also assume that all of the mass is in MACHOs (i.e., don't worry about the disk). With these assumptions, calculate ρ.

(d) Now compute the probability that a star at the center of the Milky Way is microlensed by a MACHO.

9.6. In this problem you will see why some lenses have two images and others have four. The simplest lens that can produce four images is an isothermal sphere with an external shear, whose lens equation is given by (9.27). Recall that γ is dimensionless, and we can take it to be positive.

(a) Consider a source placed on the horizontal axis in the source plane (i.e., $\beta_2 = 0$). Solve the lens equation (working with symbols) to show that:

- For $0 \leq \beta_1 < 2\gamma\theta_E/(1-\gamma)$ there are four images. Two are on the θ_1-axis and two are off the axis.
- For $2\gamma\theta_E/(1-\gamma) < \beta_1 < \theta_E$ there are two images, both on the θ_1-axis.

Give the positions of all images in both cases.

(b) Now assume $\theta_E = 1''$ and $\gamma = 0.1$, which are typical values for galaxy lenses. Sketch the image configurations for the following source positions:

- $\beta_1 = \beta_2 = 0$
- $\beta_1 = 0.15''$ and $\beta_2 = 0$
- $\beta_1 = 0.35''$ and $\beta_2 = 0$

9.7. This problem will give you a taste of how we model gravitational lens systems to measure galaxy masses. Imagine you observe a galaxy lens system with distances $D_l = 940\,\text{Mpc}$, $D_{ls} = 1{,}293\,\text{Mpc}$, and $D_s = 1{,}745\,\text{Mpc}$. One image appears at an angular position of $\theta_+ = 1.05''$ from the lens galaxy, while the other appears at an angular position $\theta_- = -0.35''$ on the opposite side of the galaxy. You may assume the lens is circularly symmetric.

(a) Assume the galaxy can be modeled as a point mass. Find the Einstein radius and mass of the lens galaxy.

(b) Now assume the galaxy can be modeled as an isothermal sphere. Again find the Einstein radius and the mass enclosed by the Einstein radius.

(c) Both models can fit the image positions, but they make different predictions for the brightnesses. Suppose the "+" image is observed to be three times brighter than the "−" image. Compute the relative magnifications of the images for your point mass and isothermal models. Which model is correct?

Hint: remember to convert between arcseconds and radians as necessary.

9.8. We know that black holes come in stellar-mass and supermassive varieties, but we do not know whether there is anything in between. In this problem we consider whether gravitational lensing could be used to look for intermediate mass black holes (IMBH) in globular clusters.

(a) Consider a globular cluster with mass M_{tot} and velocity dispersion σ. Assuming a uniform density of stars with mass m, use the virial theorem to estimate the size of the cluster and the number density of stars.

(b) Suppose there is an IMBH at the center of the cluster, and the mass is $M_\bullet$ such that $m \ll M_\bullet \ll M_{tot}$. The black hole can lens background stars that are in its "Einstein cone"—the region behind the black hole whose projected radius equals the Einstein radius.[14] Find an approximate expression for the size of the Einstein cone as a function of D_{ls}. Hints: you may assume $D_l \approx D_s$ and $D_{ls} \ll D_s$; recall that Eq. (9.7) gives the Einstein radius in angular units.

(c) Estimate the total number of stars in the Einstein cone. This is the expected number of lens systems within the globular cluster. Hint: the answer can be expressed in terms of $M_\bullet$, m, and σ.

(d) Obtain a quantitative estimate for the number of lenses by assuming that the M-σ relation for supermassive black holes (Sect. 3.2.2) can be applied to globular clusters:

$$M_\bullet = 1.35 \times 10^8 \, M_\odot \times \left(\frac{\sigma}{200 \, \text{km} \, \text{s}^{-1}} \right)^{4.02}$$

Use $\sigma \sim 10 \, \text{km} \, \text{s}^{-1}$ for a globular cluster.

(e) The same analysis can be applied to an SMBH in an elliptical galaxy. Repeat part (d) for a galaxy with $\sigma \sim 200 \, \text{km} \, \text{s}^{-1}$.

(f) Comment on our ability to detect and identify lensing of stars in a globular cluster or galaxy by a massive black hole *within* the stellar system.

9.9. Suppose you observe a binary star system consisting of a white dwarf with a radius of 6,100 km and a neutron star with a radius of 10 km. The system is 2 kpc from us and viewed edge-on. The radial velocity curves are shown below, where WD labels the white dwarf and NS labels the neutron star.

(a) What are the masses of the two stars, and the distance between them?

(b) Sketch the light curve when the neutron star passes behind the white dwarf. Hint: since the problem appears in this chapter, you can assume it involves gravitational lensing, but that is not the only phenomenon at work.

---

[14]This is not strictly a cone because the edge is not straight, but the terminology is helpful because the region does grow with distance behind the black hole.

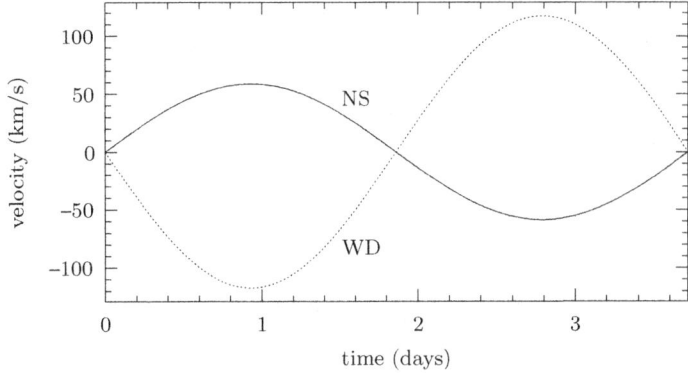

References

1. C.R. Keeton, A.O. Petters, Phys. Rev. D **72**(10), 104006 (2005)
2. V. Bozza, Gen. Relativ. Gravit. **42**, 2269 (2010)
3. F.W. Dyson, A.S. Eddington, C. Davidson, R. Soc. Lond. Philos. Trans. Ser. A **220**, 291 (1920)
4. W.W. Campbell, Publ. Astron. Soc. Pac. **35**, 11 (1923)
5. C.R. Keeton, S. Burles, P.L. Schechter, J. Wambsganss, Astrophys. J. **639**, 1 (2006)
6. S. Dong et al., Astrophys. J. **642**, 842 (2006)
7. A. Gould et al., Astrophys. J. **720**, 1073 (2010)
8. K. Griest, Astrophys. J. **366**, 412 (1991)
9. J.L. Feng, Ann. Rev. Astron. Astrophys. **48**, 495 (2010)
10. C. Alcock et al., Astrophys. J. **541**, 734 (2000)
11. C. Alcock et al., Astrophys. J. **542**, 281 (2000)
12. C. Alcock et al., Astrophys. J. **479**, 119 (1997)
13. D. Kubas et al., Astron. Astrophys. **435**, 941 (2005)
14. B.S. Gaudi, Ann. Rev. Astron. Astrophys. **50**, 411 (2012)
15. J.P. Beaulieu et al., Nature **439**, 437 (2006)
16. C.R. Keeton, C.S. Kochanek, Astrophys. J. **495**, 157 (1998)
17. P. Saha, L.L.R. Williams, Astron. J. **127**, 2604 (2004)
18. T. Treu, Ann. Rev. Astron. Astrophys. **48**, 87 (2010)
19. K.C. Wong, C.R. Keeton, K.A. Williams, I.G. Momcheva, A.I. Zabludoff, Astrophys. J. **726**, 84 (2011)
20. R.A.C. Croft, C.A. Metzler, Astrophys. J. **545**, 561 (2000)
21. A. Heavens, A. Refregier, C. Heymans, Mon. Not. R. Astron. Soc. **319**, 649 (2000)
22. J. Blazek, R. Mandelbaum, U. Seljak, R. Nakajima, J. Cosmol. Astropart. Phys. **5**, 041 (2012)
23. N. Kaiser, G. Squires, Astrophys. J. **404**, 441 (1993)
24. J.P. Kneib, P. Natarajan, Astron. Astrophys. Rev. **19**, 47 (2011)
25. D. Clowe, M. Bradač, A.H. Gonzalez, M. Markevitch, S.W. Randall, C. Jones, D. Zaritsky, Astrophys. J. Lett. **648**, L109 (2006)
26. M. Bradač, S.W. Allen, T. Treu, H. Ebeling, R. Massey, R.G. Morris, A. von der Linden, D. Applegate, Astrophys. J. **687**, 959 (2008)
27. B. Famaey, S.S. McGaugh, Living Rev. Relativ. **15**, 10 (2012)
28. A. Refregier, Ann. Rev. Astron. Astrophys. **41**, 645 (2003)
29. A. Albrecht et al., ArXiv e-prints arXiv:astro-ph/0609591 (2006)

Chapter 10
Relativity

With his special and general theories of relativity, Einstein revolutionized our understanding of space, time, gravity, and hence motion. Why, then, have we spent so much time with Newton? Newtonian physics is a good approximation when motion is slow compared with the speed of light and gravity is "weak" in a sense to be defined. A lot of astrophysics research is still carried out under these assumptions. That said, discussing our modern conception of gravity and motion opens fascinating topics such as the weirdness of spacetime around black holes and (in Chap. 11) the expanding universe.[1]

10.1 Space and Time: Classical View

All of our discussion of motion so far has relied on an implicit understanding of "space" and "time." Intuitively, we think everyone agrees on what space and time are; we imagine there are universal rods and clocks we can use to define them. Although physicists knew that only *relative* motion matters for inertial observers, they assumed that a universal, absolute reference frame does exist.

To glimpse some consequences of this assumption, let's examine the relation between two inertial reference frames that are moving relative to each other. Consider one frame (x, y, z, t), and a second frame (x', y', z', t') moving relative to the first with a constant speed u in the x-direction. If time and space are the same in both frames, the coordinates must be the same except for a translation in the x-direction:

---

[1]Parts of this presentation draw from books by Carroll and Ostlie [1] and Schutz [2].

C. Keeton, *Principles of Astrophysics: Using Gravity and Stellar Physics to Explore the Cosmos*, Undergraduate Lecture Notes in Physics, DOI 10.1007/978-1-4614-9236-8_10, © Springer Science+Business Media New York 2014

$$t = t'$$
$$x = x' + u\,t'$$
$$y = y'$$
$$z = z'$$

This is referred to as the **Galilean transformation** between reference frames. A direct corollary is

$$\frac{\mathrm{d}x}{\mathrm{d}t} = \frac{\mathrm{d}x'}{\mathrm{d}t'} + u$$

The velocity with respect to the unprimed frame is the simple sum of the velocity with respect to the primed frame, plus the velocity of the primed frame with respect to the unprimed frame. This certainly makes sense intuitively.

If we take a second derivative, we find

$$\frac{\mathrm{d}^2x}{\mathrm{d}t^2} = \frac{\mathrm{d}^2x'}{\mathrm{d}t'^2}$$

so the accelerations are the same in both frames. Then by Newton's second law the net force must likewise be the same, and the laws of physics are equally valid in either frame.

This all made sense until physicists studied electricity, magnetism, and light in the late nineteenth century. On the theoretical side, James Clerk Maxwell's theory of electrodynamics indicated that light is an electromagnetic wave traveling at speed $c = 3.0 \times 10^8\,\mathrm{m\,s^{-1}}$ *in all inertial reference frames*. On the experimental side, Albert Michelson and Edward Morley tried to measure differences in the speed of light emitted by sources moving at different speeds—*and found that there were no differences*. Physicists were stunned. Some suggested there must be a problem in Maxwell's theory. Others supposed there was some substance known as æther pervading the universe whose properties caused all inertial observers to measure the same speed of light.

10.2   Special Theory of Relativity

Albert Einstein took a different approach: he wondered whether the problem lay in misconceptions about space and time. Instead of assuming absolute space and time, he took an operational view: he described how to use a system of rigid rods and synchronized clocks to construct a coordinate system in any reference frame.[2] To Einstein, space and time could be real only to the extent that they could be measured.

That only served to define coordinates within a given reference frame. To relate different reference frames, Einstein proposed two postulates [4]:

1. The equations of motion of any (mechanical) system are the same in all inertial reference frames.
2. The speed of light is constant and universal.

The first postulate is called the **principle of relativity**, and it predated Einstein. What Einstein did was introduce the second postulate as an extension of the theory of electrodynamics, and show that together the two postulates are inconsistent with the Galilean transformation.

10.2.1 Lorentz Transformation

Einstein worked out what different observers would have to say about space and time in order for them to agree on the speed of light. He found the relations:

$$ct = \gamma \, ct' + \gamma \, \beta \, x' \tag{10.1a}$$

$$x = \gamma \, x' + \gamma \, \beta \, ct' \tag{10.1b}$$

$$y = y' \tag{10.1c}$$

$$z = z' \tag{10.1d}$$

where

$$\beta = \frac{u}{c} \quad \text{and} \quad \gamma = \left(1 - \beta^2\right)^{-1/2} \tag{10.2}$$

Note that we use ct and ct' because working with a quantity that has dimensions of length clarifies the interplay between time and space coordinates. The inverse relations are:

$$ct' = \gamma \, ct - \gamma \, \beta \, x \tag{10.3a}$$

$$x' = \gamma \, x - \gamma \, \beta \, ct \tag{10.3b}$$

$$y' = y \tag{10.3c}$$

$$z' = z \tag{10.3d}$$

---

[2]Peter Galison [3] notes that, as a clerk in the Swiss Patent Office, Einstein probably saw many patent applications for schemes to synchronize clocks. The spread of the railroad and telegraph had prompted a need for long-distance synchronization.

The relations (10.1) and (10.3) were already known as the **Lorentz transformation** after Hendrik Lorentz. They had been derived in electrodynamics as the transformation that preserves Maxwell's equations in both reference frames.[3] What Einstein offered was a sweeping new interpretation: time and space are no longer separate, absolute quantities. Rather, they are linked in a 4-dimensional structure we now call **spacetime**. Points in spacetime are referred to as **events**. (We will say more about the structure of spacetime beginning in Sect. 10.5.)

We can use the Lorentz transformation to relate velocities measured in the primed and unprimed frames. The differential version of Eq. (10.1) is

$$c\,dt = \gamma\,c\,dt' + \gamma\,\beta\,dx'$$
$$dx = \gamma\,dx' + \gamma\,\beta\,c\,dt'$$
$$dy = dy'$$
$$dz = dz'$$

Let's rewrite these relations using the components of velocity measured in the primed frame: $v'_x = dx'/dt'$, $v'_y = dy'/dt'$, and $v'_z = dz'/dt'$. We also use $\beta = u/c$. These substitutions yield

$$dt = \left(1 + \frac{uv'_x}{c^2}\right)\gamma\,dt'$$
$$dx = \left(v'_x + u\right)\gamma\,dt'$$
$$dy = v'_y\,dt'$$
$$dz = v'_z\,dt'$$

Now we can find the velocity components in the unprimed frame:

$$v_x = \frac{dx}{dt} = \frac{v'_x + u}{1 + uv'_x/c^2} \tag{10.4a}$$

$$v_y = \frac{dy}{dt} = \frac{v'_y}{\gamma(1 + uv'_x/c^2)} \tag{10.4b}$$

$$v_z = \frac{dz}{dt} = \frac{v'_z}{\gamma(1 + uv'_x/c^2)} \tag{10.4c}$$

Under the Lorentz transformation, velocity in the unprimed frame is no longer a simple sum of the velocity in the primed frame and the velocity of the frame itself. In Problem 10.2 you can see how the modified transformation explains why the speed of light does not depend on the speed of the source.

---

[3]Maxwell's equations are not invariant under the Galilean transformation.

Non-relativistic Limit

Our experience and intuition are more closely aligned with the Galilean transformation than the Lorentz transformation, but of course they are associated with motion that is slow compared with the speed of light. Let's see if we can confirm our intuition by examining the Lorentz transformation when $u/c \ll 1$. A Taylor series expansion of Eq. (10.1) yields

$$t \approx t' + \frac{ux'}{c^2} + \mathcal{O}\left(\frac{u^2}{c^2}\right) \quad \text{and} \quad x \approx x' + ut' + \mathcal{O}\left(\frac{u^2}{c^2}\right)$$

Since the speed of light is so large, ux'/c^2 is small and to a very good approximation we can write

$$t \approx t' \quad \text{and} \quad x \approx x' + ut$$

A similar analysis applied to Eq. (10.4) yields

$$v_x \approx v'_x + u \qquad v_y \approx v'_y \qquad v_z \approx v'_z$$

Thus, the Lorentz transformation does not actually invalidate the Galilean transformation (which is reassuring since the latter was the basis of all physics prior to the twentieth century). Rather, it clarifies that the Galilean transformation should be used only when motion is slow compared with the speed of light.

10.2.2 Loss of Simultaneity

When speeds are not small, however, we must use the full Lorentz transformation. Following Einstein, we can use some *gedanken* (German for "thought") experiments to uncover some consequences of the interplay between space and time. First, consider two lights that are set up to flash at the same time in the primed coordinate system (which we can take to be $t' = 0$). What are the times of the flashes in the unprimed coordinate system? Let the first event be the flash of light #1:

$$\text{event 1:} \quad (t', x') = (0, x'_1) \quad \Rightarrow \quad ct_1 = \gamma \beta x'_1$$

Let the second event be the flash of light #2:

$$\text{event 2:} \quad (t', x') = (0, x'_2) \quad \Rightarrow \quad ct_2 = \gamma \beta x'_2$$

The time between the two flashes is $\Delta t' = 0$ in the primed frame, but the time between the two flashes in the unprimed frame is

$$c \, \Delta t = \gamma \beta (x'_2 - x'_1) \neq 0$$

In other words, events that are simultaneous in one reference frame *are not simultaneous in other reference frames*. This is the first indication that there is something decidedly non-intuitive in the new way of thinking about the universe.

10.2.3 Time Dilation

Now focus on a single flashing light and consider the time between flashes. For simplicity, let's put the light at the origin of the primed frame. If we again consider two flashes as two spacetime events, we have:

$$\text{event 1:} \quad (t', x') = (0, 0) \quad \Rightarrow \quad t_1 = 0$$
$$\text{event 2:} \quad (t', x') = (\Delta t', 0) \quad \Rightarrow \quad t_2 = \gamma \, \Delta t'$$

In other words, the time between flashes as measured in the unprimed frame is

$$\Delta t = \gamma \, \Delta t' \tag{10.5}$$

Since $\gamma \geq 1$, more time passes between flashes in the unprimed frame than in the primed frame. This effect is known as **time dilation**. If we think of a flashing light as a kind of clock, we can distill this into the maxim "moving clocks run slowly."

If the measurement of time between events depends on the reference frame, how can we single out a frame to focus on when we study physical laws? The most natural quantity is the time interval measured by a clock at rest with respect to the events, which has the advantage of being the *smallest* time interval that any clock will measure. We call this the **proper time**.

Time dilation is a definite and weird prediction of relativity, so it deserves to be tested experimentally. One of the classic tests was performed in 1963, when David Frisch and James Smith [5] studied elementary particles called muons coming from space. Frisch and Smith compared the number of muons detected at the top of Mt. Washington in New Hampshire (1,907 m above sea level) with the number detected at sea level. It takes a certain amount of time Δt for muons to travel the intervening distance, but the measurements indicated that the muons "experienced" a much shorter interval $\Delta t' < \Delta t$. The experiment confirmed predictions of time dilation, as you can see in more detail in Problem 10.3. In 1971, Joseph Hafele and Richard Keating [6, 7] flew atomic clocks on airplanes to do a more controlled test of time dilation. That experiment involved gravity as well as motion, so we will consider it among tests of general relativity (Sect. 10.4.5). Today, relativistic time dilation is built into the Global Positioning System (Sect. 10.4.6).

10.2.4 Doppler Effect

In a final use of the flashing light, let's consider the times when flashes reach an observer who is stationary at the origin of the unprimed frame. This is what the observer (whether a person or an instrument) would actually measure. The first flash occurs at coordinates $(t_1', x_1') = (t_1', 0)$ in the primed frame, which correspond to coordinates $(t_1, x_1) = (\gamma\, t_1', \ \gamma\, u\, t_1')$ in the unprimed frame. In order for this flash to be observed, it must travel to the observer at the origin. The distance it must travel is $\gamma\, u\, t_1'$, and the time it takes is $\gamma\, u\, t_1'/c$. The time at which the flash is observed is therefore

$$t_{1,\text{obs}} \;=\; \gamma\, t_1' + \frac{\gamma\, u t_1'}{c} \;=\; \gamma\left(1 + \frac{u}{c}\right) t_1'$$

By similar reasoning, we find the time at which the second flash is observed to be

$$t_{2,\text{obs}} = \gamma\left(1 + \frac{u}{c}\right) t_2'$$

Thus, the time that elapses between observations of the two flashes is

$$\Delta t \;=\; t_{2,\text{obs}} - t_{1,\text{obs}} \;=\; \gamma\left(1 + \frac{u}{c}\right)\Delta t' \;=\; \left(\frac{1 + u/c}{1 - u/c}\right)^{1/2}\Delta t'$$

where $\Delta t' = t_2' - t_1'$ is the time interval between flashes in the frame in which they are emitted, and in the last step we substitute for γ using Eq. (10.2).

Now if we replace the flashes with peaks of a wave, the time between peaks is the inverse of the frequency of the wave: so $\nu_{\text{obs}} = 1/\Delta t$ in the frame of observations, and $\nu_{\text{em}} = 1/\Delta t'$ in the frame of emission. Then we have

$$\frac{\nu_{\text{obs}}}{\nu_{\text{em}}} = \left(\frac{1 - u/c}{1 + u/c}\right)^{1/2} \tag{10.6}$$

Equivalently, in terms of wavelength we can use $\lambda \propto \nu^{-1}$ to write

$$\frac{\lambda_{\text{obs}}}{\lambda_{\text{em}}} = \left(\frac{1 + u/c}{1 - u/c}\right)^{1/2} \tag{10.7}$$

This is the **relativistic Doppler effect**. It says that if a light source is moving away from the observer ($u > 0$), the observed frequency of light is lower than the emitted frequency; this corresponds to a longer wavelength and hence a redder color, so we call this a **redshift**. Conversely, if a light source is moving toward the observer ($u < 0$), the Doppler effect produces a **blueshift**.

In the non-relativistic limit, we can use a Taylor series expansion to write

$$\frac{\lambda_{obs}}{\lambda_{em}} \approx 1 + \frac{u}{c} + \mathcal{O}\left(\frac{u^2}{c^2}\right)$$

We often express the shift in terms of the change in wavelength, $\Delta\lambda = \lambda_{obs} - \lambda_{em}$:

$$\frac{\Delta\lambda}{\lambda_{em}} \approx \frac{u}{c} \tag{10.8}$$

This is the Doppler shift of light when the source of light is moving at non-relativistic speeds. It is what we use, for example, to measure the motions of stars and discover that they are orbited by planets (Sect. 4.3.1).

10.2.5 Length Contraction

Let's change gedanken tools and consider a ruler oriented along the x-axis that is at rest in the primed frame. Place one end of the ruler at $x_1' = 0$ and the other end at $x_2' = L'$, so the ruler's length in the primed frame is L'. What is the length of the ruler in the unprimed frame? It may seem backward at first, but let's use the Lorentz transformation $x_1' = \gamma x_1 - \gamma u t_1$ (and likewise for the other end). Then we have

$$x_2' - x_1' = (\gamma x_2 - \gamma u t_2) - (\gamma x_1 - \gamma u t_1) = \gamma (x_2 - x_1) - \gamma u (t_2 - t_1)$$

It is important to measure the ends of the ruler *at the same time* in the unprimed frame. Then we can put $t_2 - t_1 = 0$ and obtain

$$x_2' - x_1' = \gamma (x_2 - x_1) \quad \Rightarrow \quad L' = \gamma L \quad \Rightarrow \quad L = \frac{L'}{\gamma}$$

In other words, the moving ruler appears to have a length $L = L'/\gamma$ that is *shorter* than its length at rest. This is known as **length contraction**: moving objects appear shorter in the direction of motion. As with proper time, if we want to single out a particular length then we usually use the **proper length** measured when the object is at rest.

10.3 General Theory of Relativity

In order to deal with gravity, Einstein had to generalize his theory from inertial to accelerated reference frames. This led him to sophisticated mathematical structures including non-Euclidean geometry, manifolds, tensors, and more. We will glimpse some of the mathematical framework in Sect. 10.5, but first let's examine the physical principles that underlie general relativity.

10.3.1   Concepts of General Relativity

General relativity is a **geometric theory of gravity**: acceleration is a consequence of the curvature of spacetime. There are two key concepts governing the interaction between curvature and mass (as stated by Misner, Thorne and Wheeler [8]):

1. "Space acts on matter, telling it how to move."
2. "In turn, matter reacts back on space, telling it how to curve."

People often think of these in terms of a rubber sheet analogy. Imagine stretching a rubber sheet so it lies flat. This is a model of a 2-dimensional universe described by special relativity. Now place a bowling ball on the sheet. The bowling ball deforms the sheet; this is point #2 above. Then roll a ping-pong ball near the bowling ball. The curvature induced by the bowling ball controls how the ping-pong ball moves; this is point #1.

   It is important to understand that this is an *analogy*. It is imperfect because it describes the 2-d rubber sheet as being curved into the third dimension. For the 3-d spatial universe, we would have to think of the curvature as extending into a fourth spatial dimension. I cannot picture such a thing! Also, the analogy works only if there is *external* gravity pulling on the bowling ball to distort the rubber sheet. In general relativity, everything needs to happen *within* the theory. So the rubber sheet is useful as a pictorial analogy, but please do not take it too literally. We will be more precise about describing curvature soon.

10.3.2   Principle of Equivalence

When Einstein was trying to figure out how to describe gravity and acceleration, he had an important thought: "If a person falls freely he will not feel his own weight." [9] To be more precise, let's go back to Newton for a moment. We have often used the equation

$$\frac{GMm}{r^2} \; = \; F \; = \; ma \tag{10.9}$$

and cancelled the m's from both sides. But it is not obvious that they have to be the same. The m on the left describes how an object *feels* the force of gravity; we might call it the "gravitational mass," m_g. The m on the right describes an object's inertia and how it *responds* to a force; we might call it the "inertial mass," m_i. We really ought to rewrite Eq. (10.9) as

$$\frac{GMm_g}{r^2} \; = \; F_g \; = \; m_i a$$

which yields the acceleration

$$a = \frac{GM}{r^2} \frac{m_g}{m_i}$$

It is an experimental fact that the ratio m_g/m_i is 1. Galileo is said to have shown this by dropping balls off the Leaning Tower of Pisa. Modern versions of the experiment reveal that the difference between gravitational and inertial masses is less than 1 part in 10^{12} (e.g., [10]). Therefore we can say to high precision that the acceleration due to gravity is independent of mass—all objects fall at the same rate.[4]

If that is true, then no experiment can reveal the acceleration because the equipment will fall at the same rate as the sample. By extension, no experiment can reveal that gravity is at work! Another way to say this is that a freely falling laboratory is equivalent to a lab floating in empty space. Within such a freely falling lab, we can apply the principles of special relativity. This simplifies things quite a lot.

Strictly speaking, this reasoning holds only in a region of space in which the acceleration due to gravity is uniform. Since objects on the surface of Earth fall towards the center of the planet, objects at different positions fall in different directions; that is enough to reveal the gravity. But if we pick a region that is small enough, these effects are negligible.

Einstein turned this idea into the foundation of his theory of gravity, calling it the **principle of equivalence**:

- All local, freely falling, non-rotating frames of reference are equivalent for performing physical experiments.

This is the fundamental principle that allows us to identify some physical aspects of general relativity.

10.3.3 Curvature of Spacetime

Let's apply the principle of equivalence to some thought experiments to understand how gravity affects spacetime. Consider a lab in freefall in a gravitational field where the acceleration due to gravity is g, as depicted in Fig. 10.1. Suppose a light source on the left-hand wall is pointed toward the right. By the principle of equivalence, the lab acts as a local inertial reference frame, so an observer in the lab would see the light travel in a straight line from one side to the other.

---

[4]This is not true of other forces. Consider the acceleration created by the electric force acting on an object with mass m and charge q near another charge Q: $a = Qq/mr^2$ does depend on mass.

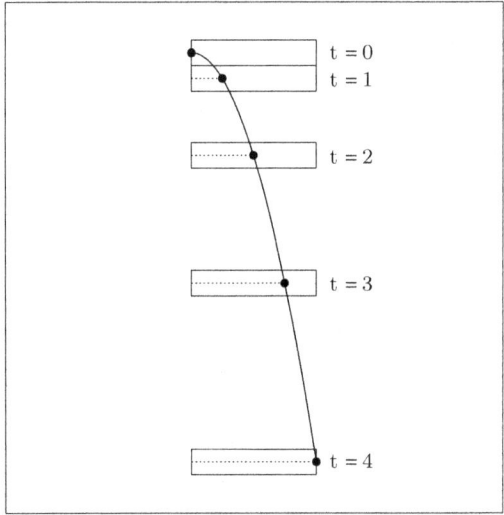

Fig. 10.1 Setup for gedanken experiment #1. A light ray travels horizontally in a lab that is in freefall with uniform acceleration. To an observer in the lab, the light ray travels straight across the room (*dotted lines*). But to an outside observer, the light ray follows a curved trajectory

What would be seen by an observer on the ground (who is stationary in the gravitational field)? As the light moves to the right in the lab, the lab accelerates downward, so the trajectory of the light looks like a parabola. Gravity has caused light to curve!

To be specific, let's write the equation of the trajectory. If the light starts at $(x, y) = (0, 0)$, its position as a function of time is

$$x = ct \quad \text{and} \quad y = -\frac{1}{2} g t^2$$

Eliminating t yields

$$y = -\frac{g x^2}{2 c^2}$$

We know this trajectory is curved, but by how much? To quantify curvature, think about a circle:

$$x^2 + y^2 = R_c^2 \quad \Rightarrow \quad y = \left(R_c^2 - x^2 \right)^{1/2}$$

If we know $y(x)$, we can extract the radius R_c as follows:

$$\frac{dy}{dx} = -\frac{x}{(R_c^2 - x^2)^{1/2}}$$

$$\frac{d^2 y}{dx^2} = -\frac{R_c^2}{(R_c^2 - x^2)^{3/2}}$$

$$\text{at } x = 0 : \quad \frac{d^2 y}{dx^2} = -\frac{1}{R_c}$$

(The minus sign means the circle is curved downward.) As a general rule, then, we can define the **radius of curvature** for a trajectory $y(x)$ as

$$R_c = \frac{1}{|d^2 y/dx^2|}$$

Heuristically, R_c is the distance over which the trajectory deviates significantly from a straight line, so a smaller value corresponds to a greater curvature. For our gedanken experiment, the radius of curvature is found as follows:

$$\frac{dy}{dx} = -\frac{gx}{c^2} \quad \Rightarrow \quad \frac{d^2 y}{dx^2} = -\frac{g}{c^2} \quad \Rightarrow \quad R_c = \frac{c^2}{g}$$

If we assume Newtonian gravity for simplicity, the gravitational acceleration at a distance r from an object of mass M is

$$g = \frac{GM}{r^2}$$

so the radius of curvature for light is

$$R_c = \frac{c^2 r^2}{GM}$$

Example: Earth

Near the surface of Earth, the acceleration due to gravity is $g = 9.80\,\text{m s}^{-2}$ so the radius of curvature for light is

$$R_c = \frac{c^2}{g} = 9.17 \times 10^{15}\,\text{m} = 0.97\,\text{ly}$$

The radius of curvature is huge—far, far bigger than the size of Earth—which means the curvature is quite small. Nevertheless, it is significant enough to produce all the familiar effects of gravity.

Example: Black Hole

Near the event horizon,

$$r = R_S = \frac{2GM}{c^2} \quad \Rightarrow \quad R_c = \frac{c^2 R_S^2}{GM} = 2R_S$$

The radius of curvature is comparable to the size of the event horizon, which means gravity is strong.

Bottom line: we have found that gravity causes light to move on a curved trajectory. A similar analysis could be done for material particles. Operationally, then, what we mean when we say spacetime is curved is that objects follow curved trajectories. To summarize:

$$\text{objects follow curved trajectories} \quad \Leftrightarrow \quad \text{space is curved}$$

10.3.4 Gravitational Redshift and Time Dilation

Consider the same freely falling lab, only now put the light source on the floor and have it shine upward. By the time the light reaches a detector in the ceiling, the lab will be moving faster because of the acceleration. If the light moves a distance h, the time elapsed is $t = h/c$ and the lab's new speed is $u = -gt = -gh/c$ (where the minus sign means downward). By Eq. (10.6), there should be a Doppler blueshift of the form

$$\Delta\nu_{\text{Doppler}} = \nu_{\text{obs}} - \nu_{\text{em}} = \nu_{\text{em}}\left[\left(\frac{1-u/c}{1+u/c}\right)^{1/2} - 1\right] \approx -\nu_{\text{em}}\frac{u}{c} \approx \nu_{\text{em}}\frac{gh}{c^2}$$

(assuming $u \ll c$). Here is the crux of this experiment: if there were a Doppler shift, we would know the lab is accelerating, and that would violate the equivalence principle. The only way out is to say that gravity causes the frequency of light to shift by just the right amount to cancel the Doppler shift. In other words, there must be a **gravitational redshift**

$$\Delta\nu_{\text{grav}} \approx -\nu_{\text{em}}\frac{gh}{c^2} \tag{10.10}$$

This actually makes sense physically: light loses energy as it moves against gravity, and since $E \propto \nu$ the frequency must decrease.

The preceding analysis assumed a constant gravitational acceleration. To deal with the general case, we can use the gravitational acceleration $g = GM/r^2$ and change the height to dr and the frequency shift to $d\nu$, obtaining

$$d\nu \approx -\nu\,\frac{GM}{c^2 r^2}\,dr$$

We can then integrate:

$$\int_{v_i}^{v_f} \frac{dv}{v} \approx - \int_{r_i}^{r_f} \frac{GM}{c^2 r^2}\, dr$$

$$\ln \frac{v_f}{v_i} \approx \frac{GM}{c^2} \left(\frac{1}{r_f} - \frac{1}{r_i} \right)$$

$$\frac{v_f}{v_i} \approx \exp\left[\frac{GM}{c^2} \left(\frac{1}{r_f} - \frac{1}{r_i} \right) \right] \approx 1 + \frac{GM}{c^2} \left(\frac{1}{r_f} - \frac{1}{r_i} \right)$$

where in the last step we use the Taylor series expansion $e^x \approx 1 + x$ for $x \ll 1$. It is convenient to take the "final" point to be at infinity, corresponding to an observer far from the object. This yields

$$\frac{v_\infty}{v(r)} \approx 1 - \frac{GM}{c^2 r}$$

The oscillations of the light act as a kind of clock, where the elapsed time is $t \propto v^{-1}$. We can therefore change the frequency equation into time,

$$\frac{t(r)}{t_\infty} \approx 1 - \frac{GM}{c^2 r}$$

This is **gravitational time dilation**: a clock in a gravitational field runs more slowly than a clock that is far away in empty space.

In this analysis we have made Taylor series approximations and computed the leading order relativistic effect. An exact analysis gives (see Sect. 10.6.1)

$$\frac{t(r)}{t_\infty} = \left(1 - \frac{2GM}{c^2 r} \right)^{1/2} \tag{10.11}$$

Gravitational time dilation becomes strong only when r gets close to $2GM/c^2$. We will see more about this when we study black holes.

Example: Surface of Earth

Clocks on the surface of Earth should run slower than clocks far away in empty space. How much slower? The difference in elapsed time is

$$\frac{\Delta t}{t_\infty} = \frac{t(r) - t_\infty}{t_\infty} \approx - \frac{GM_\oplus}{c^2 R_\oplus}$$

$$\approx - \frac{(6.67 \times 10^{-11}\, \mathrm{m^3\, kg^{-1}\, s^{-2}}) \times (5.97 \times 10^{24}\, \mathrm{kg})}{(3.0 \times 10^8\, \mathrm{m\, s^{-1}})^2 \times (6.38 \times 10^6\, \mathrm{m})}$$

$$\approx -7 \times 10^{-10}$$

To put this number in context: if your life expectancy is 100 years, you would get to live about 2 s longer on Earth than if you were in space with no gravity. Please note, though, that your experience of time is unaffected by acceleration or gravity; you would not actually have more "time" to enjoy life. You would just appear to age slowly as seen by those living in weaker gravity, while to you they would seem to age quickly.

10.4 Applications of General Relativity

In the previous section we used gedanken experiments to discover the curvature of spacetime, gravitational redshift, and gravitational time dilation. Now let's consider several real experiments that have confirmed these predictions of general relativity.

10.4.1 Mercury's Perihelion Shift (1916)

When we studied planetary motion (Chaps. 3 and 4), we said a planet follows a perfectly elliptical orbit and traces it over and over again. Strictly speaking that is true only in an ideal two-body problem. When a planet's orbit is dominated by the Sun but perturbed by another planet, the situation is only approximately two-body. The resulting orbit can be thought of as an ellipse that precesses, or rotates a little each time the planet goes around (see Fig. 10.2). We can quantify the effect by measuring the shift in the perihelion position.[5]

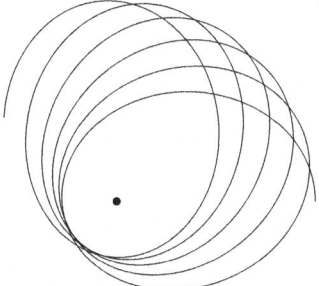

Fig. 10.2 An illustration of perihelion shift. Roughly speaking, the orbit is approximately elliptical but the ellipse rotates with time, which causes the location of perihelion to vary. The effect shown here is greatly exaggerated, with $\epsilon = 0.05$ (compared with $\epsilon = 8 \times 10^{-8}$ for Mercury)

---

[5]We could use any part of the orbit, but perihelion is distinctive.

In our Solar System, the largest perihelion shift is 560 arcsec/century for Mercury. Most of the shift can be attributed to perturbations from other planets, but after all the known planets are taken into account a shift of 43 arcsec/century remains unexplained. Historically, some people speculated that there might be another planet closer to the Sun than Mercury that caused the additional perihelion shift. The hypothetical planet was called Vulcan [11].

When Einstein considered Mercury's orbit in the context of general relativity, he discovered that it would be described by the equation of motion (see Sect. 10.6.4)

$$\frac{d^2 r}{d\tau^2} = -\frac{GM}{r^2} + \frac{\ell^2}{r^3} - \frac{3GM\ell^2}{c^2 r^4}$$

where $\ell = r^2 d\phi/d\tau$ is the specific angular momentum, which is conserved. The first term is standard Newtonian gravity, and the second term is the usual centrifugal term. The third term is new in general relativity, and it perturbs the orbit away from a pure ellipse. To see this, let's go all the way back to our analysis of Newtonian gravity in Chap. 3. Recall that we changed independent variables from time to angle, and we put $r = 1/u$. Repeating the analysis yields

$$\frac{d^2 r}{d\tau^2} = -\ell^2 u^2 \frac{d^2 u}{d\phi^2}$$

so the equation of motion becomes

$$\frac{d^2 u}{d\phi^2} + u = \frac{GM}{\ell^2} + \frac{3GM}{c^2} u^2$$

Let's define

$$\epsilon = \frac{3(GM)^2}{c^2 \ell^2} \tag{10.12}$$

and then rewrite the equation of motion as

$$\frac{d^2 u}{d\phi^2} + u = \frac{GM}{\ell^2} + \epsilon \frac{\ell^2}{GM} u^2$$

With $\epsilon = 0$ this would be Newtonian gravity and the solution would be an ellipse. For Mercury, ϵ is very small and so we can look for a solution that is perturbed away from an ellipse. The solution has the form [12]

$$u(\phi) \approx \frac{GM}{\ell^2} \left\{ 1 + e \cos[\phi(1 - \epsilon)] + \epsilon \left[1 + e^2 \left(\frac{1}{2} - \frac{1}{6} \cos 2\phi \right) \right] + \mathcal{O}\left(\epsilon^2\right) \right\}$$

Notice the second term. In order for $\cos[\phi(1 - \epsilon)]$ to complete a full cycle, ϕ has to range from 0 to $2\pi/(1 - \epsilon) \approx 2\pi(1 + \epsilon)$. Therefore it takes an extra azimuthal angle $\Delta\phi \approx 2\pi\epsilon$ for the planet to return to perihelion.

How strong is the effect? Using Eq. (3.11) we can rewrite ℓ in terms of orbital elements and obtain

$$\epsilon = \frac{3GM}{c^2 a(1 - e^2)} \tag{10.13}$$

Mercury has $a = 0.387$ AU and $e = 0.206$, yielding

$$\epsilon = \frac{3 \times 6.67 \times 10^{-11}\, \text{m}^3\, \text{kg}^{-1}\, \text{s}^{-2} \times 1.99 \times 10^{30}\, \text{kg}}{(3.0 \times 10^8\, \text{m s}^{-1})^2 \times (0.387 \times 1.50 \times 10^{11}\, \text{m}) \times (1 - 0.206^2)} = 8 \times 10^{-8}$$

The precession is an angle of about $2\pi\epsilon$ per orbit, so given Mercury's orbital period of $P = 87.97$ day the precession rate is

$$\frac{2\pi\epsilon}{P} = \frac{2\pi \times 8 \times 10^{-8}}{87.97 \times 86{,}400\, \text{s}} = 6.6 \times 10^{-14}\, \text{rad s}^{-1} = 43\, \text{arcsec/century}$$

This was the first good explanation for Mercury's perihelion shift, and it convinced Einstein that he was on the right track with his new theory of gravity.

10.4.2 Bending of Light (1919)

We have already discussed Einstein's prediction for the bending of light by the Sun, and the measurements in 1919 and 1922 that confirmed the prediction. This was the second significant test of general relativity, and the first true prediction. (Einstein's explanation of Mercury's perihelion shift was "merely" an explanation of existing data.)

10.4.3 Gravitational Redshift on Earth (1960)

In Sect. 10.3.4 we discussed gravitational time dilation on the surface of Earth. While the effect is small, it turns out that we can measure the corresponding gravitational redshift. Atomic nuclei have energy levels just like atomic electrons, so they can produce emission or absorption lines in energy spectra. The difference is that nuclear lines generally involve much higher energies and are very narrow. For example, iron-57 has a spectral line with energy $E = 14.4$ keV and line width $\delta E \sim 10^{-11}$ keV. Because the line is so narrow, we can measure energies or frequencies very precisely. In 1959 and 1960, Pound and Rebka [13, 14] realized they could use this to measure gravitational redshift. They wanted of course to have light traverse as much vertical distance as possible; in the physics department at Harvard, the best option was a height of $h = 22.6\,\text{m}$ in the stairwell. From Eq. (10.10) with $g = GM/r^2$, the fractional change in frequency as the light travels upward is

$$
\left(\frac{\Delta\nu}{\nu}\right)_{\text{up}} = -\frac{GM_{\oplus}h}{c^2 R_{\oplus}^2}
$$

$$
= -\frac{(6.67 \times 10^{-11}\,\text{m}^3\,\text{kg}^{-1}\,\text{s}^{-2}) \times (5.97 \times 10^{24}\,\text{kg}) \times (22.6\,\text{m})}{(3.0 \times 10^8\,\text{m}\,\text{s}^{-1})^2 \times (6.38 \times 10^6\,\text{m})^2}
$$

$$
= -2.5 \times 10^{-15}
$$

As the light descends there would be a shift that has the same amplitude but the opposite sign. Pound and Rebka measured the combination of the upward and downward shifts as a way to remove any non-gravitational effects, finding

$$
\left(\frac{\Delta\nu}{\nu}\right)_{\text{down}} - \left(\frac{\Delta\nu}{\nu}\right)_{\text{up}} = (5.1 \pm 0.5) \times 10^{-15}
$$

It seems astounding that we can measure relativistic effects to a few parts in 10^{15}. More recent experiments are even more precise.

10.4.4 Gravitational Redshift from a White Dwarf (1971)

The gravitational redshift on Earth is small because Earth's gravity is weak. Even on the Sun the effect is small: $\Delta\nu/\nu = -2 \times 10^{-6}$ between the surface and a point at infinity. To get a larger shift we need an object that is massive but compact. A white dwarf star is typically about as massive as the Sun but only as large as Earth (see Sect. 17.2). The nearest white dwarf is Sirius B, so named because it is in a binary with the bright star Sirius. In 1971, Greenstein et al. [15] managed to measure the gravitational redshift of light from Sirius B. (See Hetherington [16] for more about the history.) Greenstein et al. analyzed the spectrum of Sirius B to infer that the star has a radius of

$$
R = 0.0078\,R_{\odot} = 5.42 \times 10^6\,\text{m} = 0.85\,R_{\oplus}
$$

and a surface gravity of

$$
g = 4.47 \times 10^6\,\text{m}\,\text{s}^{-2}
$$

Together, these imply a mass of

$$M = \frac{gR^2}{G} = 1.97 \times 10^{30}\,\text{kg} = 0.99\,M_\odot$$

The gravitational redshift of a photon leaving the surface and traveling to infinity, expressed in terms of wavelength, is

$$\frac{\lambda(\infty)}{\lambda(r)} = \frac{\nu(r)}{\nu(\infty)} \approx 1 + \frac{GM}{c^2 r}$$

What we measure is the shift in wavelength,

$$z \equiv \frac{\lambda(\infty)}{\lambda(r)} - 1 \approx \frac{GM}{c^2 r} \approx \frac{gR}{c^2}$$

Given the properties of Sirius B, we predict a redshift of

$$z_{\text{predicted}} \approx \frac{(4.47 \times 10^6\,\text{m s}^{-2}) \times (5.42 \times 10^6\,\text{m})}{(3.0 \times 10^8\,\text{m s}^{-1})^2} \approx 2.7 \times 10^{-4}$$

The measured shifts in the spectral lines were

$$z_{\text{measured}} = (3.0 \pm 0.5) \times 10^{-4}$$

10.4.5 Flying Clocks (1971)

In October 1971, Joseph Hafele and Richard Keating [6, 7] flew atomic clocks on airplanes around the Earth. Airborne clocks experience time dilation (relative to surface clocks) for two reasons: motion, because airplanes move at different speeds than the surface of Earth; and gravity, because gravity is a little weaker at the altitudes where planes fly.

Let's consider the motion first. Both the airplane and the surface of Earth follow curved trajectories, so strictly speaking they are not inertial reference frames, but we will still use the special relativistic expression (10.5) to estimate the time dilation due to motion. We will, however, reference our measurements to the center of Earth so we can treat the surface and airplane on equal footing. Earth's rotation causes a clock at the equator to have a speed relative to the center of Earth of

$$v_S = \frac{2\pi R_\oplus}{P_{\text{rot}}} = \frac{2\pi \times (6.38 \times 10^6\,\text{m})}{86,400\,\text{s}} = 464\,\text{m s}^{-1}$$

Suppose the airplane is flying east/west with speed v_A relative to Earth's center; then write

$$v_A = v_S + u$$

so u is the speed of the airplane relative to Earth's surface. Relative to Earth's center, the surface and airplane have relativistic factors

$$\gamma_S = \left(1 - \frac{v_S^2}{c^2}\right)^{-1/2} \quad \text{and} \quad \gamma_A = \left(1 - \frac{v_A^2}{c^2}\right)^{-1/2}$$

Let t_C be the duration of the airplane flight measured in the reference frame of Earth's center. Then the durations in the surface and airplane frames are

$$t_S' = \frac{t_C}{\gamma_S} \quad \text{and} \quad t_A' = \frac{t_C}{\gamma_A}$$

The difference between the time elapsed on the airplane and the time elapsed on the surface is

$$t_A' - t_S' = \frac{t_C}{\gamma_A} - t_S' = \left(\frac{\gamma_S}{\gamma_A} - 1\right) t_S'$$

The fractional change induced by the motion is

$$\left[\frac{t_A' - t_S'}{t_S'}\right]_{\text{motion}} = \left(\frac{1 - v_A^2/c^2}{1 - v_S^2/c^2}\right)^{1/2} - 1$$

Since the speeds are small compared with the speed of light, we can do a Taylor series expansion in v_S/c and v_A/c:

$$\left[\frac{t_A' - t_S'}{t_S'}\right]_{\text{motion}} \approx \left(1 - \frac{v_A^2}{2c^2}\right)\left(1 + \frac{v_S^2}{2c^2}\right) - 1 \approx \frac{v_S^2 - v_A^2}{2c^2}$$

Now we write $v_A = v_S + u$ and simplify:

$$\left[\frac{t_A' - t_S'}{t_S'}\right]_{\text{motion}} \approx \frac{v_S^2 - (v_S^2 + 2v_S u + u^2)}{2c^2} \approx -\frac{(2v_S + u)u}{2c^2}$$

To get some specific numbers, let's suppose the clocks flew on the Concorde, which used to reach a groundspeed of about $650\,\mathrm{m\,s^{-1}}$. Then we find:

eastbound, $u = +650\,\mathrm{m\,s^{-1}}$: $\left[\dfrac{t_A' - t_S'}{t_S'}\right]_{\text{motion}} \approx -5.7 \times 10^{-12} \approx -490$ ns/day

westbound, $u = -650\,\mathrm{m\,s^{-1}}$: $\left[\dfrac{t_A' - t_S'}{t_S'}\right]_{\text{motion}} \approx 1.0 \times 10^{-12} \approx +90$ ns/day

Note that the time dilation due to motion depends on the direction in which the airborne clock flies. This is an important part of the relativistic prediction.

Now let's consider the effect of gravity. Using Eq. (10.11) but making a Taylor series expansion, we can write the gravitational time dilation as

$$\frac{t(r+h)}{t(r)} = \frac{t(r+h)/t_\infty}{t(r)/t_\infty} \approx \frac{1 - GM/c^2(r+h)}{1 - GM/c^2 r}$$

$$\approx 1 - \frac{GM}{c^2(r+h)} + \frac{GM}{c^2 r} \approx 1 + \frac{GM h}{c^2 r(r+h)}$$

To this point we have only assumed that r and $r+h$ are large compared with GM/c^2. For this experiment we can do an additional Taylor series expansion with $h \ll r$:

$$\frac{t(r+h)}{t(r)} \approx 1 + \frac{GM h}{c^2 r^2}$$

Then we identify $t(r) = t'_S$ with the surface and $t(r+h) = t'_A$ with the airplane, so we can write the time dilation induced by gravity as

$$\left[\frac{t'_A - t'_S}{t'_S}\right]_{\text{gravity}} \approx \frac{GM h}{c^2 r^2}$$

The Concorde flew at an altitude of about 20 km, so the gravitational time shift is

$$\left[\frac{t'_A - t'_S}{t'_S}\right]_{\text{gravity}} \approx 2.2 \times 10^{-12} \approx +190 \text{ ns/day}$$

This is the same for airplanes moving both east and west.

We have considered an idealized experiment (daylong Concorde flights over the equator) that captures the main ideas, but Hafele and Keating analyzed the actual flight paths. They found the following time shifts (measured in nanoseconds):

	Motion	Gravity	Net prediction	Measurement
Eastbound	-184 ± 18	144 ± 14	-40 ± 23	-59 ± 10
Westbound	96 ± 10	179 ± 18	275 ± 21	273 ± 7

(The uncertainties in the predictions include uncertainties in the flight parameters.) Notice that the east- and westbound flights have motion shifts with different signs, as we discussed. Also, they have different gravity shifts, presumably because the flights had different altitudes and/or durations. The key result is that the measurements confirm the predictions; relativistic time dilation can be measured in a controlled experiment.

10.4.6   Global Positioning System (1989)

Since 1989 we have had a widespread example of flying clocks: the Global Positioning System. GPS receivers take the time received from a satellite, compare it with the time on Earth, and use the difference (along with the known speed of light) to determine the distance to the satellite. Measuring distances to multiple satellites makes it possible to triangulate a position on Earth to high precision. The entire system rests on careful coordination between satellite and surface clocks, but relativity says they tick at different rates. Relativistic effects must therefore be taken into account for GPS to work. Let's estimate the size of those effects (see [17] for a more detailed discussion).

Each GPS satellite orbits about $h = 20{,}000\,\text{km} = 2 \times 10^7\,\text{m}$ above the surface of Earth. Its orbital speed is therefore

$$
v = \left(\frac{GM}{r+h} \right)^{1/2} = 3.9 \times 10^3\,\text{m s}^{-1}
$$

Each GPS satellite is moving faster than the surface of the Earth, so there is time dilation due to motion[6]:

$$
\left[\frac{t'_A - t'_S}{t'_S} \right]_{\text{motion}} \approx \frac{v_S^2 - v_A^2}{2c^2} \approx -8.3 \times 10^{-11} \approx -7.2\,\mu\text{s/day}
$$

where we again use $v_S = 464\,\text{m/s}$ as the velocity of the surface of Earth. There is also time dilation due to gravity (note that we no longer have $h \ll r$):

$$
\left[\frac{t'_A - t'_S}{t'_S} \right]_{\text{gravity}} \approx \frac{GMh}{c^2 r(r+h)} \approx 5.3 \times 10^{-10} \approx 45.6\,\mu\text{s/day}
$$

The net effect is that GPS satellites gain about $38\,\mu\text{s}$ per day relative to clocks on the ground. If this difference were not taken into account, the time it takes the signal to travel from the satellite would be calculated incorrectly, so the distance to the satellite would be wrong, and the triangulation would be thrown off. How badly? After 1 day the time error would be $\Delta t = 38\,\mu\text{s}$, which would translate into an error in the distance to each satellite of $\Delta \ell = c\,\Delta t = 11\,\text{km}$. This is not precisely the same as the error that a GPS receiver would make when triangulating from multiple satellites, but it does give a sense of the magnitude of the effect.

GPS is successful because the engineers who designed the system used the anticipated orbits to build clocks that would compensate for most of the relativistic effects. Also, each GPS receiver has a computer that performs relativistic calculations to determine additional corrections. Impressive, eh? General relativity at work!

---

[6]For comparison with Sect. 10.4.5, we retain the subscripts "A" for airplane and "S" for surface, respectively, even though the airplane is now a satellite.

10.5 Mathematics of Relativity

Now we turn to the mathematical framework of relativity. While we will not delve into all of the details, we want to get to the point where we can analyze motion around a black hole.[7]

10.5.1 Spacetime Interval

Fundamentally, relativity is about the geometry of spacetime. How do we quantify geometry? The first step is to measure the distance between points. In familiar Euclidean geometry, if we have two points

$$(x, y, z) \quad \text{and} \quad (x + dx, y + dy, z + dz)$$

then we define

$$d\ell^2 = dx^2 + dy^2 + dz^2$$

and say that $d\ell$ is the distance between the two points. If we have a curve, we imagine breaking it into a series of small segments, computing $d\ell$ for each segment, and adding them up (by integrating).

The distance $d\ell$ is the same in all coordinate systems—it is **invariant**. We can rotate or translate the coordinate system any way we like and still get the same distance between the points.

In the spacetime of special relativity, we add time to the mix by defining

$$ds^2 = c^2 dt^2 - (dx^2 + dy^2 + dz^2) \tag{10.14}$$

This is the "distance" between two points in spacetime, which we call the **spacetime interval**. The expression for ds^2 is known as the **metric** because it specifies how we "measure" intervals. Notice that space and time both enter the metric but with different signs. A key property of the spacetime interval is that it is invariant under the Lorentz transformation, so it is a good tool for characterizing the geometry of spacetime in special relativity.

In Euclidean geometry $d\ell^2$ is non-negative. In special relativity, by contrast, ds^2 can be positive, negative, or zero.[8] As we will see below, a light ray has $ds^2 = 0$;

---

[7]Many books do give more details; *A First Course in General Relativity* by Bernard Schutz [2] is a good example.

[8]It is tempting to think that ds is a real-valued quantity such that ds^2 must be non-negative. In relativity, ds^2 is the quantity we work with, and it can be positive, zero, or negative. We may write $\sqrt{ds^2}$ (see below), but we do not write ds by itself.

we call this a *lightlike interval*. For a clock sitting at a fixed position, the spacetime interval between any two ticks has $ds^2 = c^2 d\tau^2$ where τ is the proper time; therefore we say that any positive spacetime interval is *timelike*. Conversely, for a ruler the spacetime interval between the two ends at any given time has $ds^2 = -dL^2$ where L is the proper length; therefore we say that any negative spacetime interval is *spacelike*. To summarize:

$$ds^2 = \begin{cases} c^2 d\tau^2 > 0 & \text{timelike} \\ 0 & \text{lightlike} \\ -dL^2 < 0 & \text{spacelike} \end{cases} \quad (10.15)$$

So far we have worked in Cartesian coordinates. Since many astrophysical objects are (approximately) spherical, it is good to be able to work in spherical coordinates (r, θ, ϕ) as well. In Euclidean geometry, the distance between nearby points in spherical coordinates can be written as

$$d\ell^2 = dr^2 + r^2 d\theta^2 + r^2 \sin^2 \theta \, d\phi^2$$

The extension to the spacetime interval of special relativity just adds time:

$$ds^2 = c^2 dt^2 - (dr^2 + r^2 d\theta^2 + r^2 \sin^2 \theta \, d\phi^2) \quad (10.16)$$

We will see variants of the spatial piece several times in this chapter and the next.

Example: Straight Line

To help understand the spacetime interval, consider a light ray moving in a straight line. Suppose it moves along a line parallel to the x-axis but offset in the z-direction by an amount b. The Cartesian spacetime coordinates can be written

$$(t, x, y, z) = (t, ct, 0, b)$$

The spacetime interval for the light ray is the

$$ds^2 = c^2 dt^2 - c^2 dt^2 = 0$$

This is a lightlike interval, as it should be.

Now consider spherical coordinates. Converting from (t, x, y, z) to (t, r, θ, ϕ) yields

$$r = \left(b^2 + c^2 t^2\right)^{1/2} \qquad \theta = \tan^{-1}\left(\frac{ct}{b}\right) \qquad \phi = 0$$

which implies

$$dr = \frac{ct}{(b^2 + c^2 t^2)^{1/2}} \, c \, dt \qquad d\theta = \frac{b}{b^2 + c^2 t^2} \, c \, dt \qquad d\phi = 0$$

The spacetime interval in spherical coordinates is then

$$ds^2 = c^2 \, dt^2 - dr^2 - r^2 \, d\theta^2$$

$$= c^2 \, dt^2 - \frac{c^2 t^2}{b^2 + c^2 t^2} \, c^2 \, dt^2 - (b^2 + c^2 t^2) \, \frac{b^2}{(b^2 + c^2 t^2)^2} c^2 \, dt^2$$

$$= c^2 \, dt^2 - \left(\frac{c^2 t^2}{b^2 + c^2 t^2} + \frac{b^2}{b^2 + c^2 t^2} \right) c^2 \, dt^2$$

$$= 0$$

While spherical coordinates are less natural for this problem than Cartesian coordinates, they yield the same result. They will be more natural when we study black holes.

10.5.2 4-Vectors

We need to introduce vectors describing motion in four-dimensional spacetime. Let

$$\mathbf{X} = (ct, \mathbf{x}) \tag{10.17}$$

be a 4-d position vector that includes the time coordinate (with a factor of c so all components have dimensions of length). To compute the spacetime interval, we need to introduce a tensor that characterizes the metric. In special relativity, the tensor has the form

$$\mathbf{g} = \begin{bmatrix} 1 & 0 & 0 & 0 \\ 0 & -1 & 0 & 0 \\ 0 & 0 & -1 & 0 \\ 0 & 0 & 0 & -1 \end{bmatrix} \tag{10.18}$$

Then we can write the spacetime interval as

$$ds^2 = \sum_{\mu,\nu=1}^{4} g_{\mu\nu} \, dX_\mu \, dX_\nu$$

More generally, we use the tensor to define the dot product of any two 4-vectors:

$$\mathbf{U} \cdot \mathbf{V} = \sum_{\mu,\nu} g_{\mu\nu} \, U_\mu \, V_\nu \tag{10.19}$$

Given **X**, we can define the associated 4-velocity to be

$$\mathbf{V} = \frac{d\mathbf{X}}{d\tau} \tag{10.20}$$

Since this is defined using proper time, the spatial part of **V** is not the same as the measured velocity $\mathbf{v} = d\mathbf{x}/dt$. In a frame where the particle moves with measured velocity **v**, time dilation says the measured time is[9]

$$t = \gamma_v \tau \quad \text{where} \quad \gamma_v = \left(1 - \frac{|\mathbf{v}|^2}{c^2}\right)^{-1/2} \tag{10.21}$$

Therefore the 4-velocity can be written in terms of the measured velocity as

$$\mathbf{V} = (\gamma_v c, \gamma_v \mathbf{v}) \tag{10.22}$$

Why do we define 4-velocity in this way? We know that **X** transforms by the Lorentz transformation; then since τ is invariant, we realize that **V** must follow the Lorentz transformation as well. This clarifies the relation between reference frames[10]:

$$V_t = \gamma_u V_t' + \gamma_u \beta_u V_x' \tag{10.23a}$$

$$V_x = \gamma_u V_x' + \gamma_u \beta_u V_t' \tag{10.23b}$$

$$V_y = V_y' \tag{10.23c}$$

$$V_z = V_z' \tag{10.23d}$$

Our definition does mean that it takes a few extra steps to relate the measured velocities in different frames. In the primed frame, we can use Eq. (10.22) to write the 4-velocity in terms of the components of the measured velocity:

$$\mathbf{V}' = \frac{\{c, v_x', v_y', v_z'\}}{[1 - (v')^2/c^2]^{1/2}}$$

We can then use Eq. (10.23) to find the 4-velocity in the unprimed frame:

$$\mathbf{V} = \frac{\{\gamma_u(c + \beta_u v_x'), \gamma_u(v_x' + \beta_u c), v_y', v_z'\}}{[1 - (v')^2/c^2]^{1/2}}$$

---

[9] We put a subscript v on this γ to indicate that it is defined in terms of the particle's velocity and is not necessarily the same as the γ factor between arbitrary inertial frames (defined in Eq. 10.2).

[10] We put a subscript u on γ and β here to distinguish these factors, which relate arbitrary inertial frames, from γ_v in Eq. (10.21), which relates an arbitrary inertial frame to the particle's rest frame.

Inverting the relation between $\mathbf{v}$ and the spatial components of $\mathbf{V}$ gives

$$\{v_x, v_y, v_z\} = \frac{\{V_x, V_y, V_z\}}{[1 + (V_x^2 + V_y^2 + V_z^2)/c^2]^{1/2}}$$

Plugging in and simplifying yields

$$v_x = \frac{v_x' + u}{1 + uv_x'/c^2}$$

$$v_y = \frac{v_y'}{\gamma_u(1 + uv_x'/c^2)}$$

$$v_z = \frac{v_z'}{\gamma_u(1 + uv_x'/c^2)}$$

This is the same transformation we found by a different approach in Eq. (10.4).

10.5.3 Relativistic Momentum and Energy

We also need to generalize the concepts of energy and momentum. We define the 4-momentum to be

$$\mathbf{P} = m\mathbf{V} = (\gamma_v mc, \gamma_v m\mathbf{v}) \tag{10.24}$$

We then take the relativistic versions of energy and momentum to be the time and space parts of $\mathbf{P}$, respectively:

$$\mathbf{P} = \left(\frac{E}{c}, \mathbf{p}\right) \tag{10.25}$$

(The factor of c is included so E has dimensions of energy.) To understand what E represents, consider the dot product of $\mathbf{P}$ with itself. Using Eq. (10.24) along with the definition of the dot product in Eq. (10.19), we have

$$\mathbf{P} \cdot \mathbf{P} = \gamma_v^2 m^2 c^2 - \gamma_v^2 m^2 v^2 = \gamma_v^2 \left(1 - \frac{v^2}{c^2}\right) m^2 c^2 = m^2 c^2$$

where we use Eq. (10.21) to simplify. If instead we computed the dot product using Eq. (10.25), we would find

$$\mathbf{P} \cdot \mathbf{P} = \left(\frac{E}{c}\right)^2 - p^2$$

In other words, the relativistic energy and momentum are related by

$$\left(\frac{E}{c}\right)^2 - p^2 = m^2c^2 \quad \Rightarrow \quad E^2 = p^2c^2 + m^2c^4 \tag{10.26}$$

In the particle's rest frame, $p = 0$ so we recover the famous relation $E = mc^2$ for the rest mass energy. In the non-relativistic limit, $p \ll mc$ so we can make a Taylor series expansion:

$$E = mc^2\left(1 + \frac{p^2}{m^2c^2}\right)^{1/2} \approx mc^2 + \frac{p^2}{2m}$$

The first term is the rest-mass energy, while the second term is the Newtonian kinetic energy. The bottom line is that we can interpret E as the total energy in relativity.

There is one more useful relation we can derive. Again combining Eqs. (10.24) and (10.25), we can write

$$\frac{p}{E} = \frac{v}{c^2}$$

Using Eq. (10.26) to rewrite E yields

$$v = \frac{pc^2}{(p^2c^2 + m^2c^4)^{1/2}} \tag{10.27}$$

This is the relativistic version of the relation between momentum and velocity. In the non-relativistic limit, $p \ll mc$ so Eq. (10.27) reduces to the familiar relation $v \approx p/m$.

10.6   Black Holes

To this point we have discussed situations in which gravity is "weak" and we can make Taylor series expansions. We now move into the regime of "strong" gravity and examine a surprising and bizarre prediction of general relativity: black holes. While we are particularly interested in the strange physics near a black hole's event horizon, our analysis actually applies outside any spherical object in GR.

10.6.1   Schwarzschild Metric

To begin, we need to specify the spacetime geometry through the metric. To understand the form of the metric, recall from Eq. (10.11) the expression for gravitational time dilation,

$$\frac{\Delta t(r)}{\Delta t(\infty)} = \left(1 - \frac{2GM}{c^2 r}\right)^{1/2}$$

Presumably this factor appears in the time term of the metric. It also appears in the space term (basically from the curvature we discussed in Sect. 10.3.3). The full metric outside any spherical object of mass M is

$$ds^2 = \left(1 - \frac{2GM}{c^2 r}\right) c^2 \, dt^2 - \left(1 - \frac{2GM}{c^2 r}\right)^{-1} dr^2 - r^2 d\theta^2 - r^2 \sin^2 \theta \, d\phi^2 \quad (10.28)$$

This is called the **Schwarzschild metric** after the German mathematician Karl Schwarzschild, who discovered it as a solution of the equations of Einstein's general theory of relativity.[11]

We think of the coordinates (t, r, θ, ϕ) as quantities that would be measured by an observer far from the object, and we refer to them as "coordinate time," "coordinate radius," etc. They are different from quantities measured by an observer near the object; understanding the difference is one of our goals.

Notice that something funny happens to the metric when r approaches the **Schwarzschild radius** $R_S = 2GM/c^2$: the time term vanishes, while the radial term diverges. In the early twentieth century, all known astrophysical objects had sizes $R \gg R_S$, so Einstein and other prominent figures such as Arthur Eddington assumed the weirdness was merely a mathematical curiosity, not a physical reality. It was only later, after Subramanyan Chandrasekhar and Robert Oppenheimer showed that stars could collapse to become comparable to or even smaller than the Schwarzschild radius, that physicists began to take the strange predictions seriously.

The Schwarzschild metric deviates from the flat spacetime from special relativity (Eq. 10.16) only to the extent that R_S/r is nonzero. This allows us, finally, to specify what we mean by "weak" or "strong" gravity:

$$r \gg R_S \quad \rightarrow \quad \frac{R_S}{r} \ll 1 \quad \rightarrow \quad \text{"weak field"}$$

$$r \sim R_S \quad \rightarrow \quad \frac{R_S}{r} \sim 1 \quad \rightarrow \quad \text{"strong field"}$$

---

[11] Historical aside (drawn from *Black Holes and Time Warps* by Kip Thorne [18]): When Einstein's general theory of relativity was published on Nov. 25, 1915, Schwarzschild was serving in the German army on the Russian front in the first World War. He managed to obtain Einstein's paper, read it, apply it to stars, discover a solution to the complicated equations Einstein had derived, write a paper of his own, and send it to Einstein—all in time for Einstein to present the paper on Schwarzschild's behalf at a meeting on Jan. 13, 1916. Unfortunately, Schwarzschild died on May 11 of illness contracted during his service.

To give some examples, let's quantify the Schwarzschild radius:

$$
\begin{aligned}
R_S &= \frac{2GM_\odot}{c^2} \times \frac{M}{M_\odot} \\
&= \frac{2 \times (6.67 \times 10^{-11}\,\mathrm{m^3\,kg^{-1}\,s^{-2}}) \times (1.99 \times 10^{30}\,\mathrm{kg})}{(3.0 \times 10^8\,\mathrm{m\,s^{-1}})^2} \times \left(\frac{M}{M_\odot}\right) \\
&= 3\,\mathrm{km} \times \left(\frac{M}{M_\odot}\right)
\end{aligned}
\tag{10.29}
$$

We retain the mass dependence but express M in solar masses so we can quickly evaluate the Schwarzschild radius for different astrophysical objects. Here are typical numbers for some systems we have studied already or will encounter:

	$M/M_\odot$	R_S	R	R_S/R
Earth	3×10^{-6}	0.009 m	6.4×10^6 m	1.4×10^{-9}
Sun	1	3 km	7×10^8 m	4×10^{-6}
White dwarf	1	3 km	6×10^6 m	5×10^{-4}
Neutron star	1.4	4 km	10 km	0.4

10.6.2   Spacetime Geometry

To begin to see some of the weird properties of a black hole, consider the spacetime interval between ticks on a stationary clock. If the clock does not move then $\mathrm{d}r = \mathrm{d}\theta = \mathrm{d}\phi = 0$, so the spacetime interval is

$$
\mathrm{d}s^2 = \left(1 - \frac{R_S}{r}\right) c^2\,\mathrm{d}t^2 = \begin{cases} > 0 \text{ (timelike)} & \text{for } r > R_S \\ < 0 \text{ (spacelike)} & \text{for } r < R_S \end{cases}
$$

The spacetime interval *changes sign* at the Schwarzschild radius, switching from spacelike to timelike. This is important because no physical object can experience a spacelike interval; to do so, it would have to move faster than the speed of light. We seem to have a paradox: a stationary clock inside the Schwarzschild radius would have a spacelike interval, which is not allowed. To resolve the paradox, we conclude that *it is impossible to remain stationary inside the Schwarzschild radius.* In fact, objects inside the Schwarzschild radius are inexorably drawn to the central singularity, just as on Earth we are inexorably drawn forward in time.

10.6.3 Particle in a Circular Orbit

As we set out to study motion in general relativity, it is good to start with the simple case of circular orbits. Such an orbit stays in a plane, and we can choose our coordinates so this is the equatorial plane:

$$r = \text{constant} \qquad \theta = \frac{\pi}{2} \qquad \phi = \omega t$$

where ω is the coordinate angular speed. The period of the orbit in coordinate time is $P = 2\pi/\omega$. The spacetime interval for the orbit is:

$$ds^2 = \left(1 - \frac{R_S}{r}\right) c^2 \, dt^2 - r^2 \omega^2 \, dt^2 = \left(1 - \frac{R_S}{r} - \frac{r^2 \omega^2}{c^2}\right) c^2 \, dt^2$$

With this we can determine the proper time (see Eq. 10.15):

$$\tau_{\text{circ}} = \frac{1}{c} \int_{\text{one orbit}} \sqrt{ds^2} = \left(1 - \frac{R_S}{r} - \frac{r^2 \omega^2}{c^2}\right)^{1/2} P$$

This is the time that would be measured on a clock that is executing the circular orbit. Note that it is not a simple integral over dt; we must account for the motion using the spacetime interval.

We have not yet specified the radius. We can find it by applying Fermat's principle of least time: for a given angular speed, the particle will "choose" the radius that minimizes the proper time. Operationally, we want to find the radius that minimizes τ, so we want to solve

$$0 = \frac{d\tau}{dr} = \frac{1}{2}\left(1 - \frac{R_S}{r} - \frac{r^2 \omega^2}{c^2}\right)^{-1/2} \left(\frac{R_S}{r^2} - \frac{2r\omega^2}{c^2}\right) P$$

The solution is

$$r = \left(\frac{c^2 R_S}{2\omega^2}\right)^{1/3}$$

It is more convenient to write the relation as

$$\omega = \left(\frac{c^2 R_S}{2r^3}\right)^{1/2} = \left(\frac{GM}{r^3}\right)^{1/2}$$

The coordinate velocity is then

$$v = \omega r = \left(\frac{GM}{r}\right)^{1/2}$$

This is the same expression we had in Newtonian gravity (see Eq. 7.7). In other words, a distant observer would measure the same orbital size and orbital velocity, and hence the same orbital period, as in Newtonian gravity. But a clock following the circular orbit would measure the proper time, which is different:

$$\tau_{\text{circ}} = \left(1 - \frac{R_S}{r} - \frac{r^2\omega^2}{c^2}\right)^{1/2} P = \left(1 - \frac{3R_S}{2r}\right)^{1/2} P$$

Out of curiosity, what about a clock *at rest* at the same radius? Such a clock has $dr = d\phi = d\theta = 0$ and hence

$$ds^2 = \left(1 - \frac{R_S}{r}\right) c^2 dt^2 \quad \Rightarrow \quad \tau_{\text{rest}} = \frac{1}{c} \int \sqrt{ds^2} = \left(1 - \frac{R_S}{r}\right)^{1/2} P$$

This is identical to the gravitational time dilation for a clock at rest in a gravitational field that we examined in Sect. 10.3.4.

We see that time is complicated! The time you measure depends on where you are and how you are moving. These are both effects that we have seen already (time dilation in special and general relativity), but it is interesting to see how they manifest themselves here.

Example: Circular Orbit Around Sgr A*

Imagine we were in a spaceship orbiting the black hole at the center of the Milky Way at $r = 3 R_S$. If we take $M_{\text{bh}} = 4 \times 10^6 M_\odot$ then from Eq. (10.29) the Schwarzschild radius is $R_S = 1.18 \times 10^{10}$ m, and so the radius of the orbit is $r = 3R_S = 3.54 \times 10^{10}$ m. The orbital period as measured by a distant observer (i.e., in coordinate time) is the same as in Newtonian gravity:

$$P = \frac{2\pi}{\omega} = 2\pi \left(\frac{r^3}{GM}\right)^{1/2} = 1{,}800\,\text{s} = 30\,\text{min}$$

However, our clocks on the spaceship show the proper time, and in our frame one orbital period takes

$$\tau_{\text{circ}} = \left(1 - \frac{3R_S}{2r}\right)^{1/2} P = 1{,}290\,\text{s} = 21\,\text{min}$$

If we had friends in a space station that is sitting at a fixed spot with $r = 3 R_S$ (i.e., not orbiting but stationary), they would measure our orbital period as

$$\tau_{\text{rest}} = \left(1 - \frac{R_S}{r}\right)^{1/2} P = 1{,}490\,\text{s} = 25\,\text{min}$$

Again, time depends on where you are and how you are moving.

10.6.4 General Motion Around a Black Hole

Now we allow general motion.[12] Let's briefly review the Newtonian case as a point of reference. As we saw in Sect. 3.1, in spherical symmetry the motion is confined to a plane, which we can define to be the equatorial plane. The equation of motion for the one-body problem is then

$$\left[\frac{d^2r}{d\tau^2} - r\left(\frac{d\phi}{d\tau}\right)^2\right]\hat{\mathbf{r}} + \frac{1}{r}\frac{d}{d\tau}\left(r^2\frac{d\phi}{d\tau}\right)\hat{\boldsymbol{\phi}} = -\frac{GM}{r^2}\hat{\mathbf{r}}$$

(In general relativity, the natural time coordinate for studying motion is the proper time, so we write τ here.) The angular component of the equation of motion implies

$$r^2\frac{d\phi}{d\tau} = \text{constant} \equiv \ell \tag{10.30}$$

where ℓ is the specific angular momentum. This is conservation of angular momentum (which we have seen many times now). The radial component of the equation of motion looks like

$$\frac{d^2r}{d\tau^2} - \frac{\ell^2}{r^3} = -\frac{GM}{r^2}$$

We rewrite this as

$$\frac{d^2r}{d\tau^2} = -\frac{d\Phi_{\text{Newt}}}{dr} \tag{10.31}$$

where we define the effective potential

$$\Phi_{\text{Newt}} = -\frac{GM}{r} + \frac{\ell^2}{2r^2} + \frac{c^2}{2} \tag{10.32}$$

The first term is the familiar Newtonian gravitational potential. The second term is the centrifugal term. The last term is just a constant that we add because it will prove to be convenient in the relativistic case.

The effective potential is useful because we can think of it as a surface and use our intuition to understand what would happen to a ball on that surface. Some examples are shown by the dashed curves in Fig. 10.3. For $\ell = 0$ the ball would roll all the way down to $r = 0$. For any nonzero value of ℓ, however, the centrifugal term causes an upturn at small radius. This creates a stationary point that corresponds to a constant radius and hence a circular orbit. In the Newtonian effective potential, the

---

[12]This presentation draws from the book by Schutz [2].

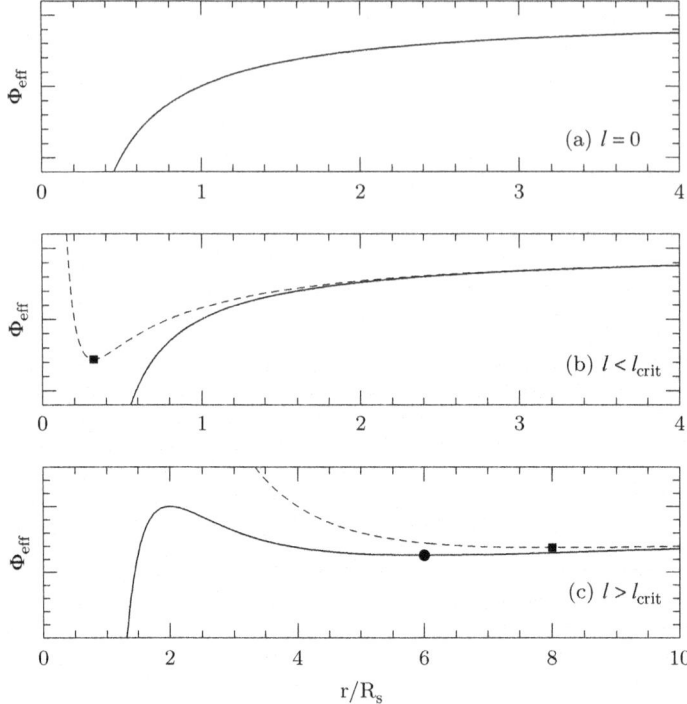

Fig. 10.3 Examples of effective potentials for a massive particle in Newtonian gravity (*dashed*) and GR (*solid*). The three panels correspond to different values of the specific angular momentum, ℓ. Note that the *bottom panel* has a different horizontal scale. Points indicate local minima (*squares* for Newtonian gravity and *circles* for GR)

stationary point is a minimum of Φ, so the orbit is stable: if you put the ball near the minimum but give it a little kick, it will oscillate around the minimum but remain confined.

In general relativity the equation of motion can be written in the form of (10.31) but with a different effective potential. To find that potential, recall that special relativity has a relation between energy, momentum, and mass: $(E/c)^2 - p^2 = m^2$ (Eq. 10.26). With the Schwarzschild metric the analogous relation has factors of $(1 - R_S/r)$:

$$\left(1 - \frac{R_S}{r}\right)^{-1} \left(\frac{m\tilde{E}}{c}\right)^2 - \left(1 - \frac{R_S}{r}\right)^{-1} \left(m\frac{dr}{d\tau}\right)^2 - \frac{m^2\ell^2}{r^2} = m^2 c^2 \quad (10.33)$$

where ℓ is the specific angular momentum and $\tilde{E} = E/m$ is the energy per unit mass, both of which are well defined only if the moving particle has a nonzero rest mass. (We consider a massless particle below.) We can divide through by m^2 and rearrange to write

$$\left(\frac{dr}{d\tau}\right)^2 = \frac{\tilde{E}^2}{c^2} - \left(1 - \frac{R_S}{r}\right)\left(c^2 + \frac{\ell^2}{r^2}\right) \tag{10.34}$$

Take the derivative $d/d\tau$, divide through by $2\,dr/d\tau$, and use $R_S = 2GM/c^2$:

$$\frac{d^2 r}{d\tau^2} = -\frac{GM}{r^2} + \frac{\ell^2}{r^3} - \frac{3GM\,\ell^2}{c^2 r^4}$$

The first two terms match the Newtonian case, but the third term is new in GR. We can capture all of the terms in the same form as Eq. (10.31) by introducing the effective potential

$$\text{(massive)} \qquad \Phi_{\text{GR}} = -\frac{GM}{r} + \frac{\ell^2}{2r^2} + \frac{c^2}{2} - \frac{GM\ell^2}{c^2 r^3}$$

$$= \frac{1}{2}\left(1 - \frac{R_S}{r}\right)\left(c^2 + \frac{\ell^2}{r^2}\right)$$

If the particle is massless (e.g., a photon), the analysis is slightly different because we cannot define the energy and angular momentum per unit mass. Nevertheless, light does carry both energy and momentum, and we can keep the same form of the equations if we define ℓ and $\tilde{E}$ to be the total angular momentum and energy, respectively. Also, we need to be careful with the derivative term in Eq. (10.33) because τ and m are both zero for photons. We can, however, define a new parameter λ that runs along the photon's trajectory in spacetime such that the derivative $dr/d\lambda$ is well defined. The upshot is that Eq. (10.33) is replaced for a massless particle by

$$\left(1 - \frac{R_S}{r}\right)^{-1}\left(\frac{\tilde{E}}{c}\right)^2 - \left(1 - \frac{R_S}{r}\right)^{-1}\left(\frac{dr}{d\lambda}\right)^2 - \frac{\ell^2}{r^2} = 0 \tag{10.35}$$

or

$$\left(\frac{dr}{d\lambda}\right)^2 = \frac{\tilde{E}^2}{c^2} - \left(1 - \frac{R_S}{r}\right)\frac{\ell^2}{r^2} \tag{10.36}$$

As before, we take the derivative $d/d\lambda$, divide through by $2\,dr/d\lambda$, and use $R_S = 2GM/c^2$:

$$\frac{d^2 r}{d\lambda^2} = \frac{\ell^2}{r^3} - \frac{3GM\,\ell^2}{c^2 r^4}$$

In this case we define the effective potential to be

$$\text{(massless)} \qquad \Phi_{\text{GR}} = \frac{1}{2}\left(1 - \frac{R_S}{r}\right)\frac{\ell^2}{r^2}$$

We can combine the expressions for the massive and massless cases into a single effective potential if we write

$$\Phi_{GR} = \frac{1}{2}\left(1 - \frac{R_S}{r}\right)\left(\tilde{m}c^2 + \frac{\ell^2}{r^2}\right) \tag{10.37}$$

and put $\tilde{m} = 1$ for a massive particle and $\tilde{m} = 0$ for the massless case.

Sample GR potentials are shown by the solid curves in Fig. 10.3. There are several important points to make:

- At large radius, the new term $GM\ell^2/c^2r^3$ from GR is small, so Newtonian gravity is a good approximation. GR effects are significant only at small radii.
- For ℓ above some critical value ℓ_{crit}, there is a minimum in the potential curve, which corresponds to a stable circular orbit. (You can find ℓ_{crit}, along with the location of the stable circular orbit, in Problem 10.7.)
- For $\ell > \ell_{crit}$, there is also a maximum in the GR potential curve. It corresponds to a second allowed circular orbit for a given angular momentum, but one that is unstable. This is new in GR.
- For $\ell < \ell_{crit}$, there is no minimum in the potential curve, and hence no stable circular orbit. Thus, there exists some *smallest* stable circular orbit in GR. This is another new feature (Newtonian gravity allows arbitrarily small circular orbits).
- The GR potential turns over at small radius, so if a particle gets too close to the black hole it cannot help but fall in. The ability to fall all the way to $r = 0$ with finite angular momentum is yet another difference from Newtonian gravity.

To learn more about the motion, let's return to Eqs. (10.34) and (10.36). We can write both in the form

$$\frac{dr}{d\tau} = \pm\left(\frac{\tilde{E}^2}{c^2} - 2\Phi\right)^{1/2} \tag{10.38}$$

This is the key equation of motion in the radial direction. We still need one more ingredient: an equation of motion for time (since t depends on position and motion). This equation involves again involves the factor $(1 - R_S/r)$:

$$\frac{dt}{d\tau} = \left(1 - \frac{R_S}{r}\right)^{-1}\frac{\tilde{E}}{c^2} \tag{10.39}$$

If we want an equation for r in terms of coordinate time (as opposed to proper time), we can combine Eqs. (10.38) and (10.39) to obtain

$$\frac{dr}{dt} = \frac{dr/d\tau}{dt/d\tau} = \pm c\left(1 - \frac{R_S}{r}\right)\frac{(\tilde{E}^2 - 2c^2\Phi)^{1/2}}{\tilde{E}} \tag{10.40}$$

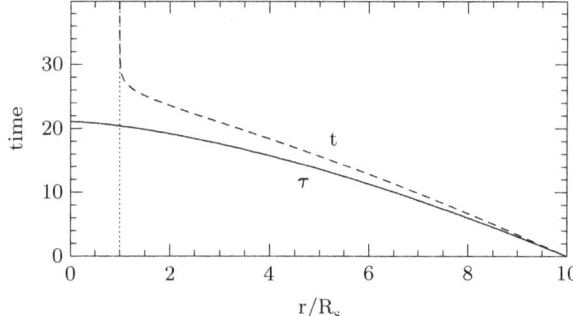

Fig. 10.4 The *dashed* and *solid lines* show how the coordinate time and proper time (respectively) flow as an object falls into a black hole. The horizontal axis is radius scaled by the Schwarzschild radius; the *dotted line* indicates the event horizon. The vertical axis is time scaled by R_S/c. In this example, time is defined to be 0 at $r/R_S = 10$ even though the particle started from rest at infinity

You can explore various aspects of motion in the Schwarzschild metric in Problems 10.6 and 10.7. Here let's consider a particle falling from rest into a black hole. We know the particle falls straight in, so one of the constants of motion is $\ell = 0$. If the particle starts from rest ($dr/d\tau = 0$) at infinity, then $\tilde{E} = c^2$. In Eq. (10.40) we choose the minus sign since we know radius must decrease with time. We can then rewrite the equation of motion as

$$c\, dt = -\left(1 - \frac{R_S}{r}\right)^{-1} \left(\frac{r}{R_S}\right)^{1/2} dr$$

Integrating both sides yields $t(r)$ as shown in Fig. 10.4. The curious result is that time goes to infinity as r approaches R_S. As seen by an observer far away, the particle never actually reaches the black hole!

This bizarre result only applies to the coordinate time. Repeating the analysis using the equation of motion (10.38) reveals that proper time—which is what you would see on your watch if you fell into a black hole—is perfectly well behaved. It *is* possible to fall into a black hole, but no one on the outside can see it.

10.6.5 Gravitational Deflection

We are now equipped to derive the relativistic deflection angle that we previously quoted in Sect. 9.1.1.[13] Consider the setup in Fig. 10.5, which is modified from Fig. 9.1 to have the particle move from right to left so the azimuthal angle ϕ

---

[13]This presentation follows the analysis given by Keeton and Petters [19].

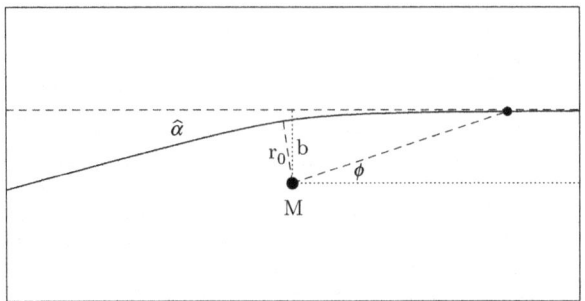

Fig. 10.5 Setup for calculating the bending angle $\hat{\alpha}$. The azimuthal angle ϕ is defined from the x-axis as usual, and the particle moves from right to left so ϕ increases monotonically from $\phi \to 0$ as $\tau \to -\infty$ to $\phi \to \pi + \hat{\alpha}$ as $\tau \to +\infty$. The impact parameter is b while the distance of closest approach is r_0

increases monotonically from an initial value of $\phi \to 0$ as $\tau \to -\infty$. If there were no deflection, ϕ would go to π for $\tau \to +\infty$; thus the angle of deflection is the amount by which $\Delta\phi$ exceeds π:

$$\hat{\alpha} = \Delta\phi - \pi$$

We need an equation of motion for ϕ. Since Eqs. (10.30) and (10.38) both have r on the right-hand side, it is useful to let r be the independent variable by writing

$$\frac{d\phi}{dr} = \frac{d\phi/d\tau}{dr/d\tau} = \pm\frac{\ell}{r^2}\left(\frac{\tilde{E}^2}{c^2} - 2\Phi\right)^{-1/2}$$

To solve this equation, we break the trajectory into two pieces. In the first half, r decreases from ∞ down to the point of closest approach r_0, and $d\phi/dr < 0$ so we take the minus sign. In the second half, r increases from r_0 out to ∞, and $d\phi/dr > 0$ so we take the plus sign. Then:

$$\hat{\alpha} = -\int_\infty^{r_0} \frac{\ell}{r^2}\left(\frac{\tilde{E}^2}{c^2} - 2\Phi\right)^{-1/2} dr + \int_{r_0}^\infty \frac{\ell}{r^2}\left(\frac{\tilde{E}^2}{c^2} - 2\Phi\right)^{-1/2} dr - \pi$$

$$= 2\int_{r_0}^\infty \frac{\ell}{r^2}\left(\frac{\tilde{E}^2}{c^2} - 2\Phi\right)^{-1/2} dr - \pi \qquad (10.41)$$

Now consider the constants of motion. At very early times the trajectory is nearly a straight line with $x = -v\tau$ and $y = b$, so the polar coordinates are

$$r = \left(b^2 + v^2\tau^2\right)^{1/2} \quad \text{and} \quad \phi = -\tan^{-1}\left(\frac{b}{v\tau}\right)$$

and the constants of motion are

$$\ell = \lim_{r \to \infty} \left[r^2 \frac{d\phi}{d\tau} \right] = bv$$

$$\frac{\tilde{E}^2}{c^2} = \lim_{r \to \infty} \left[\left(\frac{dr}{d\tau} \right)^2 + \left(1 - \frac{R_S}{r} \right) \left(\tilde{m}c^2 + \frac{\ell^2}{r^2} \right) \right] = v^2 + \tilde{m}c^2$$

Thus we can rewrite Eq. (10.41) as

$$\hat{\alpha} = 2 \int_{r_0}^\infty \frac{bv}{r^2} \left[\tilde{m}c^2 + v^2 - \left(1 - \frac{R_S}{r} \right) \left(\tilde{m}c^2 + \frac{b^2v^2}{r^2} \right) \right]^{-1/2} dr \; - \; \pi \quad (10.42)$$

There is one final ingredient: we need to relate the impact parameter b to the distance of closest approach r_0. We use the fact that $dr/d\tau = 0$ at the point of closest approach to put $\tilde{E}^2/c^2 = 2\Phi(r_0)$ or

$$\tilde{m}c^2 + v^2 = \left(1 - \frac{R_S}{r_0} \right) \left(\tilde{m}c^2 + \frac{b^2v^2}{r_0^2} \right)$$

which can be solved to find

$$b = \frac{r_0}{v} \left(\frac{v^2 + \tilde{m}c^2 R_S/r_0}{1 - R_S/r_0} \right)^{1/2} \tag{10.43}$$

Plug this into Eq. (10.42) and make the change of variables $r = r_0/w$:

$$\hat{\alpha} = 2 \int_0^1 \left[\frac{s\tilde{m}c^2 + v^2}{s\tilde{m}c^2 w(1-w)[1-s(1+w)] + v^2[1 - w^2 - s(1-w^3)]} \right]^{1/2} dw \; - \; \pi$$

where $s \equiv R_S/r_0$. This is the general expression for the deflection angle in the Schwarzschild metric, but it is not terribly enlightening. We can make more progress if the trajectory never gets very close to the central object. In that case $s \ll 1$ and we can expand the integrand as a Taylor series in s:

$$\hat{\alpha} \approx \int_0^1 \left[\frac{2}{\sqrt{1-w^2}} + \frac{\tilde{m}c^2 + v^2(1 + w + w^2)}{v^2(1+w)\sqrt{1-w^2}} s + \mathscr{O}\left(s^2 \right) \right] dw \; - \; \pi$$

$$\approx \left(2 + \frac{\tilde{m}c^2}{v^2} \right) s \; + \; \mathscr{O}\left(s^2 \right)$$

Recall that $s = R_S/r_0$, but from Eq. (10.43) we can replace this with $s \approx R_S/b \approx 2GM/c^2 b$ at the order of approximation to which we are working. This yields

$$\hat{\alpha} \approx \left(2 + \frac{\tilde{m}c^2}{v^2} \right) \frac{2GM}{c^2 b} \tag{10.44}$$

In the case of a massless particle (e.g., a photon), $\tilde{m} = 0$ and so we have

$$\hat{\alpha} \approx \frac{4GM}{c^2 b} \qquad \text{(massless)}$$

which is the result that we used in Chap. 9 to build the theory of gravitational lensing. In the case of a massive particle $\tilde{m} = 1$, and if the particle is non-relativistic then $v/c \ll 1$ and the second term in parentheses in Eq. (10.44) dominates the first term to yield

$$\hat{\alpha} \approx \frac{2GM}{v^2 b} \qquad \text{(massive and non-relativistic)}$$

which is the same result that we derived using Newtonian gravity in Sect. 9.1.1.

10.7   Other Effects

Many other aspects of relativity lie beyond the scope of this book, but two are worth mentioning briefly. First, we have studied black holes that are static and spherically symmetric, but most objects in the universe rotate. Roy Kerr [20] found a solution to Einstein's equations that describes a rotating black hole. The spin modifies spacetime in the vicinity of the black hole, which affects the motion of any matter in an accretion disk and the properties of light emitted from the disk. Observations of spectral lines from black hole accretion disks can therefore be used to measure black hole spin and probe the Kerr metric (see [21] for a review).

Second, relativity predicts that accelerating masses create ripples in spacetime that propagate as **gravitational waves**. Gravitational radiation from accelerating masses is somewhat analogous to electromagnetic radiation from accelerating charges, although the analogy is not precise. The waves are predicted to carry energy away from a binary star system and cause the stars' orbits to decay. (You can explore this process in Problem 10.8.) The energy loss scales with orbital separation as $P \propto a^{-5}$, so it mainly affects close binaries. Two particular systems show clear evidence for orbital decay due to gravitational radiation: the "binary pulsar" PSR B1913+16, discovered in 1974 by Hulse and Taylor [22] (for which they received the 1993 Nobel Prize in Physics); and the "double pulsar" J0737−3039, discovered in 2003 by Burgay et al. [23]. These systems provide strong if indirect tests of predictions for gravitational radiation [24, 25]. The next goal is to detect gravitational waves directly. Projects such as the Laser Interferometer Gravitational-Wave Observatory as well as Virgo, AURIGA, and MiniGRAIL are trying to create new ways for us to observe the effects of strong gravity in extreme events throughout the universe.

Problems

10.1. If the Sun were replaced by a black hole with the same mass, would Earth's orbit change significantly? Why or why not?

10.2. The Michelson-Morley experiment showed that the speed of light does not depend on the speed of the source. Use the velocity transformation (10.4) to explain the result. Specifically:

(a) Suppose a source moving horizontally with speed u emits a light ray going in the horizontal direction. What is the speed and direction of the light ray as measured by a stationary observer?

(b) Suppose a source moving horizontally with speed u emits a light ray going in the vertical direction (in the source's reference frame). What is the speed and direction of the light ray as measured by a stationary observer?

10.3. Muons are elementary particles produced when cosmic rays collide with atoms in Earth's upper atmosphere. Muons are unstable and decay, so the number of muons as a function of time has the form $N(t) = N_0\, e^{-t/\tau}$ where $\tau = 2.20 \times 10^{-6}\,\mathrm{s}$ and N_0 is the number at $t = 0$. In 1963, Frisch and Smith [5] put a muon detector at the top of Mt. Washington (1,907 m above sea level) and counted 563 muons per hour coming down through the atmosphere. Then they took their detector to sea level and counted 408 muons per hour. From the muon energies they inferred a speed of $0.995\,c$.

(a) If there were no time dilation, how many muons should have been measured at sea level?

(b) How does time dilation affect the experiment?

(c) Use the experimental data to determine the muons' relativistic γ factor.

10.4. Let's see how light from the Hβ transition of hydrogen (wavelength 486.13 nm) is affected by the relativistic Doppler effect and time dilation.

(a) Consider debris from a supernova moving directly toward an observer on Earth with a speed $v = 18{,}000\,\mathrm{km\,s^{-1}}$. At what wavelength would the Hβ spectral line from the debris be observed?

(b) Imagine instead the debris is moving perpendicular to our line of sight (i.e., "in the plane of the sky") with a transverse velocity $v = 18{,}000\,\mathrm{km\,s^{-1}}$. Now at what wavelength would the Hβ line be observed? Hint: you can consider the light to be a "clock" with a frequency v.

(c) What would the predicted wavelengths have been for parts (a) and (b) if we had ignored special relativity and used the "classical" Doppler formula $\Delta\lambda/\lambda = v_{\text{radial}}/c$?

10.5. The rest mass energy of a proton is about 938 MeV. The Large Hadron Collider is designed to accelerate protons to an energy of about 7 TeV. How fast do such protons move? Hint: write $v = (1 - \delta)c$ and find δ.

10.6. Suppose a probe that emits a flash of green light ($\lambda = 500\,\mathrm{nm}$) once every second is dropped into the black hole at the center of the Milky Way, starting at rest from $r = 2R_S$. Use $M = 4 \times 10^6\,M_\odot$ (see Sect. 3.2.1).

(a) In the probe's reference frame, how much time elapses as it falls from its starting point to the event horizon? From the event horizon to the center? Hint: by changing variables, you can express the integral in a form that can be evaluated using Sect. A.7.

(b) Describe qualitatively what you would see from a fixed vantage point far from the black hole as the probe falls in.

(c) What is the wavelength of the first flash, as measured by a distant observer? What about the last flash emitted by the probe before it crosses the event horizon? Hint: you will need to use a numerical root finder to solve for radius corresponding to the last integer second before the probe crosses the event horizon.

10.7. For a particle moving in the Schwarzschild metric, the effective potential is given by Eq. (10.37). Consider a circular orbit.

(a) For a massive particle ($\tilde{m} = 1$), there are two possible circular orbits. What are their radii? What is the smallest value of ℓ for which the answer is physical? What is the radius of the smallest possible circular orbit?

(b) For a photon ($\tilde{m} = 0$), show that it is possible to have a circular orbit at a particular radius, but the orbit is unstable.

10.8. In a binary star system, the accelerating masses create gravitational radiation that removes energy from the system. For stars of mass M_1 and M_2 in nearly circular orbits separated by distance a, the power emitted in gravitational radiation is [26]

$$P = \frac{32G^4}{5c^5} \frac{(M_1 M_2)^2 (M_1 + M_2)}{a^5}$$

(a) Explain conceptually what happens to the stars' orbits.

(b) Using conservation of energy, derive a differential equation for the semimajor axis as a function of time. Solve the equation to find how long it takes to go from some initial semimajor axis $a = a_0$ to $a = 0$.

(c) The binary pulsar system J0737−3039 has an orbital period of $P = 0.102\,\mathrm{day}$, and the stars move in nearly circular orbits with speeds of about $310\,\mathrm{km\,s^{-1}}$. When will the stars merge? (Assume the stars have the same mass. Problem 4.5 gives more precise parameters, but we use simpler approximations here.)

References

1. B.W. Carroll, D.A. Ostlie, *An Introduction to Modern Astrophysics*, 2nd edn. (Addison-Wesley, San Francisco, 2007)
2. B.F. Schutz, *A First Course in General Relativity* (Cambridge University Press, Cambridge, 1985)
3. P. Galison, *Einstein's Clocks, Poincaré's Maps: Empires of Time* (W.W. Norton, New York, 2003)
4. A. Einstein, *Einstein's 1912 Manuscript on the Special Theory of Relativity: A Facsimile* (George Braziller, New York, 1996)
5. D.H. Frisch, J.H. Smith, Am. J. Phys. **31**, 342 (1963)
6. J.C. Hafele, R.E. Keating, Science **177**, 166 (1972)
7. J.C. Hafele, R.E. Keating, Science **177**, 168 (1972)
8. C. Misner, K. Thorne, J. Wheeler, *Gravitation* (W.H. Freeman, San Francisco, 1973)
9. A. Pais, *Subtle Is the Lord: The Science and the Life of Albert Einstein* (Oxford University Press, Oxford/New York, 1982)
10. S. Schlamminger, K.Y. Choi, T.A. Wagner, J.H. Gundlach, E.G. Adelberger, Phys. Rev. Lett. **100**(4), 041101 (2008)
11. R. Baum, W. Sheehan, *In Search of Planet Vulcan: The Ghost in Newton's Clockwork Universe* (Basic Books, New York, 2003)
12. C. Pollack, Mercury's Perihelion (2003), http://www.math.toronto.edu/~colliand/426_03/-Papers03/C_Pollock.pdf
13. R.V. Pound, G.A. Rebka, Phys. Rev. Lett. **3**, 439 (1959)
14. R.V. Pound, G.A. Rebka, Phys. Rev. Lett. **4**, 337 (1960)
15. J.L. Greenstein, J.B. Oke, H.L. Shipman, Astrophys. J. **169**, 563 (1971)
16. N.S. Hetherington, Q. J. R. Astron. Soc. **21**, 246 (1980)
17. N. Ashby, Living Rev. Relativ. **6**, 1 (2003)
18. K.S. Thorne, *Black Holes & Time Warps: Einstein's Outrageous Legacy* (W.W. Norton, New York, 1994)
19. C.R. Keeton, A.O. Petters, Phys. Rev. D **72**(10), 104006 (2005)
20. R.P. Kerr, Phys. Rev. Lett. **11**, 237 (1963)
21. J.M. Miller, Annu. Rev. Astron. Astrophys. **45**, 441 (2007)
22. R.A. Hulse, J.H. Taylor, Astrophys. J. Lett. **195**, L51 (1975)
23. M. Burgay et al., Nature **426**, 531 (2003)
24. J.M. Weisberg, D.J. Nice, J.H. Taylor, Astrophys. J. **722**, 1030 (2010)
25. M. Kramer, et al., Science **314**, 97 (2006)
26. S.L. Shapiro, S.A. Teukolsky, *Black Holes, White Dwarfs and Neutron Stars: The Physics of Compact Objects* (Wiley, New York, 1986)

Chapter 11
Cosmology: Expanding Universe

Beyond describing the spacetime around a black hole, general relativity provides a framework for studying the universe as a whole. Some of the great discoveries in cosmology during the twentieth century related to the expansion of the universe. In this chapter we study cosmic expansion in the context of relativistic cosmology. Later, in Chap. 20, we will examine how particles and gas behaved within the expanding universe shortly after the Big Bang.

11.1 Hubble's Law and the Expanding Universe

In the early twentieth century, Vesto Slipher [1], William Wallace Campbell [2], and others measured spectra of "nebulae" (these objects were not yet known to be other galaxies) and used the Doppler effect to infer that most of them are moving away from us. For nearby galaxies, whose recession speeds are small compared with the speed of light, we can use the non-relativistic limit of the Doppler formula (10.8) to write the redshift as

$$z \equiv \frac{\Delta\lambda}{\lambda_{\mathrm{rest}}} \approx \frac{v}{c}$$

In 1929, Edwin Hubble [3] discovered a linear correlation between a galaxy's recession speed, v, and its distance from us, d, which is now known as **Hubble's law**:

$$v = H_0\, d \tag{11.1}$$

where the proportionality factor H_0 is called the **Hubble constant**. As we will see, we now understand that the original relation is limited in two ways: the Hubble "constant" actually varies with time, and the linear relation holds only for galaxies that are "nearby" on cosmic scales (within a few hundred Mpc). Hubble's discovery

C. Keeton, *Principles of Astrophysics: Using Gravity and Stellar Physics to Explore the Cosmos*, Undergraduate Lecture Notes in Physics, DOI 10.1007/978-1-4614-9236-8_11, © Springer Science+Business Media New York 2014

nevertheless played a key role in revealing the expansion of the universe. Assuming that we are not in a special place in the universe (see Sect. 11.2.1), Hubble's law implies that galaxies are receding not only from us but also from each other; all galaxies are moving apart as the universe itself expands.

After many decades of effort (and dispute), we now have a good value for the Hubble constant [4, 5],

$$H_0 = 73.8 \pm 2.4 \, \mathrm{km\,s^{-1}\,Mpc^{-1}} \qquad (11.2)$$

The controversy is not entirely gone; a recent analysis of the Cosmic Microwave Background from the Planck spacecraft (see Sect. 20.1.3) yielded a value $H_0 = 67.3 \pm 1.2 \, \mathrm{km\,s^{-1}\,Mpc^{-1}}$ [6] that is formally inconsistent with Eq. (11.2). The Planck analysis involved certain assumptions that remain to be verified, so I prefer to adopt the value in Eq. (11.2) that comes more directly from measurements of expansion. The dimensions of the Hubble constant are inverse time, but using units of $\mathrm{km\,s^{-1}\,Mpc^{-1}}$ gives us the convenience of expressing distances in Mpc and velocities in $\mathrm{km\,s^{-1}}$. For example, Eqs. (11.1) and (11.2) together indicate that a galaxy 100 Mpc away will have a cosmological recession velocity of about 7,380 km s^{-1}.

If the universe is expanding and all galaxies are moving apart, we can imagine reversing time and watching them come together at some moment in the past: the Big Bang. If the expansion occurred at a uniform rate, the age of the universe (the time since the Big Bang) would be $t_0 = H_0^{-1}$. In fact, the expansion rate has not been constant, but the quantity H_0^{-1} still sets the basic time scale for the universe, so it is known as the **Hubble time**:

$$H_0^{-1} = \frac{1}{73.8 \, \mathrm{km\,s^{-1}\,Mpc^{-1}}} \frac{1 \, \mathrm{km}}{10^3 \, \mathrm{m}} \frac{3.09 \times 10^{22} \, \mathrm{m}}{1 \, \mathrm{Mpc}} = 4.18 \times 10^{17} \, \mathrm{s} = 13.3 \, \mathrm{Gyr}$$

11.2 Relativistic Cosmology

Why is relativity important for cosmology? Recall from Sect. 10.6.1 that relativistic effects become important when the ratio $\xi = GM/c^2R$ becomes comparable to unity. In the case of black holes and other compact objects, ξ can approach 1 when R is small. In the case of the universe, by contrast, ξ can be significant when R is large. If the density of the universe is roughly uniform, a region of size R will have mass $M \propto R^3$ and the "relativity indicator" will scale as $\xi \propto R^2$. Even if ξ is small on the scale of a planetary system or galaxy, it will become appreciable as we consider larger and larger scales.

11.2.1 Robertson-Walker Metric

The first step in any relativistic analysis is to specify the spacetime geometry. It may seem daunting to describe the universe in its entirety, but certain symmetries simplify the task. According to the **cosmological principle**, no place or direction is special: the universe is homogeneous and isotropic. This is a working hypothesis, not a proven result, and it is obviously wrong on small scales; we can make a long list of ways in which the place we live differs from other places (on Earth, in the Solar System, etc.). However, as we examine larger and larger volumes of the universe, it does seem that everything begins to look the same [7].

In order for the universe to be isotropic, the angular piece of the metric must have the form $r^2 d\theta^2 + r^2 \sin^2 \theta d\phi^2$ that is familiar from the spherically-symmetric case. In order for the universe to be homogeneous, the spatial curvature must be the same everywhere. In the 1930s, Howard Percy Robertson [8] and Arthur Geoffrey Walker [9] showed that the most general metric satisfying the cosmological principle has the form

$$ds^2 = c^2 dt^2 - R(t)^2 \left[\frac{dr^2}{1 - kr^2} + r^2 d\theta^2 + r^2 \sin^2 \theta d\phi^2 \right] \tag{11.3}$$

This is now known as the **Robertson-Walker metric**. The constant k describes the spatial geometry:

$$k \begin{cases} > 0 & \text{curved like a sphere ("closed")} \\ = 0 & \text{flat} \\ < 0 & \text{curved like a saddle ("open")} \end{cases}$$

While it may be difficult to picture curved 3-d space, we can think of 2-d analogs as shown in Fig. 11.1. We describe the $k > 0$ case as "closed" because it is possible to walk in a "straight" line (i.e., always going in the same direction) and still return to the starting point, tracing out a closed curve; another way to say this is that the area of a $k > 0$ surface is finite. By contrast, a $k < 0$ surface is "open" in the sense that "straight" lines do not close on themselves, and the area is infinite (unless the surface has an edge). A $k = 0$ surface is also infinite, but it is special for being geometrically flat.

The factor $R(t)$ multiplying the spatial part of the metric is called the **scale factor**. To understand why, consider for a moment a radial spoke in a flat universe ($k = 0$). The proper distance between two points with radial coordinates r_1 and r_2 is

$$L = \sqrt{-ds^2} = R(t) |r_2 - r_1|$$

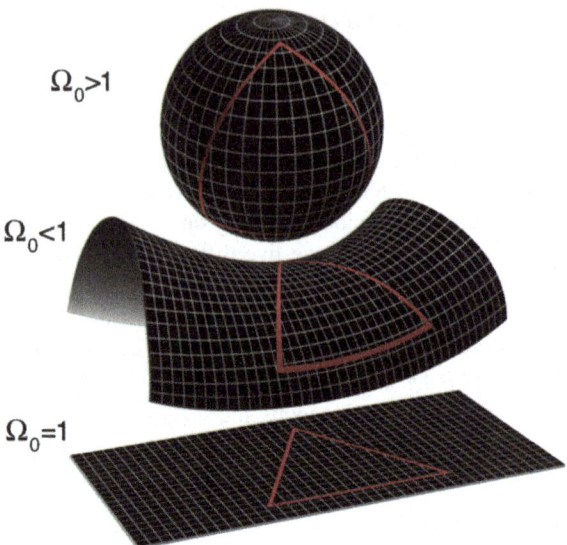

$\Omega_0 > 1$

$\Omega_0 < 1$

$\Omega_0 = 1$

Fig. 11.1 2-d curved surfaces: sphere (positive curvature), saddle (negative curvature), and flat. For a matter-filled universe, the curvature is related to the density of the universe (represented here by Ω_0; see text) (Credit: NASA / WMAP Science Team)

If $R(t)$ changes with time, the proper distance between the two points changes *even if they stay at the same coordinates*. Heuristically, we say the "size" of the universe changes with time. In what follows we take $R(t)$ to have dimensions of length, making r a dimensionless radial coordinate.

11.2.2   The Friedmann Equation

In the Robertson-Walker metric, the expansion of the universe is described by the way the scale factor changes with time. The proper way to derive an "equation of motion" for $R(t)$ involves more details of relativity than I want to go into. Several sources (e.g., [10, 11]) present a nice argument based on Newtonian gravity that, while not precisely correct, captures the key physics and yields the right equation.

Consider a sphere of radius R and density ρ, so the mass is $M = (4/3)\pi R^3 \rho$. Place a small "test" mass $m \ll M$ on the surface of the sphere. Now let the sphere expand and suppose the test mass stays right on the surface. At any time the gravitational potential energy of the test mass is

$$U = -\frac{G M m}{R} = -\frac{4\pi}{3} G \rho R^2 m$$

The test mass moves with speed $v = \mathrm{d}R/\mathrm{d}t$, so its kinetic energy is

$$K = \frac{m}{2}\left(\frac{\mathrm{d}R}{\mathrm{d}t}\right)^2$$

The total energy is then

$$\frac{m}{2}\left(\frac{\mathrm{d}R}{\mathrm{d}t}\right)^2 - \frac{4\pi}{3}G\rho R^2 m = E_{\text{tot}} \equiv -\frac{1}{2}mc^2 k$$

By conservation of energy, E_{tot} is constant and so we can define an equivalent constant k as in the last step. A full GR derivation reveals that this k is the same as the k in the Robertson-Walker metric. Dividing through by $-m/2$, and explicitly writing the t dependence for clarity, yields

$$\left(\frac{\mathrm{d}R}{\mathrm{d}t}\right)^2 - \frac{8\pi}{3}G\,\rho(t)\,R(t)^2 = -kc^2 \tag{11.4}$$

This is called the **Friedmann equation** after Alexander Friedmann, who derived it in 1922 [12]. (It was derived independently by Georges Lemaître in 1927 [13].) In general relativity, it is the equation of motion for a universe that is filled with matter that exerts gravity but has effectively no pressure.

It is customary to define some new quantities:

$$\text{expansion rate} \quad H(t) = \frac{1}{R}\frac{\mathrm{d}R}{\mathrm{d}t} \tag{11.5a}$$

$$\text{critical density} \quad \rho_{\text{crit}} = \frac{3H^2}{8\pi G} \tag{11.5b}$$

$$\text{density parameter} \quad \Omega = \frac{\rho}{\rho_{\text{crit}}} \tag{11.5c}$$

Defining the expansion rate in this way makes it scale free, so its dimension is inverse time; we use the symbol H deliberately, because it turns out that the Hubble constant H_0 is just the value of $H(t)$ today. Defining a density parameter that is dimensionless is not merely convenient but also enlightening if we use the right normalization factor. We will interpret the density normalization ρ_{crit} momentarily. Using these quantities, we can rewrite the Friedmann equation as

$$kc^2 = R^2\left[\frac{8\pi}{3}G\rho - \frac{1}{R^2}\left(\frac{\mathrm{d}R}{\mathrm{d}t}\right)^2\right]$$

$$= R^2 H^2\left[\frac{8\pi G}{3H^2}\rho - 1\right]$$

$$= R^2 H^2(\Omega - 1)$$

This equation reveals that the density of the universe is related to its geometry:

$$\Omega > 1 \leftrightarrow k > 0 \quad \text{"closed"}$$

$$\Omega = 1 \leftrightarrow k = 0 \quad \text{"flat"}$$

$$\Omega < 1 \leftrightarrow k < 0 \quad \text{"open"}$$

The transition corresponds to $\Omega = 1$, or $\rho = \rho_{\text{crit}}$. Now we understand that ρ_{crit} is the "critical" density that makes the universe flat. A density higher than the critical value causes the universe to be closed, while a density lower than the critical value makes the universe open.

As written in Eq. (11.4), the Friedmann equation involves what is effectively the velocity of expansion $(\mathrm{d}R/\mathrm{d}t)$. We can also obtain an equation for the effective acceleration. As the universe expands, conservation of mass implies

$$\rho_0 \, R_0^3 = \rho \, R^3$$

where the subscript 0 indicates a value today. Using this in Eq. (11.4) yields

$$\left(\frac{\mathrm{d}R}{\mathrm{d}t}\right)^2 - \frac{8\pi}{3} G \, \rho_0 \, \frac{R_0^3}{R} = -kc^2$$

Taking the time derivative gives

$$2 \frac{\mathrm{d}R}{\mathrm{d}t} \frac{\mathrm{d}^2 R}{\mathrm{d}t^2} + \frac{8\pi}{3} G \, \rho_0 \, \frac{R_0^3}{R^2} \frac{\mathrm{d}R}{\mathrm{d}t} = 0$$

Once again using $\rho_0 \, R_0^3 = \rho \, R^3$ finally yields

$$\frac{\mathrm{d}^2 R}{\mathrm{d}t^2} = -\frac{4\pi}{3} G \, \rho \, R \tag{11.6}$$

This **acceleration equation** is another key component of relativistic cosmology. Since ρ and R are always positive, the right-hand side is always negative. This means a universe filled with matter that exerts gravity but no pressure is *always decelerating* (and so the equation might be better termed the "deceleration equation"). No static solution is allowed; such a universe must be dynamic.

The Friedmann equation can be solved for the model discussed here (see Problem 11.5) to obtain the curves shown in Fig. 11.2. An interesting conceptual point is that for $\Omega \le 1$ the universe expands forever, while for $\Omega > 1$ the universe expands only to some finite maximum size before turning around and collapsing (see Problem 11.4). For a universe filled with pressure-less matter, the density of the universe determines its destiny.

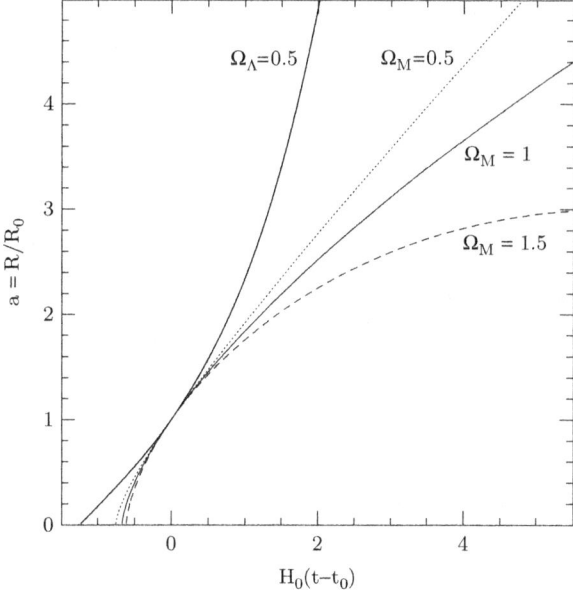

Fig. 11.2 Sample solutions of the Friedmann equation. The three concave down curves represent matter-filled universes with different values of Ω_M (and no cosmological constant, $\Omega_\Lambda = 0$). The thick, concave up curve represents a universe with a nonzero cosmological constant (and no matter, $\Omega_M = 0$). All curves are calibrated to have the same scale factor and expansion rate today

11.2.3 Einstein's Greatest Blunder

Einstein discovered in 1917 that his equations did not naturally permit a static universe. He did not believe the universe is dynamic, though, so he modified the equations. He introduced a new term so the Friedmann equation would become

$$\left(\frac{\mathrm{d}R}{\mathrm{d}t}\right)^2 - \left(\frac{8\pi}{3}\,G\,\rho + \frac{1}{3}\,\Lambda\,c^2\right) R^2 = -kc^2 \tag{11.7}$$

and the corresponding acceleration equation would be

$$\frac{\mathrm{d}^2 R}{\mathrm{d}t^2} = \left(-4\pi\,G\,\rho + \Lambda c^2\right) \frac{R}{3} \tag{11.8}$$

Here Λ is a new constant that is now known as the **cosmological constant**. With the modification, Einstein was able to find a model of the universe that was static: the two equations admit a solution with $\mathrm{d}R/\mathrm{d}t = 0$ and $\mathrm{d}^2 R/\mathrm{d}t^2 = 0$ if

$$\rho = \frac{kc^2}{4\pi GR^2} \quad \text{and} \quad \Lambda = \frac{k}{R^2} \quad \text{(static)}$$

In other words, the universe can be static—but only if it has certain values of the matter density, the cosmological constant, and the scale factor. If $\rho > 0$ then $k > 0$ as well, so the universe must have positive curvature.

After Edwin Hubble discovered the expansion of the universe in 1929, Einstein dismissed Λ as his "greatest blunder"; he had missed an opportunity to *predict* a dynamic universe. Despite this inauspicious beginning, the notion of a cosmological constant has come to play a major role in modern cosmology. Over the decades, theorists have explored the implications of Λ. One key concept is that if Λ exceeds some value then $d^2 R/dt^2$ can be positive: the expansion of the universe can accelerate! This feature began to receive broad attention in 1998, when two groups mapping the universe with supernovae announced observational evidence for accelerated expansion (see Sect. 11.3.3). Their work ultimately led to the 2011 Nobel Prize in Physics.

11.2.4 FRW Cosmology

Combining the Robertson-Walker metric with the Friedmann equation for the scale factor leads to what we call **Friedmann-Robertson-Walker (FRW) cosmology**. Let's manipulate the Friedmann equation again to put it in a form used in research today. First, define a dimensionless scale factor

$$a = \frac{R}{R_0} \tag{11.9}$$

where the subscript 0 again denotes the value today. At a general time, the Friedmann equation can be written as

$$
-kc^2 = \left(\frac{dR}{dt}\right)^2 - \frac{8\pi}{3} G \rho R^2 - \frac{1}{3} \Lambda c^2 R^2
$$
$$
= R_0^2 a^2 \left[H^2 - \frac{8\pi}{3} G \rho_0 a^{-3} - \frac{1}{3} \Lambda c^2 \right] \tag{11.10}
$$

where we use $dR/dt = RH$, $R = R_0 a$, and $\rho = \rho_0 R_0^3/R^3 = \rho_0 a^{-3}$. This equation must hold today, so

$$
-kc^2 = R_0^2 \left[H_0^2 - \frac{8\pi}{3} G \rho_0 - \frac{1}{3} \Lambda c^2 \right] \tag{11.11}
$$

Equating (11.10) and (11.11) and rearranging yields

$$
H^2 = H_0^2 a^{-2} - \frac{8\pi}{3} G \rho_0 (a^{-2} - a^{-3}) - \frac{1}{3} \Lambda c^2 (a^{-2} - 1)
$$

$$= H_0^2 \left[a^{-2} - \Omega_M (a^{-2} - a^{-3}) - \Omega_\Lambda (a^{-2} - 1) \right] \qquad (11.12)$$

where we have defined the density parameters

$$\Omega_M = \frac{8\pi G \rho_0}{3 H_0^2} \quad \text{and} \quad \Omega_\Lambda = \frac{\Lambda c^2}{3 H_0^2} \qquad (11.13)$$

(Note that Ω_M is the dimensionless density parameter today. It gets a subscript M now to distinguish it from the dimensionless version of the cosmological constant, Ω_Λ.) Going back to Eq. (11.12) and collecting terms with the same power of a yields

$$H^2 = H_0^2 \left[(1 - \Omega_M - \Omega_\Lambda) a^{-2} + \Omega_M a^{-3} + \Omega_\Lambda \right] \qquad (11.14)$$

This is the version of the Friedmann equation that is used most often in current cosmology research.

Returning to Eq. (11.11) and making the substitutions, we can write

$$k = -\frac{H_0^2 R_0^2}{c^2} (1 - \Omega_M - \Omega_\Lambda) \equiv -\frac{H_0^2 R_0^2}{c^2} \Omega_k \qquad (11.15)$$

where $\Omega_k = 1 - \Omega_M - \Omega_\Lambda$ is defined to be the "curvature density." Now we understand that in a universe with both matter and Λ, the *total* density of the universe is still related to its geometry:

$$\Omega_M + \Omega_\Lambda > 1 \leftrightarrow \text{"closed"}$$

$$\Omega_M + \Omega_\Lambda = 1 \leftrightarrow \text{"flat"}$$

$$\Omega_M + \Omega_\Lambda < 1 \leftrightarrow \text{"open"}$$

However, the total density of the universe no longer uniquely determines its destiny, because matter and Λ affect the expansion in different ways.

11.3   Observational Cosmology

With FRW theory in place, it would be lovely if we could measure $R(t)$ directly and compare it with the curves in Fig. 11.2. Unfortunately, we cannot. What we can measure is redshift as galaxies recede from us, and distance. In this section we examine redshifts and distances in FRW cosmology, and then say a few words about how we measure them and what the measurements have revealed.

11.3.1 Cosmological Redshift

The expansion of the universe carries galaxies away from us and therefore creates a **cosmological redshift**. Consider a light source sitting at coordinate r_e and emitting flashes of light. The light travels radially to Earth. Since it is light, its spacetime interval must be zero:

$$0 = c^2\,dt^2 - R(t)^2\,\frac{dr^2}{1-kr^2} \quad \Rightarrow \quad \frac{dt}{R(t)} = \pm\frac{1}{c}\,\frac{dr}{\sqrt{1-kr^2}}$$

Consider one flash emitted at (r_e, t_e) and observed at $(0, t_o)$. Since r is decreasing, we use the minus sign and integrate from r_e to 0, but then we use the sign to switch the limits of integration:

$$\int_{t_e}^{t_o} \frac{dt}{R(t)} = \frac{1}{c}\int_{0}^{r_e} \frac{dr}{\sqrt{1-kr^2}} \tag{11.16}$$

Now consider the next flash, emitted at $(r_e, t_e + \Delta t_e)$ and observed at $(0, t_o + \Delta t_o)$:

$$\int_{t_e+\Delta t_e}^{t_o+\Delta t_o} \frac{dt}{R(t)} = \frac{1}{c}\int_{0}^{r_e} \frac{dr}{\sqrt{1-kr^2}} \tag{11.17}$$

Using relations from calculus, we can rewrite the integral on the left-hand side as

$$\int_{t_e+\Delta t_e}^{t_o+\Delta t_o} \frac{dt}{R(t)} = \int_{t_e}^{t_o} \frac{dt}{R(t)} + \int_{t_o}^{t_o+\Delta t_o} \frac{dt}{R(t)} - \int_{t_e}^{t_e+\Delta t_e} \frac{dt}{R(t)}$$

If the time between flashes is small compared with the time it takes the light to travel, then Δt_e and Δt_o are small and we can approximate the second and third integrals on the right-hand side to obtain

$$\int_{t_e+\Delta t_e}^{t_o+\Delta t_o} \frac{dt}{R(t)} \approx \int_{t_e}^{t_o} \frac{dt}{R(t)} + \frac{\Delta t_o}{R(t_o)} - \frac{\Delta t_e}{R(t_e)} \tag{11.18}$$

Subtracting Eqs. (11.16) and (11.17) and using Eq. (11.18) yields

$$\frac{\Delta t_o}{R(t_o)} = \frac{\Delta t_e}{R(t_e)} \quad \Rightarrow \quad \frac{\Delta t_o}{\Delta t_e} = \frac{R(t_o)}{R(t_e)}$$

As we have done before, we can interpret small time intervals as inverse frequencies, and then relate frequencies to wavelengths, obtaining

$$\frac{R(t_o)}{R(t_e)} = \frac{\nu_e}{\nu_o} = \frac{\lambda_o}{\lambda_e}$$

Finally, we define the redshift z_e from which the light was emitted:

$$z_e \equiv \frac{\lambda_o}{\lambda_e} - 1 = \frac{R(t_o)}{R(t_e)} - 1 \tag{11.19}$$

If we are the observers then $R(t_o)$ is just the scale factor today, R_0, and the ratio on the right-hand side is the inverse of the dimensionless scale factor at the time the light was emitted. Dropping the subscript "e" (since it is implicit that we discuss the scale factor and redshift at the time the light was emitted), we have

$$z = \frac{1}{a} - 1 \quad \Leftrightarrow \quad a = \frac{1}{1+z} \tag{11.20}$$

We see that cosmological redshift is directly related to the scale factor: as the universe expands, a light ray gets stretched to longer wavelengths.

11.3.2 Cosmological Distances

We also want to discuss distance, but in a universe that is expanding and possibly curved we need to take care to specify what we mean by the notion of "distance." Let's first consider the distance measured purely in terms of coordinates. From Eq. (11.16), we can say that if light is emitted at (r_e, t_e) and observed at $(0, t_o)$, the coordinates are related by

$$\int_0^{r_e} \frac{dr}{\sqrt{1 - kr^2}} = \int_{t_e}^{t_o} \frac{c\,dt}{R(t)} = \frac{c}{R_0 H_0}\chi \tag{11.21}$$

where we have defined the t integral to be χ, with some multiplicative factors for convenience. Let's manipulate this integral:

$$\chi = R_0 H_0 \int_{t_e}^{t_o} \frac{dt}{R(t)} = R_0 H_0 \int_{R_e}^{R_0} \frac{1}{R}\frac{dt}{dR}\,dR = R_0 H_0 \int_{R_e}^{R_0} \frac{dR}{R^2 H(R)} = H_0 \int_0^{z_e} \frac{dz}{H(z)}$$

We change variables from t to R, use $dR/dt = RH$, and then change variables to $R = R_0/(1+z)$. Finally, using Eq. (11.14) for $H(z)$ yields

$$\chi(z_e) = \int_0^{z_e} \frac{dz}{[(1 - \Omega_M - \Omega_\Lambda)(1+z)^2 + \Omega_M(1+z)^3 + \Omega_\Lambda]^{1/2}} \tag{11.22}$$

Returning to Eq. (11.21), the r integral evaluates to

$$\int \frac{dr}{\sqrt{1 - kr^2}} = \begin{cases} |k|^{-1/2}\sin^{-1}(|k|^{1/2}r) & k > 0 \\ r & k = 0 \\ |k|^{-1/2}\sinh^{-1}(|k|^{1/2}r) & k < 0 \end{cases}$$

We equate this to $(c/R_0 H_0)\chi$ and then invert to isolate r. We can use Eq. (11.15) to put $k = -(H_0 R_0/c)^2 \Omega_k$ where $\Omega_k = 1 - \Omega_M - \Omega_\Lambda$ is the curvature density. This finally yields

$$R_0 r(z) = \frac{c}{H_0} \times \begin{cases} |\Omega_k|^{-1/2} \sin\left(|\Omega_k|^{1/2}\chi(z)\right) & \Omega_k < 0 \\ \chi(z) & \Omega_k = 0 \\ |\Omega_k|^{-1/2} \sinh\left(|\Omega_k|^{1/2}\chi(z)\right) & \Omega_k > 0 \end{cases} \qquad (11.23)$$

This is the general expression for the coordinate distance to redshift z. (We have dropped the subscript on z for simplicity.) The factor of R_0 appears on the left-hand side because the coordinate radius always appears in the metric in combination with the scale factor.

Like the scale factor, coordinate distances cannot be measured directly. What, then, can we measure? We can measure the apparent brightness of an object, quantified in terms of the flux F defined to be the energy received per unit time per unit area. The flux is related to the intrinsic luminosity L (energy emitted per unit time) via the inverse square law,

$$F = \frac{L}{4\pi d_L^2}$$

The distance that appears here is known as the **luminosity distance** (hence the subscript "L" for "luminosity"). As light propagates through the universe, it is subject to two effects. First, as we saw in Sect. 11.3.1, the time between photons scales as $(1 + z)$, so the rate at which photons arrive scales as $(1 + z)^{-1}$. Second, the cosmological redshift increases the wavelength, which scales the photon energy by $(1 + z)^{-1}$. These two factors mean the observed flux is related to the coordinate distance as

$$F = \frac{L}{4\pi R_0^2 r^2 (1 + z)^2}$$

Thus, luminosity distance is related to coordinate distance as

$$d_L(z) = (1 + z) R_0 r(z) \qquad (11.24)$$

This, not $R(t)$, is the relation that is actually measurable (see below). Figure 11.3 shows $d_L(z)$ curves for the FRW solutions that were shown in Fig. 11.2.

We can also measure the angular size of an object. Consider an object with proper length dL placed perpendicular to the line of sight such that the two ends have the same azimuthal angle ϕ but span a small range of polar angle, $d\theta$. We define the **angular diameter distance**, d_A, such that

$$d\theta = \frac{dL}{d_A}$$

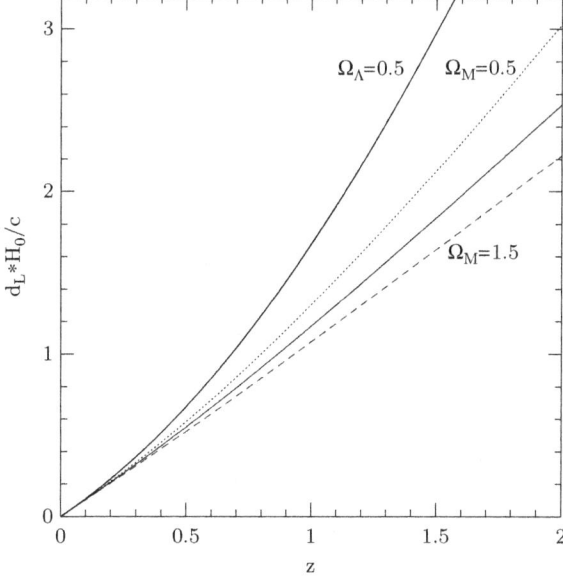

Fig. 11.3 Sample curves of luminosity distance versus redshift for the FRW solutions shown in Fig. 11.2. The *solid curve* is for $\Omega_M = 1$ and $\Omega_\Lambda = 0$. Luminosity distances are plotted in units of c/H_0

in the small-angle approximation. From the Robertson-Walker metric (11.3), we can write the proper length as

$$dL = R(t)\, r\, d\theta$$

Then using $R = R_0/(1 + z)$ from Eq. (11.20), we can write

$$d_A(z) = \frac{R_0 r(z)}{1 + z} \tag{11.25}$$

We have already encountered the angular diameter distance in the context of gravitational lensing (Chap. 9).

11.3.3 Results

We can measure luminosity distances if we can observe a set of sources whose luminosities are known. The simplest way to do this is to find **standard candles**, or sources that all have the same intrinsic luminosity. In practice, few objects are perfectly identical, but several classes of objects are "standardizable" in the sense that their observable properties allow us to infer their intrinsic luminosities with

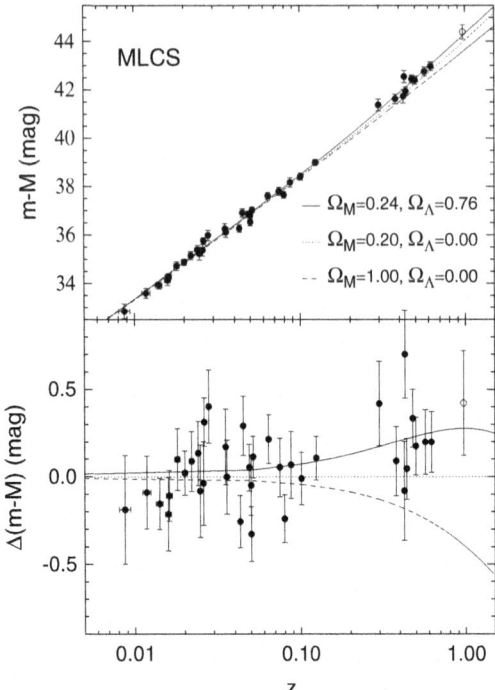

Fig. 11.4 Supernova data as of 1998. In the *top panel*, the vertical axis is $m - M = 5\log_{10} d_L/(10\,\text{pc})$. The points show the supernova data, while the curves show theoretical $d_L(z)$ curves for different values of Ω_M and Ω_Λ. In the *bottom panel*, the curve with $\Omega_M = 0.2$ and $\Omega_\Lambda = 0$ has been subtracted off to make the differences more clear (Credit: Riess et al. [14]. Reproduced by permission of the AAS. Also see Perlmutter et al. [15])

enough accuracy and precision to enable distance measurements. We will see in Chap. 18 that this includes a class of variable stars called Cepheids and a class of exploding stars called Type Ia supernovae.

The latter objects revealed the accelerated expansion of the universe. In 1998, two teams of astronomers [14, 15] reported that Type Ia supernovae at redshifts $z \gtrsim 0.1$ are farther away than expected for a universe in which matter has been causing the expansion of the universe to decelerate (see Fig. 11.4). They concluded that the universe contains some substance, dubbed **dark energy**, that causes the expansion to accelerate. For now let us assume that dark energy is just the cosmological constant, but we will briefly discuss other possibilities.

While the evidence for accelerated expansion was strong, there was some concern that a phenomenon other than dark energy might make distant supernovae appear to be farther away than they actually are (e.g., supernovae were less luminous when the universe was younger, or some unusual kind of dust absorbs supernova light without changing the color). A model with dark energy stands out from those possibilities in an important way: in the past, the density of matter was higher than

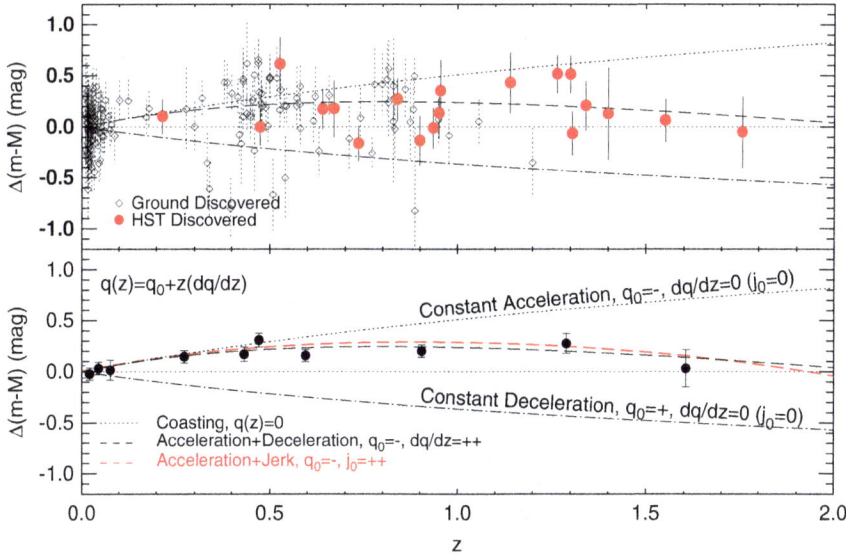

Fig. 11.5 More supernova data, showing that the universe was *decelerating* at higher redshifts (earlier times) before it switched to *accelerating* at lower redshifts (later times). That behavior is expected for a model with both matter and dark energy. The *top panel* shows data points for individual supernovae, while the *bottom panel* shows average values (Credit: Riess et al. [16]. Reproduced by permission of the AAS)

now (because the universe was smaller and mass is conserved), while the density of Λ was the same (hence cosmological "constant"). Such a universe would shift from deceleration at early times (when matter dominates) to acceleration at late times (when Λ dominates). To test this prediction, astronomers searched for more distant supernovae. Figure 11.5 shows that higher-redshift supernovae indeed reveal that the universe was decelerating in the past before it switched to accelerating more recently.

The next step is to determine what values of Ω_M and Ω_Λ are consistent with the data. Broadly speaking, the idea is to adjust these two parameters to make the predicted $d_L(z)$ curve pass through the data points. Figure 11.6 shows the constraints from supernova data (in blue), together with constraints from two other probes of the geometry of the universe: the cosmic microwave background (in orange; see Chap. 20), and baryon acoustic oscillations (in green).[1] Individually, no probe uniquely determines both Ω_M and Ω_Λ. Supernovae, for example, permit models with less mass and less Λ or more mass and more Λ, as long as the

---

[1]The fluctuations we see in the cosmic microwave background were created by sound waves, or acoustic oscillations, in the gas in the young universe. Those fluctuations left subtle imprints in the structure of normal ("baryonic") matter that are known as "baryon acoustic oscillations." BAO can be measured using the distribution of galaxies [18, 19].

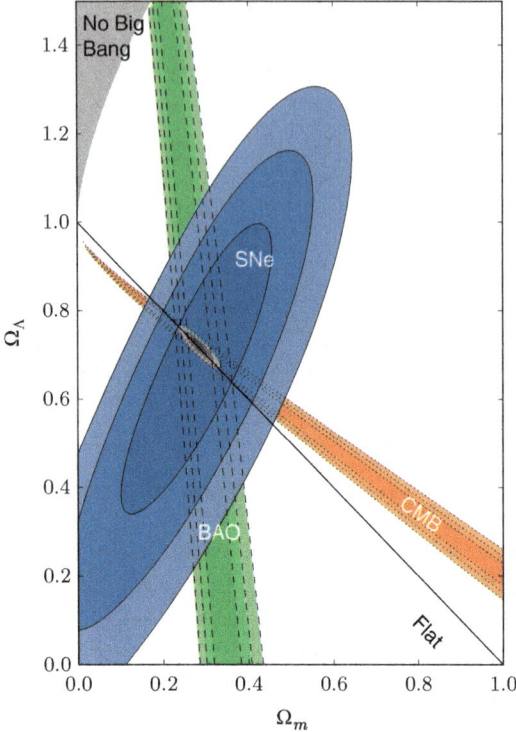

Fig. 11.6 Constraint regions for the cosmological parameters Ω_M and Ω_Λ, from Type Ia supernovae (*blue*) as well as the cosmic microwave background (CMB, *orange*; see Chap. 20) and baryon acoustic oscillations (BAO, *green*). In the upper left corner of the plot, the *gray region* corresponds to parameter values that are unphysical because there was no big bang (Credit: Suzuki et al. [17] Reproduced by permission of the AAS)

combination yields a certain net acceleration. The three sets of constraints are complementary, though, and together they determine Ω_M and Ω_Λ with small uncertainties. The conclusion is that the universe is flat or very close to it, with mass (mostly dark matter) contributing about 32% and Λ contributing about 68% of the total energy density of the universe.

The cosmological constant is weird: not only does it act as a kind of anti-gravity, but as the universe expands the energy density associated with Λ stays constant. Since the energy density associated with matter decreases (mass is constant as volume increases), dark energy will dominate more and more over matter, and the accelerated expansion will continue forever. In this scenario, everything outside our Local Group of galaxies (which includes not only the Milky Way but also the Andromeda galaxy and a few dozen small galaxies that orbit the two big ones) will eventually have recession velocities so large that their light will never reach us. The visible universe beyond the Local Group will eventually go dark (see Problem 11.7).

That scenario assumes dark energy is Einstein's cosmological constant, which is a simple assumption that is consistent with today's best data. However various theoretical concerns about Λ have led many people to hypothesize that dark energy could be more complicated and dynamic. We can characterize the repulsive effect of dark energy by saying that it has a negative pressure and then relating the pressure (P) to the density (ρ) via an equation of state of the form $P = w\rho c^2$. The cosmological constant corresponds to $w = -1$, but other models of dark energy have values of w that differ from -1 and maybe even vary with time. In such scenarios, the ultimate fate of the universe is uncertain. In some models, dark energy might be able to convert into matter or radiation and even lead to a collapse and Big Crunch. Alternatively, if $w < -1$ then the density of dark energy will *increase* with time to the point that the expansion will ultimately overcome all forces (not only gravity but also electromagnetism and nuclear forces) and tear apart all galaxies, stars, planets, atoms, and even nuclei at some finite time in the future, known as the Big Rip.

Having said all that, we still need to ask: is dark energy even the correct model? What if our assumptions about the cosmological principle are wrong, or general relativity is not the correct theory of gravity? People are thinking seriously about the possibility that Einstein's theory is an approximation to some more general theory of gravity (similar to the way that Newton's theory is an approximation to Einstein's) that can accommodate accelerated expansion without invoking an exotic substance. We will not say more about that here, except to note that the issue of dark energy versus modified gravity will ultimately be decided not by debate but by data. More and better data, from supernovae, the cosmic microwave background, and a wealth of other techniques, will provide the evidence we need to distinguish between different models for the structure, content, and eventual fate of the universe.

Problems

Where necessary, you may take the Hubble constant to be $H_0 = 74\,\mathrm{km\,s^{-1}Mpc^{-1}}$.

11.1. To visualize Hubble's law, imagine a circular rubber band with three points marked as follows: A at polar angle $\phi = 0$, B and $\phi = \pi/4$, and C at $\phi = \pi/2$.

(a) When the circle has radius R, what is the distance along the rubber band from A to B? From A to C?

(b) If the circle expands at the rate dR/dt, what is the rate of change of the distances from A to B and from A to C?

(c) Show that velocity and distance have a relation of the form $v = H_0 d$. What is H_0 in this problem?

11.2. Here is an alternate, and perhaps more intuitive, way to derive the critical density of the universe.

(a) Consider throwing a ball upward with speed v from the surface of a planet with radius R. Show that if the planet's density exceeds some critical value, the ball will go up, stop, and come back down, but if the density is lower than the critical value, the ball will go up forever. Derive an expression for the critical density of the planet in terms of v, R, and fundamental constants.

(b) Now write $v = HR$ in analog with the Hubble law, and show that your expression is equivalent to the critical density of the universe.

(c) Using the present-day value of the Hubble constant, compute the critical density of the universe in g/cm^3. How many protons per cubic meter is that?

11.3. Show that the general expressions for cosmological distances reduce to Hubble's law at low redshifts ($z \ll 1$). In this limit, the recession speed is $v \approx cz$.

11.4. For certain values of the cosmological density parameters, the expansion of the universe can change direction. For each of the following scenarios, find the value of a for the turnaround point, and indicate whether the universe changes from expansion to contraction or vice versa. Explain your analysis.

(a) $\Omega_\Lambda = 0$ and $\Omega_M > 1$
(b) $\Omega_M = 0$ and $\Omega_\Lambda > 1$

11.5. Consider how the scale factor evolves with time (see Fig. 11.2).

(a) Use the Friedmann equation to obtain an expression for $t(a)$ in terms of an integral over a.

(b) Evaluate the integral for the two cases $(\Omega_M, \Omega_\Lambda) = (1, 0)$ and $(0, 1)$. Fix the integration constant by setting $a = 1$ today. Invert the results to find $a(t)$.

(c) Now consider more general cases: (i) $\Omega_\Lambda = 0$ and $\Omega_M \neq 0$; (ii) $\Omega_M = 0$ and $\Omega_\Lambda \neq 0$. You may find it helpful to use a table of integrals or symbolic mathematics software.

11.6. The cosmological **lookback time** to redshift z is the difference between the age of the universe now and the age it had at redshift z. In a universe with $\Omega_M = 1$ and $\Omega_\Lambda = 0$, find the age of the universe today and the lookback times to $z = 1$ and $z = 2$. (Your results from Problem 11.5 should be useful.)

11.7. In a universe with accelerated expansion, the future will be lonely. Since galaxies are accelerating away from us but the speed of light is finite, there is some time in the future when light emitted by a given galaxy will no longer be able to reach us. Consider $\Omega_M = 0$ and $\Omega_\Lambda = 1$ and find the evolution of the scale factor (see Problem 11.5). Then use $ds^2 = 0$ for light to compute the time it would take light emitted from a given galaxy at a given time to reach us. Finally, find the time when light emitted by the galaxy would take an infinite amount of time to reach us. How long in the future will we lose contact with a galaxy that is presently 10 Mpc away?

References

1. V.M. Slipher, Lowell Obs. Bull. **2**, 56 (1913)
2. W.W. Campbell, G.F. Paddock, Publ. Astron. Soc. Pac. **30**, 68 (1918)
3. E. Hubble, Proc. Nat. Acad. Sci. **15**, 168 (1929)
4. W.L. Freedman, B.F. Madore, Ann. Rev. Astron. Astrophys. **48**, 673 (2010)
5. A.G. Riess, L. Macri, S. Casertano, H. Lampeitl, H.C. Ferguson, A.V. Filippenko, S.W. Jha, W. Li, R. Chornock, Astrophys. J. **730**, 119 (2011)
6. Planck Collaboration, ArXiv e-prints arXiv:1303.5076 (2013)
7. C. Marinoni, J. Bel, A. Buzzi, J. Cosmol. Astropart. Phys. **10**, 036 (2012)
8. H.P. Robertson, Astrophys. J. **82**, 284 (1935)
9. A.G. Walker, Q. J. Math. **os-6**(1), 81 (1935)
10. B.W. Carroll, D.A. Ostlie, *An Introduction to Modern Astrophysics*, 2nd edn. (Addison-Wesley, San Francisco, 2007)
11. D. Maoz, *Astrophysics in a Nutshell* (Princeton University Press, Princeton, 2007)
12. A. Friedmann, Zeitschrift fur Physik **10**, 377 (1922)
13. G. Lemaître, Annales de la Societe Scietifique de Bruxelles **47**, 49 (1927)
14. A.G. Riess et al., Astron. J. **116**, 1009 (1998)
15. S. Perlmutter et al., Astrophys. J. **517**, 565 (1999)
16. A.G. Riess et al., Astrophys. J. **607**, 665 (2004)
17. N. Suzuki et al., Astrophys. J. **746**, 85 (2012)
18. D.J. Eisenstein, New Astron. Rev. **49**, 360 (2005)
19. D.J. Eisenstein et al., Astrophys. J. **633**, 560 (2005)

Part II
Using Stellar Physics to Explore the Cosmos

Chapter 12
Planetary Atmospheres

We now shift attention from gravity to other aspects of physics that are relevant for astronomical systems. We begin with gas physics, which has two facets. **Thermodynamics** describes the bulk properties of a gas (such as temperature, density, and pressure), while **statistical mechanics** describes the microscopic motions of the particles in the gas. In this chapter we use both to study Earth's atmosphere in the context of a basic theory of planetary atmospheres.

12.1 Kinetic Theory of Gases

We can connect the macroscopic and microscopic pictures of a gas by characterizing the motions of particles as they move within a container and bump into the walls. In the simplest version of **kinetic theory**, we view gas particles as billiard balls that do not interact except for occasional collisions.

12.1.1 Temperature and the Boltzmann Distribution

There are many more particles than we can track individually, so we focus on *statistical* properties of the motion such as the distribution of velocities. Elastic collisions allow particles to exchange energy and momentum and settle into an equilibrium in which the statistical properties do not change with time. This lets us invoke a general discovery made by Ludwig Boltzmann (and also Josiah Gibbs): for a system in equilibrium with temperature T, the number of particles with energy E_i is

$$N_i \propto g_i \, e^{-E_i/kT} \tag{12.1}$$

C. Keeton, *Principles of Astrophysics: Using Gravity and Stellar Physics to Explore the Cosmos*, Undergraduate Lecture Notes in Physics, DOI 10.1007/978-1-4614-9236-8_12, © Springer Science+Business Media New York 2014

where g_i counts the number of different ways in which a particle can have energy E_i, and $k = 1.38 \times 10^{-23}$ kg m^2 s^{-2} K^{-1} is the Boltzmann constant. Equation (12.1) is known as the **Boltzmann distribution**, and it plays a fundamental role in many parts of statistical physics.

In everyday life we measure temperature using the Fahrenheit and Celsius scales, but in the Boltzmann distribution (and elsewhere in physics) it is better to use the **Kelvin scale**. The reason is that Kelvin temperature is directly related to kinetic energy in a way that we will see shortly. For reference, the Fahrenheit (T_F), Celsius (T_C), and Kelvin (T) temperatures are related as follows:

$$T = T_C + 273.15 = \frac{5}{9}(T_F + 459.67)$$

12.1.2 Maxwell-Boltzmann Distribution of Particle Speeds

We can use the Boltzmann distribution to derive the distribution of speeds for particles in a gas, at least in the following idealized scenario:

1. The motion is non-relativistic.
2. The particles interact only through collisions.
3. The particles have no significant internal structure, so collisions are elastic.

These assumptions describe an **ideal gas**, which is simplified but instructive and a good representation of many real gases. We will check their validity in Sect. 12.1.4.

Assumption #1 lets us write the kinetic energy of a single particle as $K = mv^2/2$. Assumptions #2 and #3 tell us there is no potential energy between particles or internal energy within a given particle, so the total energy of a particle is just $E = mv^2/2$. This means the Boltzmann distribution will have a factor of $\exp(-mv^2/2kT)$. What about the factor g_i? If the motion is isotropic, g_i is a constant. The Boltzmann distribution then gives the number of particles with velocity between $\mathbf{v}$ and $\mathbf{v} + d\mathbf{v}$ as

$$N(\mathbf{v})\,d\mathbf{v} = C \times e^{-m|\mathbf{v}|^2/2kT}\,d\mathbf{v}$$

The normalization constant C can be specified by setting the total number of particles equal to N_{tot}:

$$N_{\text{tot}} \equiv \int N(\mathbf{v})\,d\mathbf{v}$$

$$= C \times \int \exp\left[-\frac{m}{2kT}\left(v_x^2 + v_y^2 + v_z^2\right)\right]\,dv_x\,dv_y\,dv_z$$

$$= C \times \int e^{-mv_x^2/2kT}\,dv_x \int e^{-mv_y^2/2kT}\,dv_y \int e^{-mv_z^2/2kT}\,dv_z$$

$$= C \times \left(\frac{2\pi kT}{m}\right)^{3/2}$$

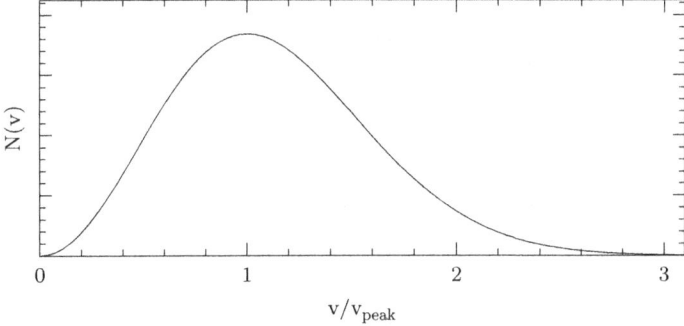

Fig. 12.1 The Maxwell-Boltzmann velocity distribution

Here we write out $|\mathbf{v}|^2 = v_x^2 + v_y^2 + v_z^2$ and then recognize that the full integral is a product of three Gaussian integrals, which can be evaluated using $\int e^{-x^2/2\sigma^2}\,dx = (2\pi)^{1/2}\sigma$ (see Sect. A.7). The upshot is that the distribution of velocities can be written as

$$N(\mathbf{v})\,d\mathbf{v} = N_{\text{tot}}\left(\frac{m}{2\pi kT}\right)^{3/2} e^{-mv^2/2kT}\,d\mathbf{v} \tag{12.2}$$

If we focus on *speed* and do not worry about direction, we can use the spherical volume element to replace $d\mathbf{v}$ with $4\pi v^2\,dv$. Then the number of particles with speed between v and $v + dv$ is

$$N(v)\,dv = N_{\text{tot}}\left(\frac{m}{2\pi kT}\right)^{3/2} e^{-mv^2/2kT}\,4\pi v^2\,dv \tag{12.3}$$

This is the **Maxwell-Boltzmann distribution** for the speeds of ideal gas particles of mass m in equilibrium at temperature T. It is shown in Fig. 12.1.

We can understand some general properties of the distribution as follows. The most common particle speed corresponds to the peak of the distribution, which can be found by solving $dN/dv = 0$:

$$\frac{dN}{dv} = N_{\text{tot}}\left(\frac{m}{2\pi kT}\right)^{3/2}\left(-\frac{mv}{kT}\,e^{-mv^2/2kT}\,4\pi v^2 + 8\pi v\,e^{-mv^2/2kT}\right)$$

$$= N_{\text{tot}}\left(\frac{m}{2\pi kT}\right)^{3/2} 8\pi v\,e^{-mv^2/2kT}\left(-\frac{mv^2}{2kT} + 1\right)$$

$$\Rightarrow \quad v_{\text{peak}} = \left(\frac{2kT}{m}\right)^{1/2} \tag{12.4}$$

What is the average velocity? We must be careful about what we mean here. If we work in a frame centered on the container of gas, there must be as many particles going left as going right (or up/down, or front/back), so the mean *velocity* must vanish, $\langle \mathbf{v} \rangle = 0$. We could instead compute the mean speed, $\langle v \rangle = \langle |\mathbf{v}| \rangle$. If we are shifting to speed, though, it actually makes more sense to compute the mean of v^2, because this is directly related to the mean kinetic energy. (Also, using v^2 has the benefit of simplifying the integrals, as we will see.) We often refer to $v_{rms} = \langle v^2 \rangle^{1/2}$ as the "root mean square" (or "RMS") velocity.

Whenever we work with a statistical distribution, the average of some quantity Q is defined to be

$$\langle Q \rangle = \frac{\int Q\, N(\mathbf{v})\, d\mathbf{v}}{\int N(\mathbf{v})\, d\mathbf{v}} = \frac{1}{N_{\text{tot}}} \int Q\, N(\mathbf{v})\, d\mathbf{v} \tag{12.5}$$

The mean of v^2 is therefore

$$
\begin{aligned}
\langle v^2 \rangle &= \left(\frac{m}{2\pi kT} \right)^{3/2} \int \left(v_x^2 + v_y^2 + v_z^2 \right) \exp\left[-\frac{m}{2kT} \left(v_x^2 + v_y^2 + v_z^2 \right) \right] dv_x\, dv_y\, dv_z \\
&= \left(\frac{m}{2\pi kT} \right)^{3/2} \int v_x^2 e^{-mv_x^2/2kT}\, dv_x \int e^{-mv_y^2/2kT}\, dv_y \int e^{-mv_z^2/2kT}\, dv_z \\
&\quad + \left(\frac{m}{2\pi kT} \right)^{3/2} \int e^{-mv_x^2/2kT}\, dv_x \int v_y^2 e^{-mv_y^2/2kT}\, dv_y \int e^{-mv_z^2/2kT}\, dv_z \\
&\quad + \left(\frac{m}{2\pi kT} \right)^{3/2} \int e^{-mv_x^2/2kT}\, dv_x \int e^{-mv_y^2/2kT}\, dv_y \int v_z^2 e^{-mv_z^2/2kT}\, dv_z
\end{aligned}
$$

We can evaluate this expression by recognizing that each term[1] has two factors of the form $\int e^{-x^2/2\sigma^2}\, dx = (2\pi)^{1/2}\sigma$ and one factor of the form $\int x^2 e^{-x^2/2\sigma^2}\, dx = (2\pi)^{1/2}\sigma^3$, where $\sigma = (kT/m)^{1/2}$. The net result is that each term evaluates to kT/m, and the fact that there are three terms means the final answer is

$$\langle v^2 \rangle = \frac{3kT}{m} \tag{12.6}$$

This result is particularly useful when we consider the average kinetic energy of a particle:

$$\langle K \rangle = \left\langle \frac{1}{2} mv^2 \right\rangle = \frac{1}{2} m \langle v^2 \rangle = \frac{3}{2} kT \tag{12.7}$$

This is the promised relation between temperature and kinetic energy. The factor of 3 entered because a particle can move in three different directions. On average there is an energy $kT/2$ associated with motion in each direction.

---

[1] Had we tried to evaluate $\langle v \rangle$, we would not have been able to separate the integral into three distinct terms as we have done here.

Example: Room Temperature

What is the typical speed of an air molecule at room temperature, $T = 293$ K? Air is mostly molecular nitrogen N_2 with $m = 28m_p$. The RMS velocity of a nitrogen molecule is

$$v_{\text{rms}} = \left[\frac{3 \times (1.38 \times 10^{-23} \, \text{kg m}^2 \, \text{s}^{-2} \, \text{K}^{-1}) \times (293 \, \text{K})}{28 \times (1.67 \times 10^{-27} \, \text{kg})} \right]^{1/2} = 509 \, \text{m s}^{-1}$$

12.1.3 Pressure and the Ideal Gas Law

The macroscopic effect we call pressure arises from particles hitting the walls of the container. We can use the microscopic picture to derive the **equation of state** relating the pressure to other physical properties of the gas.

Let's begin with motion in the x-direction and then generalize. Suppose the gas is in a container of length L and cross sectional area A, so the volume is $V = AL$. Consider a particle moving to the right with velocity v_x and momentum p_x.[2] When it hits the right-hand wall, the particle rebounds with momentum p_x in the opposite direction (assuming an elastic collision). The change in the momentum of the particle is thus $\Delta p_x = -2p_x$. By Newton's third law, this same impulse is applied to the wall of the container, but in the opposite direction. The time-averaged force on the wall is then $|\Delta p_x|$ divided by the time between collisions with the right-hand wall, which is $\Delta t = 2L/v_x$ (the particle must traverse the length of the container twice, going first left and then back right). Thus, the time-averaged force on the right wall from this one particle is

$$f_x = \frac{\Delta p_x}{\Delta t} = \frac{v_x p_x}{L}$$

The average force on the wall from *all* particles is

$$F = \frac{1}{L} \int v_x \, p_x \, N(p_x) \, dp_x$$

where $N(p_x) \, dp_x$ is the number of particles whose x-component of momentum is between p_x and $p_x + dp_x$.[3] Pressure is then force per unit area:

---

[2] Note that velocity and momentum enter this analysis in different ways. While the two quantities are related, keeping them separate lets the framework describe either classical or relativistic motion.

[3] We could equivalently write the integral in terms of velocity using $N(v_x) \, dv_x$.

$$P = \frac{F}{A} = \frac{1}{AL} \int v_x \, p_x \, N(p_x) \, dp_x = \int v_x \, p_x \, n(p_x) \, dp_x$$

where we recognize AL as the volume of the container so $N/(AL)$ is the number density of particles, n. Since the total number density of particles can be written as $n_{tot} = \int n(p_x) \, dp_x$, we can write

$$P = n_{tot} \frac{\int v_x \, p_x \, n(p_x) \, dp_x}{\int n(p_x) \, dp_x} = n_{tot} \langle v_x \, p_x \rangle \tag{12.8}$$

where we use Eq. (12.5) to recognize the average of $v_x \, p_x$.

Now we generalize to motion in all directions. For a large collection of particles moving randomly, the motion is isotropic and the averages in different directions are the same:

$$\langle v_x \, p_x \rangle = \langle v_y \, p_y \rangle = \langle v_z \, p_z \rangle$$

This implies

$$\langle vp \rangle = \langle \mathbf{v} \cdot \mathbf{p} \rangle = \langle v_x \, p_x + v_y \, p_y + v_z \, p_z \rangle = 3 \langle v_x \, p_x \rangle \tag{12.9}$$

so we can rewrite Eq. (12.8) as

$$P = \frac{1}{3} \int v \, p \, n(p) \, dp \tag{12.10}$$

This **pressure integral** specifies how to compute the pressure from any collection of particles with a known momentum distribution $n(p)$. It can handle either classical or relativistic motion as long as we use the appropriate relation between v and p.

In particular, for a non-relativistic ideal gas we can use $p = mv$. Then the pressure integral yields

$$P = \frac{1}{3m} \int p^2 \, n(p) \, dp = \frac{n_{tot} \langle p^2 \rangle}{3m} = \frac{n_{tot} m \langle v^2 \rangle}{3}$$

where we use Eq. (12.5). Then using Eq. (12.6) for $\langle v^2 \rangle$, and dropping the subscript "tot" for simplicity, we obtain

$$P = nkT \tag{12.11}$$

This is the famous **ideal gas law**. You may have seen the law in different forms, but they are equivalent to this one. For example, since $n = N/V$ we can write

$$PV = NkT$$

If we specify the number of moles (rather than the number of particles), we can put $\mathcal{N} = N/N_A$ where N_A is Avogadro's number and then write

$$PV = \mathcal{N}RT$$

where the "gas constant" is $R = N_A k$. In this book we use Eq. (12.11) so we can work with *local* quantities (e.g., number density) and avoid having to specify the global size and shape of the container holding the gas.

Example: Gas Density on Earth

What is the number density of gas molecules at the surface of Earth? Room temperature is about $T = 293$ K, and standard air pressure at sea level is about 101 kPa where a Pascal is a unit of pressure: Pa $=$ kg m^{-1} s^{-2}. Treating the atmosphere as an ideal gas yields

$$n = \frac{P}{kT} = \frac{1.01 \times 10^5 \text{ kg m}^{-1} \text{ s}^{-2}}{(1.38 \times 10^{-23} \text{ kg m}^2 \text{ s}^{-2} \text{ K}^{-1}) \times (293 \text{ K})} = 2.5 \times 10^{25} \text{ m}^{-3}$$

12.1.4 Assumptions in the Ideal Gas Law

Is it reasonable to treat Earth's atmosphere as an ideal gas? Let's check the assumptions stated in Sect. 12.1.2. We have already seen that molecules in Earth's atmosphere have speeds of a few hundred meters per second, so the motion is non-relativistic. Most of the atoms and molecules are electrically neutral, so there is little electromagnetic interaction between particles. Nuclear forces operate only within the nucleus, so they are not important across a single atom let alone between atoms. Gravity between particles is insignificant compared with the gravity from Earth. (We will see how gravity affects gas in Sect. 12.2.)

That leaves the question of internal structure. As we will discuss in Chap. 14, all gas particles have energy levels associated with electron excitation, and molecules can have additional energy associated with vibrational and rotational motion. The kinetic energy scale at room temperature, $kT = (8.62 \times 10^{-5} \text{ eV K}^{-1}) \times (293 \text{ K}) = 0.025$ eV, is too low for collisions to excite electron transitions (which typically require energies of a few eV), but it may allow collisions to set molecules rotating. Internal structure does not greatly affect the distribution of particle speeds, though. In equilibrium, any type of motion that can be excited already will be. Rotational modes therefore absorb energy in some collisions but give it back in others, so the overall set of speeds remains close to the Maxwell-Boltzmann distribution.[4] On the whole, then, the assumptions of an ideal gas are not too bad for Earth's atmosphere.

---

[4]Internal modes play a more significant role in the specific heat of a gas, which quantifies the amount of energy required to raise the temperature by a certain amount (see Sect. 16.1.2).

To enhance our intuitive picture, let's estimate some characteristic properties of the gas on a microscopic scale. First, how densely or sparsely are the particles distributed? More specifically, what fraction of the volume is occupied by the particles? For simplicity, let's treat each particle as a sphere of radius R, where the typical size of atoms and molecules is around an Angstrom ($1 \text{ Å} = 10^{-10}$ m).[5] The total volume occupied by all particles is $\sim N_{tot} \times (4/3)\pi R^3$, so the fraction of the overall volume that is taken up by particles is

$$f_V \sim \frac{N_{tot}}{V} \times \frac{4}{3}\pi R^3 \sim \frac{4}{3}\pi R^3 n$$

Using $R \sim 3 \times 10^{-10}$ m as an estimate for molecular nitrogen, along with $n \sim 2.5 \times 10^{25}$ m^{-3} from above, we estimate $f_V \sim 3 \times 10^{-3}$. The gas around us is mostly empty space!

Second, how far do particles typically travel between collisions? This is a quantity known as the **mean free path**. With our simple spherical model,[6] we say that each particle has a cross sectional area $\sigma \sim \pi R^2$. As a particle travels distance ℓ, its cross section sweeps out a volume $\ell\sigma$, and the number of other particles it encounters can be estimated as $N = \ell\sigma n$. The mean free path is the distance ℓ such that $N \sim 1$:

$$\ell = \frac{1}{n\sigma} \sim \frac{1}{\pi R^2 n} \tag{12.12}$$

For Earth's atmosphere, the numbers give

$$\ell \sim \frac{1}{\pi \times (3 \times 10^{-10} \text{ m})^2 \times (2.5 \times 10^{25} \text{ m}^{-3})} \sim 1.4 \times 10^{-7} \text{ m}$$

While this may seem like a small number, what matters is how it compares with the size of a particle:

$$\frac{\ell}{R} \sim \frac{1}{\pi R^3 n}$$

Interestingly, this is (within a factor of order unity) just the inverse of the volume fraction, f_V. For the numbers we have been using, $\ell/R \sim 500$ and we see that particles do travel a fairly long way, at least compared with their own size, between collisions.

---

[5]While atoms and molecules do not have sharp edges and need not be spheres, the simple assumption is adequate for rough estimates.

[6]This is appropriate if we think of particles as billiard balls that literally hit one another, but the concept of cross section can be generalized to other interactions (see Sect. 15.2.3).

Finally, what is the time between collisions?

$$t \sim \frac{\ell}{v_{rms}} \sim \frac{1.4 \times 10^{-7}\,m}{509\,m\,s^{-1}} \sim 3 \times 10^{-10}\,s$$

On a human time scale, collisions happen incredibly often. That is what allows the gas to be in thermodynamic equilibrium.

12.2 Hydrostatic Equilibrium

Before we can analyze atmospheres we need one more piece of physics: the effect of gravity on gas. On Earth, why doesn't gravity pull all the gas particles down to the surface? As gravity tries to squeeze the gas, the pressure rises until it counteracts the gravity. Here we seek to specify how gravity and pressure can achieve the balance that allows gas to be in equilibrium.

Let's imagine dividing the atmosphere into a number of rectangular volume elements[7] with radial thickness Δr and cross sectional area A. Let the mass density be ρ, so the mass of this volume element is $m = \rho A \Delta r$. The force from gravity is $F_{grav} = -mg(r)$ where $g(r)$ is the local acceleration due to gravity, and the minus sign indicates that the force is downward. Pressure on the bottom of the volume element creates an upward force $F_{bottom} = AP(r - \Delta r/2)$, while pressure on the top creates a downward force $F_{top} = -AP(r + \Delta r/2)$. In order for the atmosphere to be in equilibrium, the net vertical force must vanish[8]:

$$AP(r - \Delta r/2) - AP(r + \Delta r/2) - mg(r) = 0$$

Using $m = \rho A \Delta r$ and rearranging yields

$$\frac{P(r + \Delta r/2) - P(r - \Delta r/2)}{\Delta r} = -g\rho$$

If Δr is sufficiently small, the left-hand side is the derivative[9] dP/dr, so we have

$$\frac{dP}{dr} = -g\rho \tag{12.13}$$

---

[7]We could divide space in different ways and still obtain the same result.

[8]The net horizontal force vanishes by symmetry: the pressure on the "left" side of the volume element is balanced by the pressure on the "right" side, and likewise for the "front" and "back."

[9]You may be more familiar with the derivative written in terms of $[P(r + \Delta r) - P(r)]/\Delta r$, but in the limit $\Delta r \to 0$ it is equivalent to use $[P(r + \Delta r/2) - P(r - \Delta r/2)]/\Delta r$. By introducing a derivative, we are assuming that P is a continuous function. While gas is made of discrete particles on a microscopic scale, the sheer number of particles allows us to treat pressure as effectively continuous on a macroscopic scale.

This is the equation of **hydrostatic equilibrium,** and it is the fundamental equation for describing gas in a gravitational field. Physically, it says that equilibrium requires not just pressure but a *pressure gradient* to offset gravity.

12.3 Planetary Atmospheres

As we apply gas physics and gravity to planetary atmospheres, several questions come to mind: Can we understand the structure of Earth's atmosphere? Should we be surprised that Earth has an atmosphere at all? Why do some bodies (like Earth) have atmospheres, while others (like the Moon) do not? These are important questions because life as we know it depends on having an atmosphere.

12.3.1 Density Profile

Let's combine the ideal gas law and the equation of hydrostatic equilibrium to construct a model for an atmosphere. We will make two simplifying assumptions:

- The gas is **isothermal,** meaning the temperature is the same throughout the atmosphere.
- The acceleration due to gravity is constant throughout the atmosphere.

Strictly speaking, these conditions do not apply to Earth: temperature tends to decrease with altitude (think of going to the top of a mountain or up in an airplane), and gravity weakens with height. However, Earth's atmosphere is thin enough compared with the size of the planet that the assumptions are not too bad. Plus, making the assumptions lets us obtain a toy model that is instructive.

Let's replace the general radius r with height above the planet's surface by writing $r = R + h$. Let $n(h)$ be the number density of gas particles at height h, so the mass density is $\rho(h) = mn(h)$ where m is the particle mass. The equation of hydrostatic equilibrium then gives

$$\frac{\mathrm{d}P}{\mathrm{d}h} = -mgn(h)$$

With the ideal gas law, this becomes

$$\frac{\mathrm{d}P}{\mathrm{d}h} = -\frac{mg}{kT}\,P = -\frac{P}{H} \tag{12.14}$$

where we define

$$H \equiv \frac{kT}{mg} \tag{12.15}$$

This quantity has dimensions of length and a physical interpretation that will soon be apparent. If H is independent of height, we can solve Eq. (12.14) by rearranging and integrating:

$$\int \frac{\mathrm{d}P}{P} = -\int \frac{\mathrm{d}h}{H}$$

$$\ln P = -\frac{h}{H} + \text{constant} \tag{12.16}$$

$$P = e^{-h/H} \times \text{constant}$$

Now we see that H characterizes **scale height** over which the pressure changes substantially. We can determine the constant by saying the pressure is P_0 at the surface:

$$P(h) = P_0\, e^{-h/H} \tag{12.17}$$

Then from the ideal gas law the number density has the same form:

$$n(h) = n_0\, e^{-h/H} \tag{12.18}$$

where n_0 is the number density at the surface. The model we have derived here, known as the **exponential atmosphere**, is admittedly idealized but still useful for understanding a lot of the basic properties of Earth's atmosphere.

Aside. If we return to Eq. (12.16) and allow H to vary with height (e.g., because T changes), we can write the solution in the form

$$P \propto e^{-\int H^{-1} dh}$$

This is a generalized version of the atmosphere model.

Example: How Thick Is Earth's Atmosphere?

We take the scale height to give the characteristic thickness of the atmosphere; in the exponential model, about 63 % of the gas lies between $h = 0$ and $h = H$. On Earth, the average temperature near the surface is around $T = 288$ K, and the acceleration due to gravity is $g = 9.80 \,\mathrm{m\,s^{-2}}$. The main components of Earth's atmosphere are molecular nitrogen and molecular oxygen. Plugging in numbers yields:

nitrogen: $\quad m = 28\,m_p$

$$H = \frac{(1.38 \times 10^{-23}\,\mathrm{kg\,m^2\,s^{-2}\,K^{-1}}) \times (288\,\mathrm{K})}{(28 \times 1.67 \times 10^{-27}\,\mathrm{kg}) \times (9.80\,\mathrm{m\,s^{-2}})} = 8.7\,\mathrm{km}$$

oxygen: $\quad m = 32\,m_p$

$$H = \frac{(1.38 \times 10^{-23} \text{ kg m}^2 \text{ s}^{-2} \text{ K}^{-1}) \times (288 \text{ K})}{(32 \times 1.67 \times 10^{-27} \text{ kg}) \times (9.80 \text{ m s}^{-2})} = 7.6 \text{ km}$$

This is not very thick compared to the radius of the Earth, $R_{\oplus} = 6{,}378$ km. Just imagine: all life as we know it exists in a very thin shell near the surface of the Earth. (Note: since $H \ll R_{\oplus}$ it is reasonable to treat the acceleration due to gravity as constant throughout the atmosphere.)

Example: Why Are Airplanes Pressurized?

Airplanes fly at an altitude of around 10 km or 33,000 ft. At that altitude in an exponential atmosphere, the pressure is

$$P(10 \text{ km}) = P_0 \, e^{-(10 \text{ km})/H} = \begin{cases} 0.32 \, P_0 & \text{nitrogen} \\ 0.27 \, P_0 & \text{oxygen} \end{cases}$$

In other words, the ambient pressure is too low to breathe or operate comfortably.

12.3.2 Exosphere

Where does Earth's atmosphere end? While there is no sharp "edge" to an exponential atmosphere, we can consider an effective boundary to be the place where the mean free path becomes long enough that a particle could escape to infinity without experiencing a collision. We call this the **exosphere**. Recall that the mean free path is $\ell = (n\sigma)^{-1}$ where n is the number density and σ is the collision cross section. If a particle moves a distance dh, the average number of collisions is $n \, \sigma \, dh$. The total number of collisions when traveling from height h to infinity is therefore

$$p(h) \sim \sigma \int_h^{\infty} n(h') \, dh' \tag{12.19}$$

If this number is much less than 1, we can interpret it as the probability for a collision (which is why we write it as p). Using the exponential atmosphere, we can compute

$$p(h) \sim n_0 \sigma \int_h^{\infty} e^{-h'/H} \, dh' \sim n_0 \sigma H \, e^{-h/H} \sim \frac{H}{\ell_0} e^{-h/H}$$

where $\ell_0 \sim 1/(n_0 \, \sigma)$ is the mean free path at Earth's surface. We can invert to find the height at which the collision probability has some value p:

$$h_{exo} \sim H \ln \left(\frac{H}{\ell_0 p} \right) \qquad (12.20)$$

In truth, the exosphere is not sharply defined; we should really consider it to be a broad region from which air molecules can escape into space. So please do not take our analysis too literally. But do use it as a guide for understanding the physical picture of air molecules leaking out of the atmosphere.

Example: Earth

In Sect. 12.1.4 we estimated $\ell_0 \sim 1.4 \times 10^{-7}$ m, and in Sect. 12.3.1 we obtained $H = 8.7$ km for nitrogen. Suppose we consider the exosphere to be the place where the probability of a collision is 1 %, or $p = 0.01$. Then we estimate the height of the exosphere to be

$$h_{exo} \sim 260 \, \text{km}$$

Note that p appears in a logarithm, so changing p would not change the answer very much. For example, if we took $p = 0.001$ we would get $h_{exo} \sim 280$ km. Similarly, the result is not very sensitive to the specific value of ℓ_0. Increasing the mean free path by a factor of 10 would change h_{exo} to 240 km.

12.3.3 Evaporation

In discussing the exosphere, we only considered whether particles would be held in place by collisions. But there is another factor: gravity. A particle can escape only if it overcomes gravity by exceeding the planet's escape velocity Eq. (2.17). The Maxwell-Boltzmann distribution extends to high speeds, though, so some particles can in fact escape. We can make a simple estimate of the temperature required to have a reasonable fraction of particles escape, and a somewhat more detailed estimate of the time scale for an atmosphere to evaporate.[10]

The first thing we might do is compare the escape speed (v_{esc}) and the typical speed of particles in the Maxwell-Boltzmann distribution (which we take to be v_{rms}). If $v_{rms} \ll v_{esc}$ then the fraction of particles that can escape will be very small. Conversely, if $v_{rms} \gtrsim v_{esc}$ then the fraction will be much larger. As a rough estimate, we might expect that evaporation can occur if

$$v_{esc} \lesssim 6 v_{rms}$$

---

[10]This analysis draws from the book by Carroll and Ostlie [1].

The factor of 6 ensures that we count only particles able to escape and roughly accounts for other restraining effects like collisions and geometry (particles moving downward will not escape even if they exceed v_{esc}); however, it is somewhat arbitrary so the following numerical values should be taken as indicative, not precise. Using v_{esc} from Eq. (2.17) and v_{rms} from Eq. (12.6), we can write the evaporation condition as

$$\left(\frac{2GM}{R}\right)^{1/2} \lesssim 6\left(\frac{3kT}{m}\right)^{1/2}$$

or

$$T \gtrsim T_{esc} \quad \text{where} \quad T_{esc} \equiv \frac{1}{54}\frac{GMm}{kR} \tag{12.21}$$

This represents a simple criterion that we can use to estimate whether a planet's atmosphere is hot enough to evaporate.

Example: Earth and Moon

Let's consider molecular nitrogen since it is the main component of Earth's atmosphere (see below for other molecules). For Earth, the relevant numbers are

$$M = 5.97 \times 10^{24}\,\text{kg}$$

$$R = 6.38 \times 10^{6}\,\text{m}$$

$$m = 28\,m_p$$

so the escape temperature is

$$T_{esc} = \frac{(6.67 \times 10^{-11}\,\text{m}^3\,\text{kg}^{-1}\,\text{s}^{-2}) \times (5.97 \times 10^{24}\,\text{kg}) \times (28 \times 1.67 \times 10^{-27}\,\text{kg})}{54 \times (1.38 \times 10^{-23}\,\text{kg}\,\text{m}^2\,\text{s}^{-2}\,\text{K}^{-1}) \times (6.38 \times 10^{6}\,\text{m})}$$

$$= 3{,}900\,\text{K}$$

For comparison, the numbers for the Moon are:

$$M = 7.35 \times 10^{22}\,\text{kg}$$

$$R = 1.74 \times 10^{6}\,\text{m}$$

$$\Rightarrow T_{esc} = \frac{(6.67 \times 10^{-11}\,\text{m}^3\,\text{kg}^{-1}\,\text{s}^{-2}) \times (7.35 \times 10^{22}\,\text{kg}) \times (28 \times 1.67 \times 10^{-27}\,\text{kg})}{54 \times (1.38 \times 10^{-23}\,\text{kg}\,\text{m}^2\,\text{s}^{-2}\,\text{K}^{-1}) \times (1.74 \times 10^{6}\,\text{m})}$$

$$= 177\,\text{K}$$

Earth's gravity is strong enough to hold onto gas particles, but the Moon's gravity is not.

We can go a step further and estimate the rate at which particles escape. Specifically, we imagine that once a particle crosses the exosphere moving upwards at a speed faster than the escape velocity, it is effectively gone. Therefore the number of particles ΔN that escape from area A in time Δt is given by counting all particles that have speed $v > v_{esc}$ and lie in a layer of thickness $v_z \Delta t$ below the exosphere, where v_z is the z-component of the velocity. If $n(h, \mathbf{v}) \, d\mathbf{v}$ is the number density of particles at height h with velocity between $\mathbf{v}$ and $\mathbf{v} + d\mathbf{v}$ (i.e., the density version of Eq. 12.2), then we can write:

$$
\begin{aligned}
\Delta N &= \int_{v > v_{esc}} d\mathbf{v} \int_{h_{exo} - v_z \Delta t}^{h_{exo}} dh \, A \, n(h, \mathbf{v}) \\
&= A \int_{v > v_{esc}} d^3 v \int_{h_{exo} - v_z \Delta t}^{h_{exo}} dh \, n_0 \, e^{-h/H} \left(\frac{m}{2\pi kT} \right)^{3/2} e^{-mv^2/2kT} \\
&= A \left(\frac{m}{2\pi kT} \right)^{3/2} n_0 \int_{v > v_{esc}} d^3 v \, e^{-mv^2/2kT} \times \left(-H e^{-h/H} \right) \Big|_{h = h_{exo} - v_z \Delta t}^{h = h_{exo}} \\
&= A \left(\frac{m}{2\pi kT} \right)^{3/2} n_0 \int_{v > v_{esc}} d^3 v \, e^{-mv^2/2kT} \times H e^{-h_{exo}/H} \left(e^{v_z \Delta t / H} - 1 \right) \\
&\approx A \left(\frac{m}{2\pi kT} \right)^{3/2} n_0 \, e^{-h_{exo}/H} \Delta t \int_{v > v_{esc}} d^3 v \, v_z \, e^{-mv^2/2kT}
\end{aligned}
$$

In the second line we use the Maxwell-Boltzmann distribution but write $d^3 v$ (instead of the spherical volume element $4\pi v^2 \, dv$) because direction is important. Also, we use the exponential atmosphere model. In the third and fourth lines we evaluate the h integral, and in the fifth line we use the approximation $\exp(v_z \Delta t / H) \approx 1 + v_z \Delta t / H$ when Δt is small.

The next step is to evaluate the velocity integral. In spherical coordinates, $d^3 v = v^2 \sin\theta \, dv \, d\theta \, d\phi$ where θ is the polar angle of the velocity vector, measured from vertical, and ϕ is the azimuthal angle of the velocity vector. We integrate over all ϕ, but only over $0 < \theta < \pi/2$ because we want upward velocities. The z-component of the velocity vector is $v_z = v \cos\theta$. Putting the pieces together yields

$$
\begin{aligned}
\frac{\Delta N}{\Delta t} &= A \left(\frac{m}{2\pi kT} \right)^{3/2} n_0 \, e^{-h_{exo}/H} \int_{v_{esc}}^{\infty} dv \, v^3 e^{-mv^2/2kT} \int_0^{2\pi} d\phi \int_0^{\pi/2} d\theta \, \sin\theta \cos\theta \\
&= A \left(\frac{kT}{2\pi m} \right)^{1/2} n_0 \, e^{-h_{exo}/H} \int_{x_{esc}}^{\infty} dx \, x \, e^{-x} \\
&= A \left(\frac{kT}{2\pi m} \right)^{1/2} n_0 \, e^{-h_{exo}/H} \left(1 + x_{esc} \right) e^{-x_{esc}}
\end{aligned}
$$

In the second line we evaluate the angular integrals and then change variables to $x = mv^2/2kT$. We also define

$$x_{\text{esc}} = \frac{mv_{\text{esc}}^2}{2kT} = \frac{v_{\text{esc}}^2}{v_{\text{peak}}^2} \tag{12.22}$$

using Eq. (12.4). In the third line we integrate by parts.[11]

The last step of the analysis is to determine the total number of particles in the "reservoir" that extends from the surface up to the exosphere above the area A. We do this by integrating the exponential atmosphere model over h:

$$N = A \int_0^{h_{\text{exo}}} dh\, n_0 e^{-h/H} = AHn_0 \left(1 - e^{-h_{\text{exo}}/H}\right)$$

Now we can estimate the time it would take for the entire atmosphere to evaporate. If the evaporation rate were constant, the time to deplete the reservoir would be

$$t_{\text{evap}} \sim \frac{N}{\Delta N/\Delta t} \sim H \left(\frac{2\pi m}{kT}\right)^{1/2} \left(e^{h_{\text{exo}}/H} - 1\right) \frac{e^{x_{\text{esc}}}}{1 + x_{\text{esc}}} \tag{12.23}$$

Even if the rate varies, this is still a useful order-of-magnitude estimate of the evaporation time.

Example: Earth

For Earth's nitrogen-dominated atmosphere, we have already estimated the scale height and the exosphere. The exosphere is actually warmer than the atmosphere near the surface because the upper atmosphere is heated by ultraviolet radiation from the Sun; let's use $T \sim 1,000$ K, which might seem to exaggerate the effects of evaporation but is actually a conservative choice (as we will see). The escape velocity at the exosphere is $v_{\text{esc}} = 11$ km s^{-1} (which is not very much smaller than the escape velocity from the surface). In the following table, Column 2 is the particle mass in units of the proton mass, Column 3 is the peak speed in the Maxwell-Boltzmann distribution, Column 4 is the dimensionless escape parameter x_{esc} from Eq. (12.22), and finally Column 5 is our estimate of the evaporation time scale.

For everything except monatomic hydrogen, the evaporation time is longer than the age of Earth (about 4.5 billion years). Earth can and will hold onto its atmosphere for a long time.[12] (Had we used a lower value of temperature, the evaporation times would have been longer; that is why using $T \sim 1,000$ K is a conservative choice.)

---

[11] Recall from calculus: $\int u\, dv = uv - \int v\, du$.

[12] Whether or not a particular molecule is abundant depends on whether any was present in the first place; that, in turn, depends on how planets formed (see Chap. 19) and how life subsequently

Molecule	m/m_p	v_{peak} (km s^{-1})	x_{esc}	t_{evap} (yr)
Hydrogen, H	1	4.1	7.3	4×10^8
Hydrogen, H$_2$	2	2.9	14.6	4×10^{11}
Helium, He	4	2.0	29.1	7×10^{17}
Nitrogen, N$_2$	28	0.8	203.7	2×10^{93}
Oxygen, O$_2$	32	0.7	232.8	7×10^{105}

Example: Moon

For comparison, the numbers for the Moon are as follows. Because the Moon has no atmosphere to act as an insulating blanket, the temperature varies quite dramatically from day to night; the dayside temperature can reach $T \sim 373$ K. The weak gravity leads to a low escape velocity (2.4 km s^{-1} from the surface, or 2.3 km s^{-1} from an altitude of 100 km) and to a large scale height ($H = 68$ km for molecular nitrogen). With no actual atmosphere it is not clear how to define an exosphere in a meaningful way; let's set $h_{\text{exo}} \sim 100$ km as something rather arbitrary but reasonable for an estimate (and conservative, as we will see).

Molecule	m/m_p	v_{peak} (km s^{-1})	x_{esc}	t_{evap} (yr)
Hydrogen, H	1	2.5	0.9	10^{-5}
Hydrogen, H$_2$	2	1.8	1.7	3×10^{-5}
Helium, He	4	1.2	3.5	10^{-4}
Nitrogen, N$_2$	28	0.5	24.2	7×10^4
Oxygen, O$_2$	32	0.4	27.7	2×10^6

While the gas speeds are lower than on Earth, the escape velocity is so much lower that the evaporation time scales are quite short. Even relatively heavy gases like molecular nitrogen and molecular oxygen would evaporate in a time much shorter than the age of the Moon. (Had we used a smaller value of h_{exo}, the evaporation times would have been even shorter.) Now we understand why Earth has an atmosphere but the Moon does not.

Problems

Here are planetary data that may be relevant for some problems. T is the mean temperature; for gaseous planets it is quoted at a depth where the atmospheric pressure is comparable to Earth's pressure at sea level.

modified Earth's atmospheric composition. What this analysis tells us is that if a certain gas is present, it will stick around.

Planet	$M/M_\oplus$	$R/R_\oplus$	T (K)
Mercury	0.06	0.38	440
Venus	0.82	0.95	740
Earth	1.00	1.00	288
Mars	0.11	0.53	210
Jupiter	317.89	11.19	165
Saturn	95.18	9.46	134
Uranus	14.54	4.01	76
Neptune	17.13	3.81	72

12.1. Suppose gas in a closed cylinder is kept at constant temperature while being compressed to half its original volume. How does the compression affect the pressure on the walls of the cylinder? How would you explain this effect in terms of the microscopic picture of kinetic theory?

12.2. When you ride on an elevator in a high-rise building, you may feel your ears "pop" from a change in pressure. If this happens when you go to the top of a skyscraper that is 400 m tall, what is the (fractional) change in pressure to which your ears are sensitive?

12.3. The atmosphere of Mars is mostly carbon dioxide, and the typical pressure is about $600\,\mathrm{kg\,m^{-1}\,s^{-2}}$.

(a) What is the typical speed of a gas molecule near the surface?
(b) What is the scale height of the atmosphere?

12.4. Rewrite the escape temperature Eq. (12.21) in terms of an object's mass and the mean mass density $\bar{\rho} = 3M/(4\pi R^3)$. Among astronomical objects that have roughly the same mean density, would you expect hydrogen to be more common in high-mass or low-mass objects?

12.5. Given the data above, which planets would you expect to have hydrogen atmospheres? Explain, and be quantitative.

12.6. Consider a hypothetical planet of radius R, whose density ρ is uniform. The planet is composed of a classical ideal gas of ionized hydrogen and is in hydrostatic equilibrium.

(a) What is the pressure as a function of radius in the planet, $P(r)$? You may take the pressure at the surface to be zero, $P(R) = 0$.
(b) Consider the planet to have the same size and mass as Jupiter. What is the temperature at the center? How fast are the protons and electrons moving there?

12.7. We can use the equation of hydrostatic equilibrium to place interesting bounds on conditions at the center of a star, even if we do not solve the equation in detail. Recall: for a spherical star with density $\rho(r)$, the mass enclosed by radius r is

$$M(r) = 4\pi \int_0^r r'^2 \, \rho(r') \, dr' \quad \text{or} \quad \rho(r) = \frac{1}{4\pi r^2} \frac{dM}{dr}$$

and the acceleration due to gravity at r is $g(r) = GM(r)/r^2$.

(a) Show that in hydrostatic equilibrium, the function

$$P(r) + \frac{GM(r)^2}{8\pi r^4}$$

must decrease as r increases. Show that this condition implies that the pressure at the center of the star must satisfy the inequality

$$P(0) > \frac{GM^2}{8\pi R^4} \tag{12.24}$$

where R is the radius of the star and M is the total mass. Hint: assume the pressure is zero at the surface of the star, $P(R) = 0$.

(b) Compute the lower bound (12.24) on the pressure at the center of the Sun.

(c) Combine your result from (b) with the ideal gas law to compute a lower bound on the temperature at the center of the Sun. For simplicity, use the average density and assume the Sun is made entirely of hydrogen.

Reference

1. B.W. Carroll, D.A. Ostlie, *An Introduction to Modern Astrophysics*, 2nd edn. (Addison-Wesley, San Francisco, 2007)

Chapter 13
Planetary Temperatures

In Chap. 12 we saw how a planet's temperature and gravity combine to determine whether the planet has an atmosphere. In this chapter we study the physical processes that determine the temperature in the first place. Intuitively, we expect a planet close to the Sun to be warmer than a planet farther away, but we seek to quantify that effect. We also consider ways in which a planet's atmosphere can act as a blanket to trap heat. The physical phenomena that play a role here are blackbody radiation and the interaction of light with matter.

13.1 Blackbody Radiation

A **blackbody** is a hypothetical object that absorbs all light incident upon it. Since light carries energy, the object must either heat up or get rid of the excess energy. Heating a solid object causes it to glow, and a blackbody emits a characteristic spectrum that depends only on the temperature (not on the size or composition). While real astrophysical objects are not perfect blackbodies, the blackbody spectrum nevertheless provides a good starting point for describing the light emitted by stars and planets.

13.1.1 Luminosity

The relationship between temperature and the total amount of energy emitted by a blackbody was determined empirically by Josef Stefan, and then explained theoretically by Ludwig Boltzmann. (We will consider the theory in Sect. 13.1.2.) The **Stefan-Boltzmann law** for luminosity is

$$L = A \sigma T^4 \tag{13.1}$$

C. Keeton, *Principles of Astrophysics: Using Gravity and Stellar Physics to Explore the Cosmos*, Undergraduate Lecture Notes in Physics, DOI 10.1007/978-1-4614-9236-8__13, © Springer Science+Business Media New York 2014

where A is the surface area of the blackbody, T is the temperature, and σ is a constant now known as the **Stefan-Boltzmann constant**:

$$\sigma = 5.67 \times 10^{-8} \, \text{kg} \, \text{s}^{-3} \, \text{K}^{-4}$$

Example: Sun

The Sun's luminosity is $L = 3.84 \times 10^{26} \, \text{J} \, \text{s}^{-1}$ and radius is $R = 6.96 \times 10^{8}$ m. What is its surface temperature? Since the Sun is not a perfect blackbody, we need to be more precise about the question: What is the temperature of a blackbody that has the same size and luminosity as the Sun? This is what we define to be the **effective temperature** of the Sun, and while it may not be precisely the same as the physical temperature in the outer layers of the star it is close enough to be very useful. Using $A = 4\pi R^2$ and inverting the Stefan-Boltzmann law yields

$$T_{\text{eff}} = \left(\frac{L}{4\pi R^2 \sigma} \right)^{1/4} \tag{13.2}$$

The numbers for the Sun give

$$T_{\odot} = \left[\frac{3.84 \times 10^{26} \, \text{J} \, \text{s}^{-1}}{4\pi \times (6.96 \times 10^{8} \, \text{m})^2 \times (5.67 \times 10^{-8} \, \text{kg} \, \text{s}^{-3} \, \text{K}^{-4})} \right]^{1/4} = 5{,}780 \, \text{K} \tag{13.3}$$

13.1.2 Spectrum

Moving beyond the total luminosity, Max Planck discovered a formula that describes the full spectrum of blackbody radiation. First, let's think about how it is expressed. We define the wavelength spectrum to be $B_\lambda(\lambda; T)$ such that the luminosity can be written as

$$L = \int_0^\infty d\lambda \int dA \int_0^{2\pi} d\phi \int_0^{\pi/2} d\theta \, \cos\theta \sin\theta \, B_\lambda(\lambda; T) \tag{13.4}$$

This expression involves two surfaces. The integral over A covers the entire surface from which light is *emitted*. The integrals over θ and ϕ cover an imaginary surface around the blackbody at which light is *received*. For each point on the emitting surface, we can define the polar angles relative to the vector perpendicular to dA, and we consider only $0 \leq \theta \leq \pi/2$ because the light must travel outward. The integrand contains the standard spherical volume element $\sin\theta \, d\theta \, d\phi$ along with an extra factor of $\cos\theta$ for geometric reasons: an observer viewing from angle θ would see the surface element subtend a *projected* area of $dA \cos \theta$. As defined through the integral, $B_\lambda(\lambda; T)$ has dimensions of energy per unit area per unit time per unit wavelength per unit solid angle,[1] and standard units[2] of $\mathrm{J\,m^{-2}\,s^{-1}\,m^{-1}\,sr^{-1}}$.

Planck considered a box filled with electromagnetic radiation that has a small hole in one side through which light can escape. He postulated that light energy can exist only in discrete packets, called "quanta" (plural of "quantum"),[3] with energy

$$E = h\nu = \frac{hc}{\lambda} \tag{13.5}$$

where ν and λ are the frequency and wavelength of light, respectively, and

$$h = 6.63 \times 10^{-34} \,\mathrm{J\,s} \tag{13.6}$$

is a new constant of nature now known as **Planck's constant**. We sometimes rewrite it as

$$\hbar = \frac{h}{2\pi} = 1.05 \times 10^{-34} \,\mathrm{J\,s} \tag{13.7}$$

With the quantum assumption, Planck derived the spectrum

$$B_\lambda(\lambda; T) = \frac{2\,h\,c^2}{\lambda^5} \frac{1}{e^{hc/\lambda kT} - 1} \tag{13.8}$$

This function, which is shown in Fig. 13.1, agrees very well with observed blackbody spectra. The most precise blackbody spectrum ever measured actually comes from astrophysics: the Cosmic Microwave Background (CMB) radiation, which was produced by hot glowing gas in the early universe. As the universe has expanded, the radiation has effectively cooled such that the spectrum today is accurately described by a Planck spectrum with a temperature of 2.73 K. (We will study the CMB in Chap. 20.)

The Planck spectrum can be expressed in terms of frequency if we rewrite Eq. (13.4) as

$$L = \int_0^\infty d\nu \int dA \int_0^{2\pi} d\phi \int_0^{\pi/2} d\theta \, \cos\theta \sin\theta \, B_\nu(\nu; T) \tag{13.9}$$

Thinking of this as a change of integration variables, we can identify the frequency spectrum as

---

[1] Solid angle is like an angular area, $d\Omega = \sin\theta \, d\theta \, d\phi$, and it is measured in steradians. There are 4π steradians on a sphere.

[2] In principle, this could be reduced to $\mathrm{kg\,m^{-1}\,s^{-3}\,sr^{-1}}$, but that would make the physical meaning much less clear.

[3] We now use "quantum" as a general term, and "photon" when speaking specifically of light.

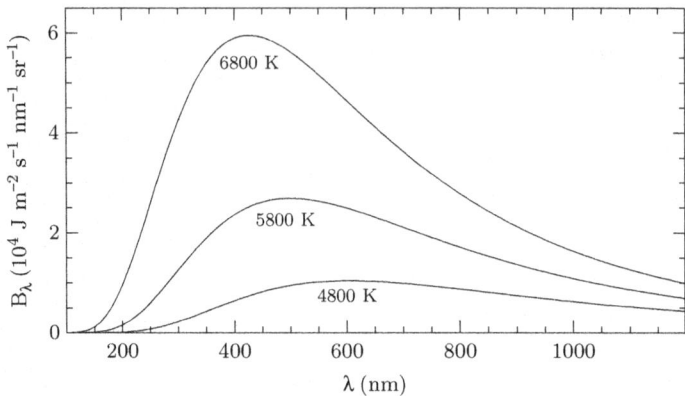

Fig. 13.1 Examples of the Planck spectrum for different temperatures

$$B_\nu(\nu; T) = B_\lambda \left| \frac{d\lambda}{d\nu} \right|_{\lambda = c/\nu} = \frac{2\,h\,\nu^3}{c^2} \frac{1}{e^{h\nu/kT} - 1} \qquad (13.10)$$

We can use the Planck spectrum to derive the Stefan-Boltzmann constant σ in terms of fundamental constants. In Eq. (13.9), the integrals over A, θ, and ϕ can be evaluated:

$$\int dA = A \qquad \int_0^{2\pi} d\phi = 2\pi \qquad \int_0^{\pi/2} \cos\theta \sin\theta \, d\theta = \frac{1}{2}$$

The ν integral can be computed by changing variables to $x = h\nu/kT$ and using the result

$$\int_0^\infty \frac{x^3}{e^x - 1} \, dx = \frac{\pi^4}{15} \qquad (13.11)$$

Thus, we find

$$L = A \times \frac{2\pi^5 k^4}{15\,c^2\,h^3} \times T^4$$

Comparing with Eq. (13.1) reveals that

$$\sigma = \frac{2\pi^5 k^4}{15\,c^2\,h^3} = \frac{\pi^2 k^4}{60\,c^2\,\hbar^3} \qquad (13.12)$$

13.1.3 *Color*

Early studies of blackbody radiation noted that an object's color shifts from red to orange to yellow (and so on) as the temperature rises. While the blackbody spectrum contains a mixture of wavelengths (the function is continuous), we can get a sense of the color by considering the location of the peak in the spectrum. To find the peak, let's begin by computing the derivative:

$$\frac{dB_\lambda}{d\lambda} = -\frac{10\,h\,c^2}{\lambda^6}\frac{1}{e^{hc/\lambda kT}-1} + \frac{2\,h^2\,c^3}{\lambda^7\,k\,T}\frac{e^{hc/\lambda kT}}{(e^{hc/\lambda kT}-1)^2}$$

Set $dB/d\lambda = 0$, substitute $x = hc/(\lambda_{\text{peak}}kT)$, and simplify:

$$0 = \frac{2\,k^6\,T^6\,x^6}{h^5\,c^4}\frac{5-5\,e^x + x\,e^x}{(e^x-1)^2} \qquad \Rightarrow \qquad \frac{x\,e^x}{e^x-1} = 5$$

This equation is solved for $x = 4.965$, or

$$\lambda_{\text{peak}}\,T = \frac{h\,c}{4.965\,k}$$

$$= \frac{(6.626 \times 10^{-34}\,\text{kg m}^2\,\text{s}^{-1}) \times (3.0 \times 10^8\,\text{m s}^{-1})}{4.965 \times (1.38 \times 10^{-23}\,\text{kg m}^2\,\text{s}^{-2}\,\text{K}^{-1})}$$

$$= 2.90 \times 10^{-3}\,\text{m K}$$

Equivalently, we can write

$$\lambda_{\text{peak}} = \frac{2.90 \times 10^{-3}\,\text{m K}}{T} \tag{13.13}$$

This relation is known as **Wien's displacement law** after Wilhelm Wien. It quantifies the connection between temperature and color: as temperature increases, the peak wavelength decreases, so the spectrum shifts toward bluer colors. Notice that we did not specify anything about the size or shape of the object; Wien's law is universal for blackbodies.

Example: What Is the Peak Wavelength of Sunlight?

Using the effective temperature of the Sun from Eq. (13.3) gives

$$\lambda_{\text{peak}} = \frac{2.90 \times 10^{-3}\,\text{m K}}{5{,}780\,\text{K}} = 5.02 \times 10^{-7}\,\text{m} = 502\,\text{nm} = 5{,}020\,\text{Å}$$

This is green light. It is no accident that our eyes are most sensitive to light around these wavelengths; they evolved to operate in the portion of the electromagnetic spectrum in which the Sun emits most of its light.

Example: In What Portion of the Spectrum Does Earth Radiate?

Using $T = 288$ K as the average temperature of Earth's surface yields

$$\lambda_{peak} = \frac{2.90 \times 10^{-3} \, \text{m K}}{288 \, \text{K}} = 10^{-5} \, \text{m} = 10 \, \mu\text{m}$$

This is in the infrared region of the electromagnetic spectrum.

13.1.4 Pressure

According to Planck, each photon carries energy $E = h\nu = hc/\lambda$. According to relativity, then, we can think of a photon as a massless particle with momentum $p = E/c$ (see Eq. 10.26), or equivalently $p = h\nu/c = h/\lambda$. Particles that carry momentum can exert pressure (recall Sect. 12.1.3), so we infer that there must be some pressure associated with light.

We can compute the pressure by thinking of blackbody radiation as a "gas" of photons. The number density of photons with momentum between p and $p + \mathrm{d}p$ is

$$n(p)\,\mathrm{d}p = \frac{2}{h^3} \frac{1}{e^{pc/kT} - 1} \, 4\pi p^2 \,\mathrm{d}p$$

This basically comes from expressing the Planck spectrum (13.10) in terms of momentum, except that we have switched to number density; note that the factor of $4\pi p^2 \,\mathrm{d}p$ is the spherical volume element. Using the pressure integral (Eq. 12.10), we can write the pressure of the photon gas as

$$P = \frac{1}{3} \int c \, p \, n(p) \,\mathrm{d}p = \frac{8\pi c}{3h^3} \int_0^\infty \frac{p^3}{e^{pc/kT} - 1} \,\mathrm{d}p = \frac{8\pi k^4}{3c^3 h^3} T^4 \int_0^\infty \frac{x^3}{e^x - 1} \,\mathrm{d}x$$

where we use $v = c$ for photons, and we change variables to $x = pc/kT$. Again using the integral (13.11), we find

$$P = \frac{8\pi^5 k^4}{45 c^3 h^3} T^4 = \frac{4\sigma}{3c} T^4 \tag{13.14}$$

where we use Eq. (13.12) to replace some of the constants with the Stefan-Boltzmann constant. Plugging in numbers yields

$$P = 2.52 \times 10^{-16} \, \text{kg} \, \text{m}^{-1} \, \text{s}^{-2} \times \left(\frac{T}{\text{K}}\right)^4$$

Comparing Eqs. (13.14) and (13.1) shows that photon pressure is proportional to luminosity, which makes sense.

13.2 Predicting Planet Temperatures

We can now use the properties of blackbody radiation to make a simple model for the temperature of a planet. The logic is as follows: light from the Sun heats the planet, causing the planet to radiate. The planet heats up until it reaches an equilibrium state in which the energy it radiates exactly balances the energy it receives from the Sun. The model is simplistic because it assumes the temperature and composition are uniform across the planet, which we know is not true, but using average quantities yields a model that works surprisingly well. Plus, the model illuminates the basic physics, which is ultimately the purpose of a simple model.

In Sect. 13.1.1 we already defined the effective temperature of the Sun, $T_\odot$, through the relation $L_\odot = 4\pi R_\odot^2 \sigma T_\odot^4$. By the inverse square law, the energy flux (energy per unit area per unit time) at a distance D from the Sun is

$$f = \frac{L_\odot}{4\pi D^2} = \frac{\sigma R_\odot^2 T_\odot^4}{D^2}$$

The total power (energy per unit time) incident on a planet of radius R_p is the flux times the area of the planet. From the perspective of the Sun, the planet appears to subtend a circle of area πR_p^2. Planets are not perfect blackbodies; they reflect part of the incident light. We define the **albedo**, a, to be the fraction of the incident light that is reflected. Thus the fraction of the incident power that is *absorbed* is $(1 - a)$, and the total power that acts to heat the planet is[4]

$$P_{\text{abs}} = (1 - a) \times \pi R_p^2 \times \frac{\sigma R_\odot^2 T_\odot^4}{D^2}$$

If the planet has (effective) temperature T_p, then from the Stefan-Boltzmann law the total luminosity (energy per unit time) it emits is

$$L_{\text{em}} = 4\pi R_p^2 \sigma T_p^4$$

---

[4] Here P denotes power, not pressure.

As we said above, the planet will reach an equilibrium in which $P_{\text{abs}} = L_{\text{em}}$, or

$$(1-a)\pi R_p^2 \frac{\sigma R_\odot^2 T_\odot^4}{D^2} = 4\pi R_p^2 \sigma T_p^4$$

Solving for T_p gives

$$T_p = T_\odot (1-a)^{1/4} \left(\frac{R_\odot}{2D}\right)^{1/2} \tag{13.15}$$

Notice that the radius of the planet has dropped out; in this simple model, a planet's temperature depends only on its distance from the Sun and its albedo (and the properties of the Sun, of course).

We have constructed the model using the average temperature and albedo of the planet. In reality, the absorption and emission properties vary with latitude and from the day-side to the night-side (not to mention smaller-scale features), and the variation depends on whether there is much atmosphere to retain and circulate warmth. We do not attempt to incorporate such details into the model, though. At this point we just want to see whether this simple model of planet heating can help us gain a basic understanding of why some planets are hot and others are cold.

Example: Earth

Earth lies $D = 1\,\text{AU} = 1.50 \times 10^{11}$ m from the Sun and has an average albedo of $a = 0.306$. What does this model predict for its average temperature?

$$T_\oplus = 5{,}780\,\text{K} \times (1-0.306)^{1/4} \times \left(\frac{6.96 \times 10^8\,\text{m}}{2 \times 1.50 \times 10^{11}\,\text{m}}\right)^{1/2} = 254\,\text{K} = -19\,\text{deg\,C} = -2\,\text{deg\,F}$$

Not bad for a simple model, but a bit colder than reality. Why? Our analysis actually applies to the outermost layer of a planet; the situation may be somewhat different on the surface, if the planet has an atmosphere.

13.3 Atmospheric Heating

In order to predict the temperature on the *surface* of a planet, we need to account for the fact that an atmosphere can trap heat near the surface. In Sect. 13.4 we will consider the physical processes by which light can interact with gas in an atmosphere; for now, let's estimate how much atmospheric heating can warm the surface.[5]

---

[5]This analysis is inspired by Problems 19.13 and 20.7 in the book by Carroll and Ostlie [1].

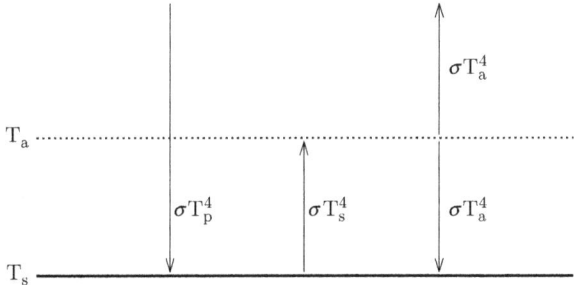

Fig. 13.2 Setup for the one-layer model of atmospheric heating. The atmosphere (*dotted line*) is treated as a thin layer at temperature T_a, while the surface (*solid line*) has temperature T_s. Energy flux (in units of energy per unit time per unit area) is denoted with *arrows*. The incident flux is quantified with T_p from Eq. (13.15). The atmosphere is transparent to the visible light incident from the Sun, but it absorbs the infrared radiation emitted by the surface and then re-radiates that energy both upward and downward

13.3.1 One Layer

To start, let's make a toy model in which the atmosphere is a thin layer hovering above the surface of the planet. Suppose the atmosphere is transparent to the visible light that comes from the Sun, but it absorbs 100 % of the infrared radiation emitted by the planet and then re-radiates that energy. (Again, we will see below how this happens at the level of atoms and molecules.) Consider the energy flow, quantified in terms of the energy flux (power per unit area), as sketched in Fig. 13.2:

- The incident flux from the Sun comes down through the atmosphere with no effect, and then hits the surface. We can use the analysis from Sect. 13.2 to quantify the incident flux as σT_p^4 where T_p is given by Eq. (13.15).
- The surface has temperature T_s (which is to be determined), so it radiates a flux σT_s^4 upward. That flux is absorbed by the atmosphere.
- The atmosphere has temperature T_a (which is to be determined), so it radiates a flux σT_a^4 both upward and downward.

The net flux incident on Earth is σT_p^4, while the net flux leaving Earth is σT_a^4. In equilibrium, we must therefore have $T_a = T_p$. Thus, what Eq. (13.15) predicts is the temperature of the *atmosphere*.

What about the surface? In order for the incident and emitted radiation to balance, we must have

$$\sigma T_s^4 = \sigma T_p^4 + \sigma T_a^4 = 2\sigma T_p^4 \quad \Rightarrow \quad T_s = 2^{1/4} T_p \tag{13.16}$$

In this simple model, the atmosphere raises the surface temperature by a factor of $2^{1/4} = 1.19$. An increase of 19 % may not seem like a lot, but remember that it applies to the Kelvin temperature.

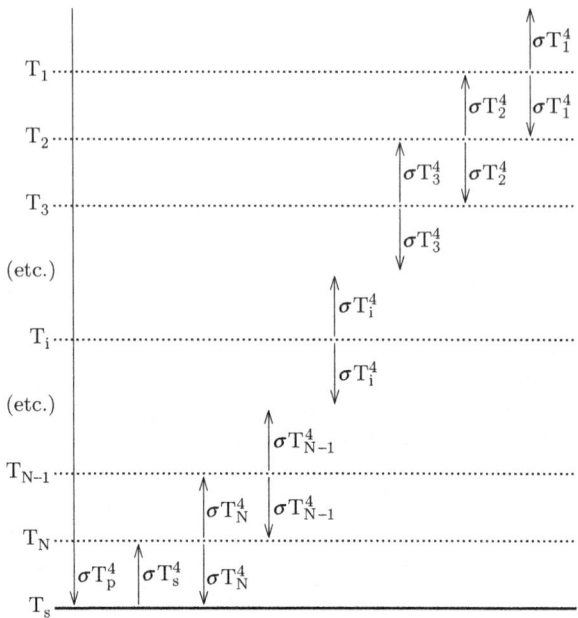

Fig. 13.3 Setup for the many-layer model of atmospheric heating. Atmospheric layers are labeled $1, \ldots, N$ from top down (most intermediate layers are not shown here). Energy flux is denoted with *arrows*. The layers are transparent to the visible light incident from the Sun, but each layer absorbs 100 % of the infrared radiation incident upon it and then re-radiates that energy both upward and downward

Example: Earth

Using $T_p = 254$ K as above, the one-layer model would predict a surface temperature of $T_s = 302$ K. This is closer to Earth's average surface temperature, but a little high. The problem, presumably, is the assumption that 100 % of the infrared light emitted by the surface is trapped by the atmosphere.

13.3.2 Many Layers

To improve the model we need to handle what is effectively a partial layer. While it may seem counterintuitive, our best bet is actually to understand what happens when we add more layers, and then to generalize the notion of layers, so we can finally circle back to the case of a partial layer.

Now suppose the atmosphere is made of N layers that transmit visible light but absorb infrared light. Number the layers from top to bottom and then consider the energy flow as sketched in Fig. 13.3. Again, we write the incident flux as σT_p^4. Now consider the energy balance for each layer; on the left of each equation is the output flux, and on the right is the input[6]:

$$2T_1^4 = T_2^4$$
$$2T_2^4 = T_1^4 + T_3^4$$
$$2T_3^4 = T_2^4 + T_4^4$$

$$\vdots$$

$$2T_i^4 = T_{i-1}^4 + T_{i+1}^4$$

$$\vdots$$

$$2T_{N-1}^4 = T_{N-2}^4 + T_N^4$$
$$2T_N^4 = T_{N-1}^4 + T_s^4$$
$$T_s^4 = T_N^4 + T_p^4$$

We also have the condition that the net input to the planet must equal the net output from the planet. This yields:

$$T_1^4 = T_p^4$$

Now that we know T_1, we can work from the top down:

$$T_1^4 = T_p^4$$
$$T_2^4 = 2T_1^4 = 2T_p^4$$
$$T_3^4 = 2T_2^4 - T_1^4 = 3T_p^4$$

$$\vdots$$

The pattern is clear: the temperature of layer i satisfies

$$T_i^4 = i\,T_p^4 \quad\Rightarrow\quad T_i = i^{1/4}\,T_p$$

Finally, the last energy balance equation tells us what happens at the surface:

$$T_s^4 = (1+N)T_p^4 \quad\Rightarrow\quad T_s = (1+N)^{1/4}\,T_p \qquad (13.17)$$

Adding more layers creates more heating at the surface,[7] with the scaling $(1+N)^{1/4}$.

---

[6]To get true energy flux we need a factor of σ multiplying each T^4, but those all factor out.

[7]This is why we wear more layers of clothing in the winter.

13.3.3   Optical Depth

The preceding analysis becomes more applicable to Earth when we generalize the notion of "layers" in the atmosphere. What actually matters is the number of times a photon is absorbed and reemitted between the surface and free space. If the mean free path of light in the atmosphere is $\ell = (n\sigma)^{-1}$, then the number of interactions over some distance dx is $n\,\sigma\,dx$. The total number of interactions as light travels through the atmosphere is then (cf. Eq. 12.19)

$$\tau = \int n\,\sigma\,dx \tag{13.18}$$

We call this the **optical depth** because it gives a sense of how far light penetrates into a gas: roughly speaking, we can see fairly clearly up until the point where $\tau \sim 1$.

Since τ counts the number of interactions, we can interpret it as the effective number of layers in the atmosphere. Then we can write our model for atmospheric heating as

$$T_s = (1 + \tau)^{1/4}\, T_p \tag{13.19}$$

This is a nice generalization because τ no longer needs to be an integer. In particular, we can use a value $\tau < 1$ to model an atmosphere that absorbs infrared radiation partially but not perfectly. In Problem 13.7 you can see how to apply this model to Earth's atmosphere.

13.4   Interaction of Light with Matter

In the previous section we postulated that Earth's atmosphere can absorb infrared light. How does this happen on a microscopic level? More generally, how do light and matter interact? At the moment we are most interested in how matter *absorbs* light, and there are four phenomena that may be relevant:

- Photoionization: a photon can give enough energy to an electron to knock it out of its atom/molecule.
- Electron excitation: a photon can give energy to an electron and cause it to jump to a higher energy level.
- Molecular vibration: a photon can cause the atoms in a molecule to vibrate relative to one another.
- Molecular rotation: a photon can cause an entire molecule to spin.

Let's consider each in turn. For now we want to figure out which phenomena affect infrared light, but we will encounter all of them in various contexts in coming chapters.

13.4.1 Photoionization

To analyze photoionization, we need to determine what wavelengths of light have enough energy to unbind an electron. Let's use the Bohr model to estimate the energy levels of electrons in atoms. In this model we picture the electron in a classical circular orbit, but we say the angular momentum must be quantized. While the physical picture is not strictly correct, the resulting energy levels turn out to be accurate for hydrogen and reasonable for some other elements.

Consider a single electron orbiting a nucleus with atomic number Z. The electric force between the nucleus and electron is Ze^2/r^2, while the force needed to keep the electron in a circular orbit with speed v is $m_e v^2/r$. Equating these lets us determine the orbital speed v and angular momentum L:

$$v = \left(\frac{Ze^2}{m_e r}\right)^{1/2} \quad \text{and} \quad L = m_e r v = \left(Ze^2 m_e r\right)^{1/2}$$

We then quantize the angular momentum by setting $L_n = n\hbar$ where n is an integer. This yields the orbital radius for level n:

$$r_n = \frac{n^2 \hbar^2}{Ze^2 m_e}$$

The total energy of the electron in this orbit is then

$$E_n = -\frac{Ze^2}{r_n} + \frac{1}{2}m_e v_n^2 = -\frac{Ze^2}{2r_n} = -\frac{Z^2 e^4 m_e}{2n^2 \hbar^2} \tag{13.20}$$

In order to eject such an electron, a photon must have energy $E > |E_n|$ or wavelength

$$\lambda < \lambda_n \quad \text{where} \quad \lambda_n = \frac{hc}{E_n} = \frac{4\pi \hbar^3 c n^2}{Z^2 e^4 m_e}$$

Plugging in numbers yields

$$\lambda_n = 911 \, \text{Å} \times \frac{n^2}{Z^2} \tag{13.21}$$

Photoionization mainly involves ultraviolet light.

13.4.2 Electron Excitation

If the photon does not have sufficient energy to unbind the electron, it might still be able to excite the electron into a higher energy level. In the Bohr model, the energy required to raise an electron from level n to level m is

$$\Delta E_{nm} = E_m - E_n = \frac{Z^2 e^4 m_e}{2\hbar^2} \left(\frac{1}{n^2} - \frac{1}{m^2} \right)$$

The corresponding wavelength is

$$\lambda_{nm} = \frac{4\pi \hbar^3 c}{Z^2 e^4 m_e} \left(\frac{1}{n^2} - \frac{1}{m^2} \right)^{-1} = \frac{911\,\text{Å}}{Z^2} \left(\frac{1}{n^2} - \frac{1}{m^2} \right)^{-1} \tag{13.22}$$

As an example, in hydrogen the $1 \rightarrow 2$ transition has $\lambda_{12} = 1,216\,\text{Å}$, while the $2 \rightarrow 3$ transition has $\lambda_{23} = 6,563\,\text{Å}$. These spectral lines will play an important role when we study stars in Chap. 14. For now, the key point is that electron excitation occurs at discrete wavelengths of visible and ultraviolet light.

13.4.3 Molecular Vibration

In a molecule, light can cause the chemical bonds to vibrate. We can use our tools of dimensional analysis and toy models to estimate the range of wavelengths that can excite vibrational motion.

Let's begin with dimensional analysis. The force involved is the electric force, so we ought to use e. Since the nuclei themselves move, the mass scale is that of a proton or neutron. The important length scale is the typical distance r_0 between atoms in the molecule. Note that r_0 encodes the physics (including quantum mechanics) that governs the chemical bond and determines the size of the molecule; we can avoid those details by working with r_0 directly. Let's write

$$\omega \sim e^\alpha\, m_p^\beta\, r_0^\gamma$$

$$[T^{-1}] \sim [M^{\alpha/2} L^{3\alpha/2} T^{-\alpha} \times M^\beta \times L^\gamma]$$

This is solved with $\alpha = 1$, $\beta = -1/2$, and $\gamma = -3/2$, yielding

$$\omega \sim \frac{e}{(m_p\, r_0^3)^{1/2}}$$

The angular frequency ω corresponds to a linear frequency $\nu = \omega/(2\pi)$. The corresponding wavelength of light is

$$\lambda = \frac{c}{\nu} \sim \frac{2\pi c}{e} \left(m_p\, r_0^3\right)^{1/2}$$

The most important vibrating molecule in Earth's atmosphere is water, which has $r_0 \approx 10^{-10}$ m, and hence

$$\lambda \sim \frac{2\pi \times (3.0 \times 10^8 \, \mathrm{m\,s^{-1}})}{1.52 \times 10^{-14} \, \mathrm{kg^{1/2}\,m^{3/2}\,s^{-1}}} \left[(1.67 \times 10^{-27} \, \mathrm{kg}) \times (10^{-10} \, \mathrm{m})^3\right]^{1/2} \sim 5\,\mu\mathrm{m}$$

This is in the infrared portion of the electromagnetic spectrum.

To make a fully realistic model, we would need to consider the structure of the bond between atoms, handle quantum mechanics properly, and account for different vibrational modes. Without getting into all of the details, we can go one step further by thinking about the bond. In water, the hydrogen and oxygen share a covalent bond: each atom contributes one electron to a sort of cloud surrounding the two nuclei.[8] Let's make a toy model in which the oxygen and hydrogen ions (each with charge $+e$) are enclosed in a spherical cloud with total charge $-2e$ and radius R. The oxygen ion is heavier so we imagine it sits motionless at the center of the cloud while the hydrogen ion moves. Let's suppose the hydrogen oscillates radially. What is its equation of motion? There is a repulsive force between the two positive ions of e^2/r^2. There is also an attractive force due to the portion of the electron cloud interior to the hydrogen ion's position.[9] If the electron cloud has a uniform charge density, the charge contained within r is $-2er^3/R^3$. Thus, the equation of motion is

$$m\frac{\mathrm{d}^2 r}{\mathrm{d}t^2} = \frac{e^2}{r^2} - \frac{2e^2 r}{R^3}$$

How do we analyze this equation? Consider: there is some equilibrium position r_0 where the attractive and repulsive forces exactly balance. This is the place where $\mathrm{d}^2 r/\mathrm{d}t^2 = 0$, or

$$r_0 = \frac{R}{2^{1/3}} \tag{13.23}$$

Now we imagine the ion makes small excursions around this position. Let's write

$$r = r_0 + \delta r = r_0\left(1 + \frac{\delta r}{r_0}\right)$$

and imagine that $\delta r \ll r_0$. Then we can write the equation of motion as

---

[8] We can picture a cloud for two reasons: according to quantum mechanics the electron wavefunctions are shells; even in classical mechanics, if we took a long-exposure photograph the electrons would look smeared out due to their motion.

[9] In analogy with gravity, the portion of the cloud outside the hydrogen ion's position produces no net force.

$$m\frac{\mathrm{d}^2(\delta r)}{\mathrm{d}t^2} = \frac{e^2}{r_0^2}\left(1+\frac{\delta r}{r_0}\right)^{-2} - \frac{e^2}{r_0^2}\left(1+\frac{\delta r}{r_0}\right)$$

In the second term we use Eq. (13.23) to substitute for R. What do we do now? If $\delta r/r_0 \ll 1$ then we can make a Taylor series expansion of the factor in the first term: $(1 + \delta r/r_0)^{-2} \approx (1 - 2\delta r/r_0)$. This gives

$$m\frac{\mathrm{d}^2(\delta r)}{\mathrm{d}t^2} \approx \frac{e^2}{r_0^2}\left(1-2\frac{\delta r}{r_0}\right) - \frac{e^2}{r_0^2}\left(1+\frac{\delta r}{r_0}\right)$$

$$\approx -\frac{3e^2}{r_0^3}\,\delta r$$

This is the equation for simple harmonic motion. Thus, in our cloud model the hydrogen ion will experience sinusoidal oscillations with angular frequency

$$\omega = \left(\frac{3e^2}{mr_0^3}\right)^{1/2}$$

This matches what we obtained from dimensional analysis, up to a factor of $\sqrt{3}$. Now, the numerical factor is not necessarily precise, because our toy model does not account for all the details of the real chemical bond. Nevertheless, the model does contain some real physics (even if simplified), so perhaps it helps you believe the dimensional analysis.

13.4.4   Molecular Rotation

A molecule can also rotate, but perhaps at a different frequency than it vibrates. Here, dimensional analysis will not involve the electric force, but it will explicitly involve $\hbar$ because that is the natural scale of angular momentum for atoms and molecules. In this case, the usual dimensional analysis yields

$$\omega \sim \frac{\hbar}{mr_0^2} \quad \Rightarrow \quad \lambda \sim 2\pi\,\frac{cmr_0^2}{\hbar}$$

Again the most important molecule in Earth's atmosphere is water, which has $m = 18m_p$ and $r_0 \approx 10^{-10}$ m, so the wavelength scale is

$$\lambda \sim 2\pi \times \frac{(3.0\times10^8\,\mathrm{m\,s^{-1}}) \times (18\times1.67\times10^{-27}\,\mathrm{kg}) \times (10^{-10}\,\mathrm{m})^2}{1.05\times10^{-34}\,\mathrm{kg\,m^2\,s^{-1}}} \sim 5\,\mathrm{mm}$$

This is in the microwave portion of the electromagnetic spectrum.[10]

---

[10]Incidentally, microwave ovens operate using molecular rotation. Microwave radiation induces water molecules in food to rotate; friction then disperses the rotational energy as heat.

13.4.5   Recap

As a rule of thumb, we can say that the four phenomena we have considered are important in different parts of the electromagnetic spectrum:

- Photoionization: ultraviolet
- Electron excitation: near-UV and visible
- Molecular vibration: infrared
- Molecular rotation: microwave

Molecular vibration is the main phenomenon that drives atmospheric absorption of infrared radiation, with molecular rotation kicking in at the long-wavelength end.

13.5   Greenhouse Effect and Climate Change

Atmospheric heating is commonly known as the **greenhouse effect**. We hear a lot about the greenhouse effect in connection with global warming and climate change, but the three phenomena are not exactly identical. The greenhouse effect is a well-understood physical effect that we know takes place on Earth. (As we have seen, Earth's surface would be measurably cooler without it.) Global warming occurs if the greenhouse effect strengthens with time, especially due to changes caused by humans. An increase in (average) temperature is part of a broader set of changes to the climate that can develop when the greenhouse effect is increased. Notwithstanding any political controversy about how we should respond to climate change, there is no doubt that the greenhouse effect is real.

13.5.1   Earth

Detailed studies of molecular rotation lead to the absorption spectra shown in Fig. 13.4. In Earth's atmosphere, water is responsible for most of the absorption between 1–$8\,\mu m$ and beyond about $20\,\mu m$. Carbon dioxide is important beyond $14\,\mu m$, and in a band around $4\,\mu m$. There is little natural greenhouse effect in an "atmospheric window" at 8–$14\,\mu m$. Using information like this, we can make a table showing how much different gases contribute to the greenhouse effect on Earth, and the degree to which people are concerned about their role in climate change. (Here ppm=parts per million) [2]

Notice that the contribution to the greenhouse effect is not dictated by abundance alone: molecular nitrogen and oxygen are very common, but they are poor absorbers at infrared wavelengths, so they contribute little to the greenhouse effect.

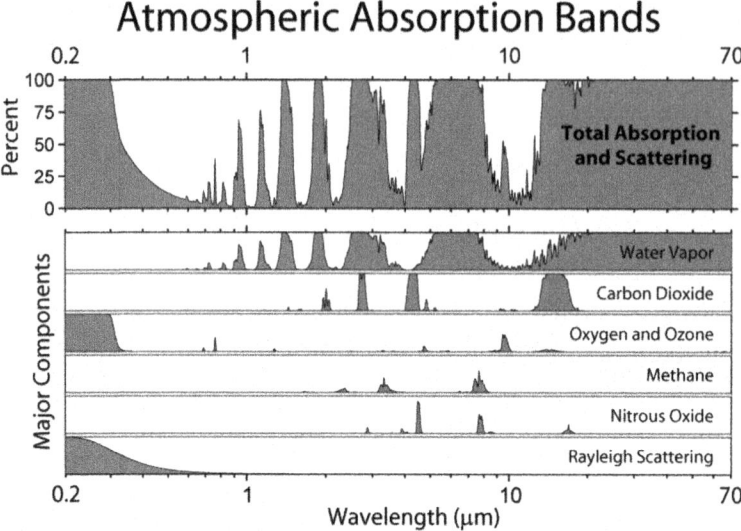

Fig. 13.4 The *top panel* shows the percentage of light that is absorbed by Earth's atmosphere as a function of wavelength, running from the visible into the infrared. The *bottom panels* show contributions from different gases (Credit: Robert A. Rohde/Global Warming Art)

Molecule	Abundance in Earth's atmosphere	Contribution to greenhouse effect	Concern?
N_2	78.1 %	–	No
O_2	20.1 %	–	No
Ar	0.9 %	–	No
H_2O	<1 %	60 %	Indirect
CO_2	0.04 %	26 % ↗	Yes
CH_4, N_2O	2 ppm	6 %	Yes
O_3	0.1 ppm	8 %	Yes

Public discussion of climate change focuses on carbon dioxide, whose atmospheric abundance is increasing due to human activity. The fact that atmospheric CO_2 is within our control means we can and should be concerned about it. Industrial gases containing carbon and fluorine and/or chlorine are also important. While they are not very abundant, they are efficient absorbers in the 8–14 μm atmospheric window so even trace amounts can strengthen the greenhouse effect.

What about water vapor? It contributes the majority of the greenhouse effect, but we have little direct control over its atmospheric abundance because of natural evaporation from the oceans. Water vapor plays a complicated role in climate change because of **feedback loops**. If the temperature rises, more water can evaporate, which would strengthen the greenhouse effect and thus raise the temperature even further. This would be an example of *positive feedback*, which reinforces any change that takes place. But there may be another effect: evaporating water can form more clouds, which can reflect more sunlight, which would increase the albedo and thereby reduce the net energy absorbed from the Sun. That would create *negative feedback*, which acts against the prevailing trend and tries to regulate the temperature.

Additional factors include snow and ice (whose formation removes water vapor from the atmosphere and increases the albedo), and forests (which consume carbon dioxide and have low albedo), to name just two. In detail, climate change is a complicated process with many connected pieces that climatologists are working hard to understand fully. Nevertheless, there is solid evidence that the average temperature of Earth's surface and the amount of carbon dioxide in the atmosphere are on the rise. Where there are legitimate scientific questions, they mainly involve detailed predictions about exactly how the climate will change in the next century. Even there, though, the question is not *whether* the climate is changing; the question is really, "how much?"

13.5.2   *Venus*

Venus offers an example of what can happen when an atmosphere has strong positive feedback. Today the planet has a thick atmosphere dominated by carbon dioxide that warms the surface to about 740 K, far above the temperature of 185 K that we would expect if there were no atmospheric heating.[11] Also, the atmospheric pressure at the surface of Venus is about 90 times higher than at the surface of Earth (and comparable to the water pressure almost a kilometer down in the oceans on Earth).

Yet Venus and Earth were probably alike when they were young: the planets are similar in size, and they presumably formed by similar processes (see Sect. 19.4.2). What happened? Venus is closer to the Sun, so young Venus would have been a little warmer than young Earth. That would have caused any water on Venus to evaporate quickly, loading the atmosphere with water vapor and creating a strong greenhouse effect. As the temperature rose, carbon dioxide began to "bake out" of rocks,[12] which created yet more positive feedback and led to a **runaway greenhouse effect**. As the atmosphere filled with CO_2, the lighter H_2O molecules rose to the upper atmosphere, where they could be dissociated by ultraviolet light from the Sun. Then the light hydrogen atoms escaped from the atmosphere (see Sect. 12.3.3), leaving mostly carbon dioxide.

A tangential aspect of this story makes a testable prediction. A small fraction of the hydrogen in Venus's atmosphere was actually the isotope deuterium (D, whose nucleus contains one proton and one neutron). Being heavier, deuterium would evaporate less quickly than hydrogen, so the ratio of deuterium to hydrogen would

---

[11] Venus's cloud cover creates a high albedo, so the predicted temperature is actually lower than Earth's even though Venus is closer to the Sun.

[12] On Earth, most of the carbon is locked up in the crust.

rise with time. Today the D/H ratio is about 100 times higher on Venus than on Earth, supporting the notion that Venus was once wet but the runaway greenhouse effect has made it such an inhospitable place [3,4].

Problems

In several problems, the albedo is relevant but not specified. Recall it must be between 0 and 1, and think about whether the specific value affects your conclusions. If it does, explain any assumptions you make.

13.1. Is a blackbody actually black? Explain. How would you determine if a given object emits electromagnetic radiation like a blackbody?

13.2. In common usage, we sometimes call things "red hot" or "white hot." Which is hotter? Explain.

13.3. Rank the following stars by luminosity:

Star	R ($R_\odot$)	T ($T_\odot$)
A	0.5	0.5
B	1.0	2.0
C	1.5	1.0
D	20.0	1.0

13.4. In the Sirius binary system (see Problem 4.4b), star A has a luminosity of $25.4\,L_\odot$ and the peak in its spectrum is at 292 nm, while star B has a luminosity of $0.026\,L_\odot$ with peak emission at 115 nm. Treating these stars as blackbodies, what physical properties of the stars can you determine from the information given? Calculate at least two properties for each star.

13.5. The star HD 209458 has mass $M_s = 1.13\,M_\odot$, luminosity $L_s = 1.61\,L_\odot$, and surface temperature $T = 6{,}000$ K. It is orbited by a planet with mass $M_p = 0.69\,M_J$ and radius $R_p = 1.35\,R_J$ that lies 0.045 AU from the star (see Sect. 4.3.2). The size and mass of the planet suggest it is a gaseous planet like Jupiter, but it is so close to its star that it must be hotter than Jupiter. Would you expect hydrogen to be able to evaporate from the planet? Explain, and be quantitative.

13.6. Consider the planet orbiting the star HD 209458 (see data in Problem 13.5). Treating both objects as blackbodies, compute the brightness of the planet relative to the star at the following wavelengths: $\lambda = 450$ nm (blue light), $\lambda = 700$ nm (red light), and $\lambda = 2.2\,\mu$m (infrared light). Consider both starlight reflected off the planet and blackbody light emitted by the planet itself. Explain your reasoning. If you want to detect light from this exoplanet, which of these three wavelengths is the best choice?

13.7. If Earth were much farther from the Sun, it would be too cold for water to remain liquid on the surface. If Earth were much closer, it would be too hot. The region in which water can remain liquid is called the "habitable zone."

(a) Taking Earth's mean surface temperature to be 288 K, estimate the optical depth of Earth's atmosphere. (Don't worry about the detailed absorption spectrum; use our simple atmospheric heating model to obtain the average optical depth.)

(b) Assuming the optical depth and albedo remain fixed, compute the inner and outer edges of the habitable zone for Earth around the Sun.

13.8. In 2005 astronomers discovered an object in the outer Solar System now known as the dwarf planet Eris (which played a major role in Pluto's demotion to dwarf planet status). Here is a way to understand how Eris's size was first determined.

(a) Using the model from Sect. 13.2, find the luminosity L_p reflected from a planet with radius R_p at a distance D from the Sun.

(b) When astronomers first discovered Eris they found it to be $D = 97\,\text{AU}$ from the Sun and measured its reflected luminosity to be $L_p = 5.8 \times 10^{11}\,\text{J s}^{-1}$. Use this information to derive a bound on Eris's size.

(c) More recently, astronomers measured the infrared light emitted by Eris and found that its spectrum peaks at a wavelength of 116 microns. Use this information to calculate the effective temperature, and state any assumptions that you have to make.

(d) Use your results from (b) and (c) to determine the size of Eris. How does Eris compare in size with Pluto ($R_{\text{Pluto}} = 1{,}153\,\text{km}$)?

References

1. B.W. Carroll, D.A. Ostlie, *An Introduction to Modern Astrophysics*, 2nd edn. (Addison-Wesley, San Francisco, 2007)
2. J.T. Kiehl, K.E. Trenberth, Bull. Am. Meteorol. Soc. **78**, 197 (1997)
3. T.M. Donahue, J.H. Hoffman, R.R. Hodges, A.J. Watson, Science **216**, 630 (1982)
4. C. de Bergh, B. Bezard, T. Owen, D. Crisp, J.P. Maillard, B.L. Lutz, Science **251**, 547 (1991)

Chapter 14
Stellar Atmospheres

The phenomena that affect the transmission of light through planetary atmospheres also operate in stellar atmospheres. For planets, our interest in infrared light led us to focus on molecular vibrations. For stars, we are more interested in visible light so our attention shifts to electron excitation and ionization. In this chapter we study how those processes affect the absorption lines that appear in stellar spectra.

14.1 Atomic Excitation and Ionization

Observed spectra of stars resemble blackbody spectra modified by absorption bands at specific wavelengths (see Figs. 14.1 and 14.2). The absorption lines are produced when light that originates from some modest depth passes through the star's outer layers.[1] The gas in stars is predominantly hydrogen, so we might expect to see strong hydrogen absorption lines. Recall from Eq. (13.22) that the transition between hydrogen energy levels n and m corresponds to wavelength

$$\lambda_{nm} = 911 \,\text{Å} \left(\frac{1}{n^2} - \frac{1}{m^2} \right)^{-1}$$

Thus the locations of hydrogen lines are as follows (with wavelengths in Å):

	$m = 2$	3	4	5	
$n = 1$	1,216	1,026	973	949	"Lyman series"
2		6,563	4,861	4,340	"Balmer series"
3			18,750	12,818	"Paschen series"

---

[1] A star's photosphere, or apparent "surface," corresponds to an optical depth of order unity (see Sect. 13.3.3).

C. Keeton, *Principles of Astrophysics: Using Gravity and Stellar Physics to Explore the Cosmos*, Undergraduate Lecture Notes in Physics, DOI 10.1007/978-1-4614-9236-8_14, © Springer Science+Business Media New York 2014

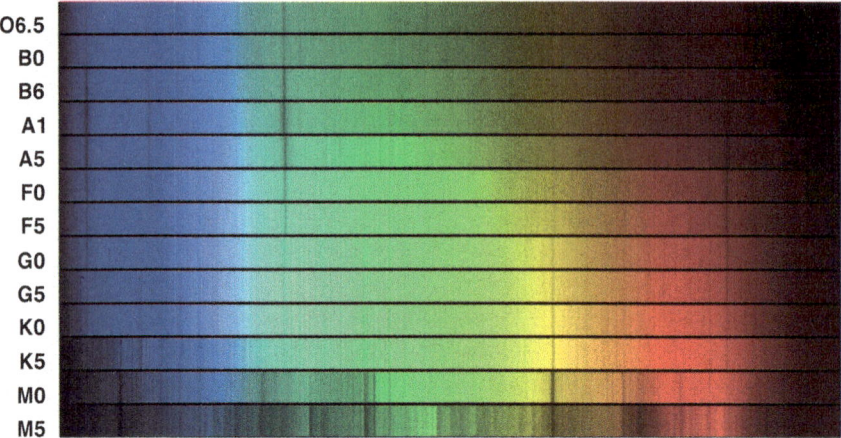

Fig. 14.1 Examples of absorption line spectra from different types of stars. The codes on the left refer to the spectral classification scheme discussed in Sect. 14.2 (the numbers indicate subcategories). The strong lines in the spectrum labeled A1 correspond to hydrogen. Moving down the sequence, the peak of the spectrum shifts to longer wavelengths and the hydrogen lines become less prominent (Credit: NOAO/AURA/NSF)

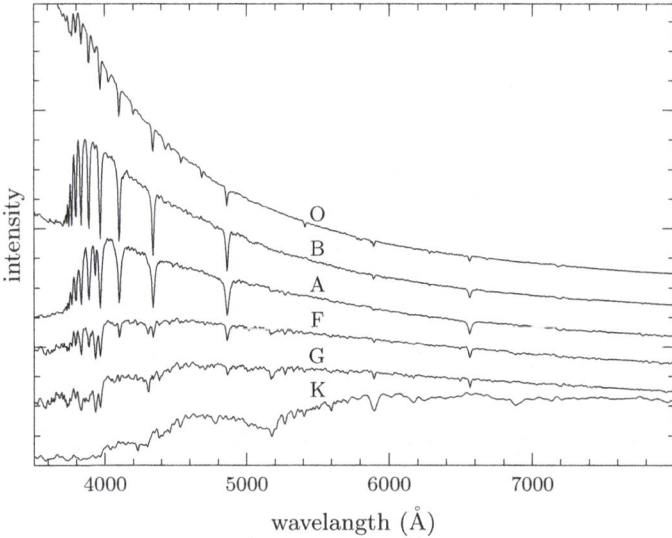

Fig. 14.2 A different view of spectra for different types of stars, from O (*top*) to K (*bottom*). M stars are not shown here. The curves show intensity versus wavelength, and are offset vertically for clarity (Data from Pickles [1])

The Balmer lines are the ones that appear at optical wavelengths. We do see them in some stellar spectra (e.g., the top half of Fig. 14.1), but not in others. Can we understand why?

14.1.1 Energy Level Occupation

Consider an absorption line corresponding to an atomic transition from energy level n to energy level $m > n$. In order for this line to be strong, two conditions must hold:

1. The atom in question must be abundant.
2. In a non-negligible fraction of atoms, level n must be occupied but level m must not be full.

Consider point #2. In isolation, an atom generally has all of its electrons in the lowest possible energy levels. In a gas, however, atoms occasionally bump into each other, and some of the kinetic energy can be transferred to internal energy. The temperature and associated kinetic energy in stellar atmospheres are high enough that collisions can excite electrons into higher energy levels.

The amount of collisional excitation depends on the kinetic energy, which in turn depends on the temperature. To quantify this effect, recall from Eq. (12.1) that the probability for level n to be occupied is

$$P_n \propto g_n\, e^{-E_n/kT}$$

where E_n is the energy and g_n is the **statistical weight**, which in this case counts the number distinct quantum states in the energy level. The ratio of the number of atoms with level m occupied to the number with level n occupied is then

$$\frac{N_m}{N_n} = \frac{g_m}{g_n}\, e^{-(E_m-E_n)/kT} \tag{14.1}$$

This **Boltzmann equation** allows us to determine which energy levels are significantly occupied and analyze whether absorption lines will be strong or weak.

14.1.2 Ionization Stages

A collision with sufficient energy can kick an electron out of an atom altogether. This influences absorption lines because electrons that are not in atoms cannot produce *atomic* transitions, and the loss of an electron may modify the energy levels for any remaining electrons in an atom. It is conventional to label different **ionization stages** with Roman numerals:

I. Neutral

II. Singly-ionized

III. Doubly-ionized

and so on. Consider a transition that starts with a neutral atom (stage I) whose most loosely bound electron has energy $E_I = -\chi_I$, where $\chi_I > 0$ is the **ionization energy**, and ends with a stage II ion plus a free electron with speed v_e or energy $E_{II} = m_e v^2/2$. The energy difference between the two states is $\Delta E = E_{II} - E_I = \chi_I + m_e v^2/2$, so the relative abundance of ionized and neutral atoms has a Boltzmann factor of the form

$$\frac{N_{II}}{N_I} \sim \exp\left(-\frac{\chi_I + m_e v^2/2}{kT}\right)$$

(We will deal with statistical weights in a moment.) We do not actually care what the final speed of the electron is, so we can integrate over all velocities:

$$\frac{N_{II}}{N_I} \propto \int \exp\left(-\frac{\chi_I + m_e v^2/2}{kT}\right) d^3v$$

$$\propto e^{-\chi_I/kT} \int_0^\infty e^{-m_e v^2/2kT} \, 4\pi v^2 \, dv$$

$$\propto \left(\frac{2\pi kT}{m_e}\right)^{3/2} e^{-\chi_I/kT}$$

The key factors here are $e^{-\chi_I/kT}$ and $(kT)^{3/2}$.

There are a few more details that enter a complete analysis. First, the number density of free electrons must play a role. It appears in the denominator because free electrons can combine with ions and return them to a neutral stage. Second, where we used statistical weights to count states for the Boltzmann equation, we must now use a more general counting that is done with the **partition function,**[2]

$$Z = \sum_{n=1}^\infty g_n \, e^{-(E_n - E_1)/kT} \tag{14.2}$$

where E_1 is the energy of the ground state. This is basically the sum of the number of ways the atom can arrange its electrons, with more energetic (and therefore less likely) configurations receiving less weight from the Boltzmann factor. Careful counting reveals that the ratio of the number of atoms in ionization stage II to the number in stage I is

---

[2]Partition functions are often studied in courses on statistical mechanics.

$$\frac{N_{II}}{N_I} = \frac{2Z_{II}}{n_e \, Z_I} \left(\frac{m_e kT}{2\pi\hbar^2}\right)^{3/2} e^{-\chi_I/kT} \tag{14.3}$$

This is called the **Saha equation** after Meghnad Saha.

Hydrogen only has two ionization stages: neutral (H I) and ionized (H II). For heavier elements, higher ionization stages are possible, and the Saha equation can be applied to them as well. For example, the ratio of doubly ionized atoms (like He III) to singly ionized atoms (like He II) would be described by an equation like (14.3) but with the second ionization energy χ_{II} and the partition functions Z_{III} and Z_{II} (see Problem 14.1).

14.1.3   Application to Hydrogen

Let's use these ideas to study hydrogen Balmer lines. These occur when an electron jumps from level $n = 2$ up to level $m > 2$, so they are strong only if a reasonable fraction of hydrogen atoms have electrons in the $n = 2$ level. Before we can use the Boltzmann equation to compute that fraction, we need to specify the statistical weight for hydrogen:

$$g_n = 2n^2$$

The n^2 comes from the orbital quantum numbers, while the 2 comes from spin. Using $T_\odot = 5{,}780$ K for the Sun yields the following numbers:

$$kT = 7.98 \times 10^{-20}\,\text{J} \; = \; 0.498\,\text{eV}$$
$$E_1 = -13.6\,\text{eV}$$
$$E_2 = -3.4\,\text{eV}$$
$$g_1 = 2$$
$$g_2 = 8$$
$$\Rightarrow \quad \frac{n_2}{n_1} = \frac{g_2}{g_1}\, e^{-(E_2-E_1)/kT} \; = \; 5 \times 10^{-9} \tag{14.4}$$

Only a tiny fraction of hydrogen atoms in the outer layers of the Sun are excited to $n = 2$, which is why Balmer lines are not very prominent in the spectrum of the Sun. The excitation fraction increases with temperature, as shown in Fig. 14.3, so in general we would expect Balmer lines to be more prominent in hotter stars.

However, as the temperature increases, so too does the ionization fraction. At some point ionization must overcome excitation and prevent hot stars from producing any hydrogen absorption lines. We can investigate this using the Saha equation, but first we need to determine the partition functions. The starting point is neutral hydrogen, whose partition function is

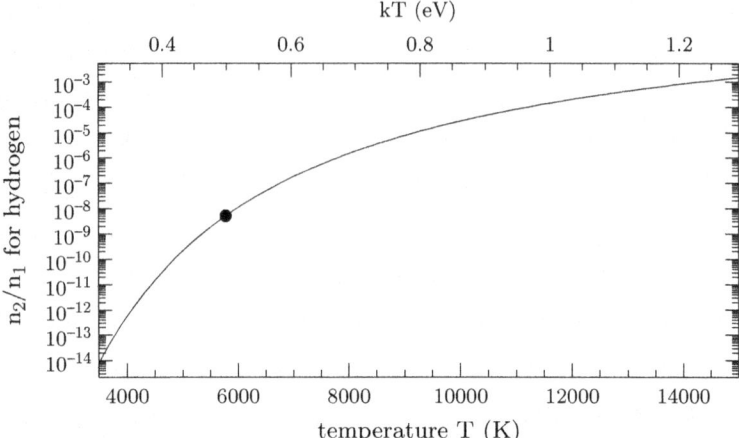

Fig. 14.3 Excitation ratio n_2/n_1 for hydrogen, as a function of temperature. The value for the Sun is marked

$$Z_I = g_1 + g_2\,e^{-(E_2-E_1)/kT} + g_3\,e^{-(E_3-E_1)/kT} + \dots$$

$$= g_1\left(1 + \frac{g_2}{g_1}\,e^{-(E_2-E_1)/kT} + \frac{g_3}{g_1}\,e^{-(E_3-E_1)/kT} + \dots\right)$$

$$= g_1\left(1 + \frac{n_2}{n_1} + \frac{n_3}{n_1} + \dots\right)$$

We just found that at temperatures relevant for the surface of stars, most of the hydrogen atoms are in the ground state. Thus n_2/n_1 is small, n_3/n_1 is even smaller, and we have $Z_I \approx g_1 = 2$. The ending point is a bare proton, which has only one possible state so $Z_{II} = 1$.

We want determine the ionization fraction, X. Let the number densities of hydrogen ions and neutral atoms be n_{II} and n_I, respectively, and the total number density of hydrogen be $n_{\text{tot}} = n_I + n_{II}$. By the definition of the ionization fraction, $n_{II} = X n_{\text{tot}}$ and $n_I = (1 - X)n_{\text{tot}}$. Then the left-hand side of the Saha equation (14.3) becomes $n_{II}/n_I = X/(1 - X)$. The right-hand side has a factor of n_e, and by charge conservation $n_e = n_{II} = X n_{\text{tot}}$ for hydrogen. Collecting all factors of X on the left-hand side then yields

$$\frac{X^2}{1 - X} = \frac{1}{n_{\text{tot}}}\left(\frac{m_e kT}{2\pi\hbar^2}\right)^{3/2} e^{-\chi_I/kT} \tag{14.5}$$

Now we need to deal with n_{tot} on the right-hand side. NASA's Sun fact sheet gives the pressure and temperature at the bottom and top of the photosphere (see Table 14.1).

We can compute a number density from the ideal gas law, $n = P/kT$ (column 3). This includes contributions from all three constituents (neutral atoms,

Table 14.1 Physical conditions at the bottom and top of the Sun's photosphere, from NASA's Sun fact sheet.

	P (kg m^{-1} s^{-2})	T (K)	n (m^{-3})	ρ (kg m^{-3})
Bottom	1.25×10^4	6,600	1.37×10^{23}	2.29×10^{-4}
Top	8.68×10^1	4,400	1.43×10^{21}	2.39×10^{-6}

ions, and electrons), so it depends on the ionization fraction. The *mass* density, by contrast, is much less sensitive to X (because electrons contribute so little mass). As we will see shortly, the ionization fraction in the Sun's photosphere is quite low so to a good approximation we can take the mean particle mass to be the mass of a neutral hydrogen atom, $\bar{m} \approx m_p$, and then compute the mass density as $\rho = \bar{m} P / kT$ (column 4). Then we can use $\rho = m_p n_{\mathrm{tot}}$ (again, the electrons are negligible in mass) to rewrite Eq. (14.5) as

$$\frac{X^2}{1-X} = \frac{m_p}{\rho} \left(\frac{m_e kT}{2\pi\hbar^2} \right)^{3/2} e^{-\chi_I/kT} \tag{14.6}$$

This is a quadratic equation that we can solve for X. Plugging in numbers for the bottom and top of the photosphere yields

$$X = \begin{cases} 6.2 \times 10^{-4} & \text{bottom} \\ 1.1 \times 10^{-5} & \text{top} \end{cases}$$

Very few hydrogen atoms in the outer layer of the Sun are ionized.

That would change if the temperature increased, of course. We can use Eq. (14.6) to estimate how the ionization fraction would increase if we assume the mass density remains fixed as we vary the temperature. This is not quite correct because the star would adjust its hydrostatic equilibrium for a different temperature, but it lets us obtain a useful estimate without getting too bogged down in details. Plugging numbers into Eq. (14.6) lets us write the temperature dependence as

$$\frac{X^2}{1-X} = \begin{cases} 2.20 \times 10^4 \left(\dfrac{kT}{\mathrm{eV}} \right)^{3/2} e^{-13.6\,\mathrm{eV}/kT} & \text{bottom} \\[3mm] 2.11 \times 10^6 \left(\dfrac{kT}{\mathrm{eV}} \right)^{3/2} e^{-13.6\,\mathrm{eV}/kT} & \text{top} \end{cases}$$

Solving for X yields the curves shown in Fig. 14.4. At the top of the photosphere, the temperature would need to reach 10^4 K or more for the ionization fraction to become substantial. At the bottom of the photosphere, an even higher temperature would be required because the higher density makes it easier for electrons and ions to recombine into neutral atoms, which reduces the equilibrium ionization fraction at a given temperature.

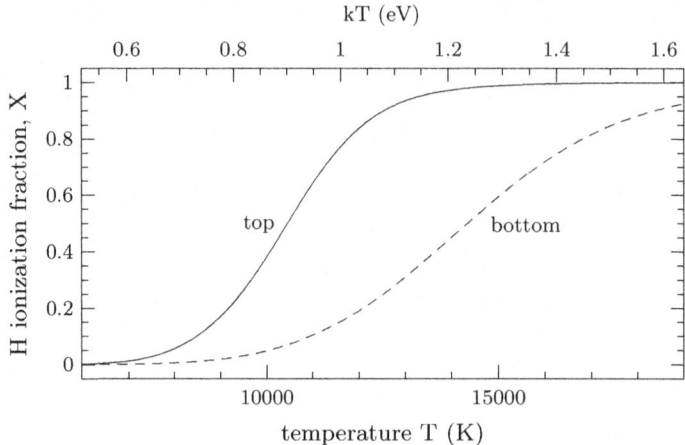

Fig. 14.4 Ionization fraction X for hydrogen, as a function of temperature, assuming the pressure at the top (*solid*) or bottom (*dashed*) of the Sun's photosphere

We can combine the Boltzmann and Saha analyses to determine the fraction of hydrogen atoms that are neutral and in the first excited state. If we assume that very few neutral atoms are excited beyond the first excited state, we can use $n_I \approx n_1 + n_2$ to write

$$\frac{n_2}{n_{\text{tot}}} = \frac{n_2}{n_I} \frac{n_I}{n_{\text{tot}}} \approx \frac{n_2}{n_1 + n_2} \left(1 - \frac{n_{II}}{n_{\text{tot}}}\right) \approx \frac{n_2/n_1}{1 + n_2/n_1} (1 - X)$$

The first factor involves n_2/n_1 from Fig. 14.3 while the second factor has X from Fig. 14.4. Putting them together yields Fig. 14.5, where we focus on the top of the photosphere because more absorption lines are produced there than at the bottom. For the assumed mass density, the excitation fraction n_2/n_{tot} peaks for a photospheric temperature of around 12,000 K. We need to be a little careful when interpreting this result because the mass density would not necessarily remain fixed as the temperature varies, and the density we have assumed for the Sun would not necessarily apply to stars with different masses. Nevertheless, our simplified analysis suggests that hydrogen Balmer lines will be most prominent in stars with photospheric temperatures in the ballpark of 10,000–15,000 K, which is in fact what we see. And we understand why: at cooler temperatures, too little hydrogen is excited into the state that can produce Balmer absorption lines; while at hotter temperatures, too much of the hydrogen is ionized.

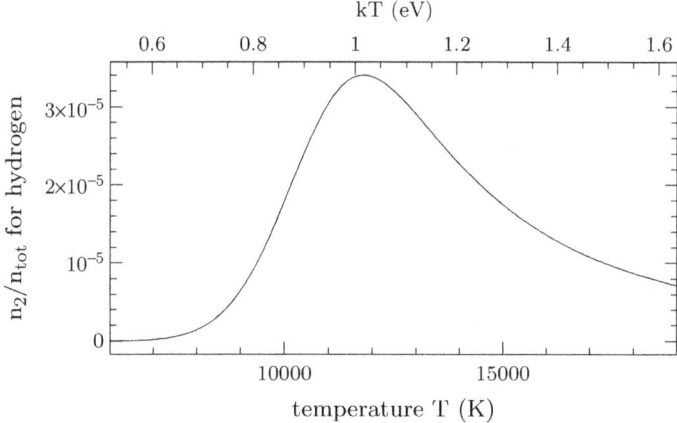

Fig. 14.5 Expected fraction of hydrogen atoms that are neutral and in the first excited state, as a function of temperature. We use the mass density at the top of the Sun's photosphere

14.2 Stellar Spectral Classification

Now we understand that the appearance of spectral lines is governed not only by the composition of stellar atmospheres, which is to be expected, but also by the temperature, which is less obvious but no less important. Using the physics of atomic excitation and ionization, we can comprehend the key patterns observed in stellar spectra.

In the 1890s, astronomers at Harvard College Observatory amassed a large collection of stellar spectra. The Observatory employed many women (who were known at the time as "computers") to analyze astronomical data. Williamina Fleming developed a taxonomy in which spectra were classified based on the strength of their hydrogen absorption lines. A stars had the strongest lines, B stars the next strongest, and so forth. Annie Jump Cannon then consolidated the spectral classes and discovered that the order "O B A F G K M" corresponds to a sequence in *temperature*, running from hot to cool.[3] Figures 14.1 and 14.2 show that the peak of the spectrum shifts with temperature (Wien's law) and the set of absorption lines varies as listed in Table 14.2.[4]

One additional pattern in star properties was discovered in the early 1900s by Ejnar Hertzsprung and Henry Norris Russell. Working independently, they compared the spectral classes and luminosities of different stars in a plot now called the **Hertzsprung-Russell (HR) diagram** (see Fig. 14.6). Stars do not scatter across

---

[3]People have invented a variety of mnemonics to remember the sequence. What's yours?

[4]Recently the spectral sequence has been extended to include types L, T, and Y for low-mass stars known as brown dwarfs whose cores are not hot enough for normal hydrogen fusion to occur (see Problem 16.5).

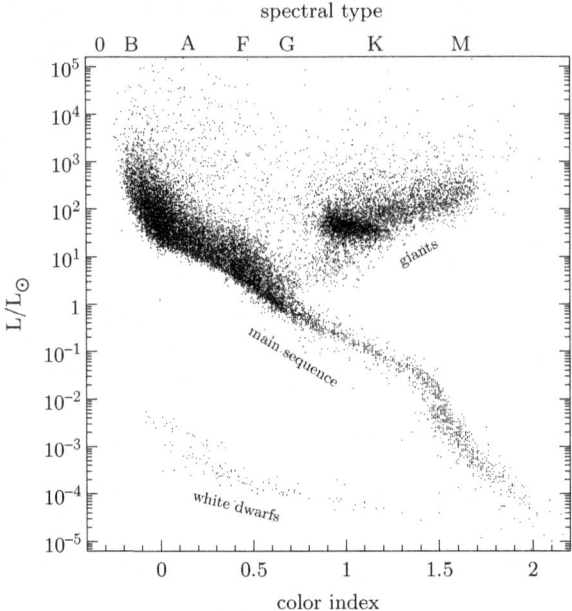

Fig. 14.6 Hertzsprung-Russell (HR) diagram, based on data from the Hipparcos and Tycho catalogs [2] and the Gliese-Jahreiss catalog [3]. The horizontal axis is color running from *blue* on the left to *red* on the right; the bottom axis indicates an index astronomers use to quantify color, while the top axis indicates the corresponding spectral type [4]. The vertical axis is luminosity relative to the Sun (measured in the *V* filter)

Table 14.2 Typical temperature ranges and spectral lines for different types of stars.

Type	T (K)	Spectral lines
O	>25,000	Neutral and ionized He lines
B	11,000–25,000	Neutral He lines, some H
A	7,500–11,000	Strong H lines; some ionized metal lines (Ca II, Mg II)
F	6,000–7,500	Weaker H lines; ionized metal lines
G	5,000–6,000	Ionized and neutral metal lines
K	3,500–5,000	Strong metal lines
M	<3,500	Molecular lines (TiO)

the entire plot but rather fall into three main groupings. The most prominent one runs from the upper left (hot, bright stars) to the lower right (cool, dim stars); this is known as the **main sequence**.

Another grouping of stars is seen in the upper right portion of the HR diagram. What can we say about these stars, in comparison with main sequence stars? Stars of a given spectral type have the same temperature, so how can they have such different luminosities? Going back to the Stefan-Boltzmann law (Eq. 13.1), recall that luminosity and (effective) temperature are related by

$$L = 4\pi R^2 \sigma T_{\text{eff}}^4$$

Consider the ratio of luminosities for two stars:

$$\frac{L_2}{L_1} = \frac{4\pi R_2^2 \sigma T_2^4}{4\pi R_1^2 \sigma T_1^4} = \frac{R_2^2}{R_1^2} \frac{T_2^4}{T_1^4}$$

If the stars have the same temperature, the only way they can have different luminosities is to have different sizes:

$$\frac{R_2}{R_1} = \left(\frac{L_2}{L_1}\right)^{1/2}$$

Compared to main sequence stars, stars in the upper right are more luminous, so they must be larger. This is the region of **giant stars**. Most of these stars have K and M spectral types, which correspond to cool temperatures and red colors, so they are **red giants**. Red giants can be 100–10,000 times brighter than main sequence stars with similar temperatures, so they must be 10–100 times bigger.

Finally, a third grouping is seen in the lower left of the HR diagram. They are less luminous than main sequence stars of the same spectral type, so they must be smaller. Because these stars are generally hot and white, they are called **white dwarfs**.

One of our goals for the coming chapters is to understand what physical processes create the patterns seen in the HR diagram.

Problems

14.1. In the text we analyzed hydrogen to understand the presence of Balmer lines in stellar spectra. Now consider helium. The presence of two electrons allows more configurations; here are the statistical weights and energies of the first five energy levels of neutral helium:

n	g_n	E_n (eV)
1	1	−24.59
2	3	−4.77
3	1	−3.97
4	9	−3.63
5	3	−3.37

Helium has three ionization stages: I is neutral, II is singly-ionized (one electron removed), and III is doubly-ionized (both electrons removed). The ionization energy to go from I to II is $\chi_I = 24.6$ eV, while to go from II to III it is $\chi_{II} = 54.4$ eV. The partition functions for the three stages are $Z_I = 1$, $Z_{II} = 2$, and $Z_{III} = 1$.

(a) Neutral helium is observed to have absorption lines at wavelengths of 447.1, 438.7, and 402.6 nm (to name a few). Do these absorption lines correspond to transitions between any of the energy levels listed above? Explain.

(b) These absorption lines correspond to transitions in which an electron jumps from level 4 to a higher level. This can happen only if atoms are excited to level $n = 4$. Use the Boltzmann equation to compute N_4/N_1 for stars with surface temperature $T = 25,000$ K (an O star), $T = 14,000$ K (a B star), and $T = 9,000$ K (an A star).

(c) Why have we focused on transitions starting from level 4, rather than level 3? Compute N_4/N_3 for our three sample stars (O, B, and A). Explain why level 4 has a higher occupation despite having a higher energy.

(d) Now compute the helium ionization fraction for our three sample stars. Since stars have more hydrogen than helium, assume the factor of n_e is dominated by electrons from ionized hydrogen, drawing on our analysis in Sect. 14.1.3.

(e) Using your results from parts (b) and (d), explain whether you would expect to see neutral helium lines, ionized helium lines, or no helium lines at all in O, B, and A stars.

14.2. As we will see later (Sect. 16.3), a planetary nebula is an expanding shell of gas expelled by a low-mass star near the end of its life. Planetary nebulae are observed to have a prominent emission line from oxygen[5] at 5,007 Å, which is produced when an excited atom decays to a lower energy state. What can you say about the temperature of the gas from knowing that a significant fraction of the oxygen atoms are collisionally excited? You may assume the lower and higher energy states have the same statistical weight.

14.3. In Problem 12.6 we modeled an object with a uniform density of hydrogen gas in hydrostatic equilibrium. We assumed the gas was fully ionized; now we can check that assumption.

(a) Apply the model to a planet with the same size and mass as Jupiter, and compute the ionization fraction at the center.

(b) Apply the model to a star with the same size and mass as the Sun, and compute the ionization fraction at the center.

(c) For the star model, plot the ionization fraction as a function of radius.

14.4. We will study the degenerate gas in white dwarf stars later (Chap. 17). For now, let's consider the possibility that the degenerate core could be surrounded by a thin non-degenerate atmosphere. The spectrum of a white dwarf called EG157 indicates a surface temperature of about 30,000 K and also shows absorption lines due to neutral hydrogen.

---

[5]In fact, doubly-ionized oxygen, although that is not important for this problem.

(a) In Sect. 14.1.3 we found that hydrogen in the Sun's atmosphere ionizes at around 10,000 K. Explain conceptually how EG157's hydrogen atmosphere can be predominantly neutral despite being so hot.
(b) What does the fact that the hydrogen is predominantly neutral imply for the physical conditions in the star's atmosphere? State your answer in terms of a lower or upper bound on some interesting property of the gas.

14.5. The star Betelgeuse (in the constellation Orion) is a red supergiant star with a surface temperature of around $T = 3,450$ K, a luminosity of about $L = 55,000 L_\odot$, and a mass of about $M = 19 M_\odot$. Compute its radius and mean mass density.

References

1. A.J. Pickles, Publ. Astron. Soc. Pac. **110**, 863 (1998)
2. M.A.C. Perryman, ESA (eds.), *The HIPPARCOS and TYCHO Catalogues. Astrometric and Photometric Star Catalogues Derived from the ESA HIPPARCOS Space Astrometry Mission*, vol. 1200 (ESA Special Publication, Noordwijk, 1997)
3. W. Gliese, H. Jahreiss (eds.), *Preliminary Version of the Third Catalogue of Nearby Stars* (Astronomisches Rechen-Institut, Heidelberg, 1991)
4. M.P. Fitzgerald, Astron. Astrophys. **4**, 234 (1970)

Chapter 15
Nuclear Fusion

To this point we have viewed star temperatures and luminosities as empirical quantities, but now we seek to understand them in terms of physical processes that occur deep within stars. Nuclear fusion provides a potent power source, and studying fusion reveals a link between properties of astrophysical objects and reactions that occurs on scales more than 20 orders of magnitude smaller.

15.1 What Powers the Sun?

Before delving into the physics of fusion, it is worthwhile to consider why that is the only source of energy that could power the Sun. At stake is the total energy emitted by the Sun during its lifetime. The Sun's present luminosity is $3.84 \times 10^{26}\,\mathrm{J\,s^{-1}}$, and the age of the Solar System inferred from radiometric dating of rocks and meteorites is about $4.5\,\mathrm{Gyr} \approx 1.4 \times 10^{17}\,\mathrm{s}$. Even if the Sun was a little dimmer in the past (see Sect. 16.3.1), we can safely estimate that the Sun has already released more than $10^{43}\,\mathrm{J}$ in light. What is the source of that energy?

Gravitational energy? The gravitational potential energy of an object of mass M and radius R is $U \sim -GM^2/R$ (see Sect. 8.2.1). If the object shrinks without losing mass then U decreases (becomes more negative). The process of releasing energy by gravitational contraction is known as the **Kelvin-Helmholtz mechanism**. If the Sun was initially large enough that its gravitational potential energy was roughly zero, the amount of gravitational energy it could have released by now is

$$\Delta U \sim \frac{GM_\odot^2}{R_\odot} \sim 4 \times 10^{41}\,\mathrm{J}$$

There is a dimensionless factor of order unity that depends on how the density changes with radius, and an additional factor of $1/2$ because, according to the virial theorem (see Sect. 8.1.3) only half of the energy can be radiated (the other half

C. Keeton, *Principles of Astrophysics: Using Gravity and Stellar Physics to Explore the Cosmos*, Undergraduate Lecture Notes in Physics, DOI 10.1007/978-1-4614-9236-8_15, © Springer Science+Business Media New York 2014

goes into kinetic energy). Thus, as an order-of-magnitude estimate we can say that something like $E_{tot} \sim 10^{41}$ J could have been released by gravitational collapse. At its current luminosity the Sun would radiate this amount of energy in a time

$$t \sim \frac{E_{tot}}{L_\odot} \sim \frac{10^{41} \text{ J}}{3.84 \times 10^{26} \text{ J s}^{-1}} \sim 3 \times 10^{14} \text{ s} \sim 8 \times 10^6 \text{ yr}$$

This is far too short compared with the age of the Solar System. The Sun cannot be powered by gravitational energy alone.

Chemical energy? Chemical reactions can release energy by rearranging electrons in atoms and molecules. The energy scale is set by the difference between electron energy levels. If we optimistically assume that each atom can release $E_1 = 10$ eV, the energy available from all atoms is $E_{tot} = NE_1$. If we assume the Sun is pure hydrogen, the total number of atoms is $N \sim M_\odot/m_p \sim 1.2 \times 10^{57}$. Then

$$E_{tot} \sim 1.2 \times 10^{58} \text{ eV} \times \frac{1.60 \times 10^{-19} \text{ J}}{1 \text{ eV}} \sim 1.9 \times 10^{39} \text{ J}$$

With this total energy, the Sun could shine at its current rate for

$$t \sim \frac{E_{tot}}{L_\odot} \sim \frac{1.9 \times 10^{39} \text{ J}}{3.84 \times 10^{26} \text{ J s}^{-1}} \sim 5.0 \times 10^{12} \text{ s} \sim 160,000 \text{ yr}$$

The Sun cannot be powered by chemical energy, either.

Nuclear energy? Nuclear reactions involve energies in the range of MeV, and thus provide something like a million times more energy than chemical reactions. Specifically, nuclear fusion involves the conversion of mass into energy via $E = mc^2$. If the entire mass of the Sun were converted to energy in this way, the total energy released would be

$$E_{tot} \sim M_\odot c^2 \sim (1.99 \times 10^{30} \text{ kg}) \times (3.0 \times 10^8 \text{ m s}^{-1})^2 \sim 2 \times 10^{47} \text{ J}$$

which would correspond to a lifetime of

$$t \sim \frac{2 \times 10^{47} \text{ J}}{3.84 \times 10^{26} \text{ J s}^{-1}} \sim 5 \times 10^{20} \text{ s} \sim 10^{13} \text{ yr}$$

In practice, the total energy available from fusion is less than this, because only a small fraction of mass is converted into energy in each reaction, and only a portion of matter in the Sun can undergo fusion. (We quantify these fractions below.) Even so, nuclear fusion provides ample energy to power the Sun.

15.2 Physics of Fusion

Fusion occurs when two lighter nuclei combine to create a heavier nucleus. The process can release energy because the mass of the final nucleus may be less than the combined mass of the starting nuclei; the "missing" mass gets converted to energy (such that the total mass/energy is conserved). In this section we study the physics of fusion to understand the conditions under which fusion can occur and the rate at which energy is released.[1]

15.2.1 Mass and Energy Scales

To discuss fusion we need to be more precise than we have been about the masses of particles and nuclei. Given the focus on energy, it is common to quote masses in terms of equivalent energies using $m = E/c^2$; the corresponding unit is MeV/c^2. The masses of the three familiar fundamental particles are:

Proton	$m_p = 938.272\,\text{MeV}/c^2$
Neutron	$m_n = 939.565\,\text{MeV}/c^2$
Electron	$m_e = 0.511\,\text{MeV}/c^2$

We will see in Sect. 15.3 that the reaction powering the Sun is the fusion of 4 hydrogen nuclei into a helium nucleus. The mass involved are[2]

$$4\ \text{hydrogen}\quad 4m_\text{H} = 3{,}753.09\,\text{MeV}/c^2$$

$$\text{helium}\quad m_\text{He} = 3{,}727.38\,\text{MeV}/c^2$$

The amount of mass that gets converted to energy is[3]

$$\Delta m = 25.71\,\text{MeV}/c^2$$

The fraction of the original mass that goes into energy is sometimes called the **efficiency** of a nuclear reaction,

$$\epsilon = \frac{m_\text{start} - m_\text{end}}{m_\text{start}} = \frac{\Delta m}{m_\text{start}}$$

---

[1] Parts of this presentation draw on the books by Carroll and Ostlie [1] and Maoz [2].

[2] These are masses of nuclei; they do not include electrons. In the core of the Sun where fusion occurs, atoms are ionized (see Sect. 16.2.2).

[3] There may be a little additional energy released when electrons annihilate with anti-electrons (see Sect. 15.3.3), but we are focusing on nuclear masses.

With this definition we can write the energy released when some mass M undergoes fusion as

$$E = \epsilon M c^2$$

For hydrogen fusing into helium, the efficiency is $\epsilon = 0.007$.

How much fusion energy is available in the Sun? A first estimate is:

$$E \sim 0.007 \times M_\odot \, c^2 \sim 1.3 \times 10^{45} \, \text{J}$$

This could power the Sun for a lifetime of

$$t \sim \frac{1.3 \times 10^{45} \, \text{J}}{3.84 \times 10^{26} \, \text{J s}^{-1}} \sim 3.3 \times 10^{18} \, \text{s} \sim 10^{11} \, \text{yr}$$

In practice, only about 10 % of the Sun's mass can undergo fusion, so the actual energy and lifetime are about a factor of 10 smaller.

15.2.2   Requirements for Fusion

Why does only a portion of the Sun's mass ever participate in fusion? Fusion requires high temperatures and densities that occur only in the core of the Sun. To understand why, note that nuclei have positive charges and thus repel each other through the Coulomb force. Fusion can occur only if the nuclei get close enough for the **strong nuclear force** to take over and create an attractive force that binds nuclear particles together (see Fig. 15.1). The strong force operates over scales of femtometers (1 fm $= 10^{-15}$ m), so we need to consider the conditions under which nuclei are able to overcome Coulomb repulsion and get close enough for the strong force to come into play.

Classical Analysis

The center of the Sun is hot, so nuclei are zipping around and bumping into one another. Is this enough to overcome the **Coulomb barrier** between nuclei? Thinking in terms of classical physics, we would say that fusion can occur only if the kinetic energy is above the height of the Coulomb barrier. For a Maxwell-Boltzmann distribution of velocities (see Sect. 12.1.2), the typical kinetic energy is

$$\text{kinetic energy} \sim \frac{3}{2} kT$$

If r_s is the scale on which the strong nuclear force acts, the height of the Coulomb barrier is

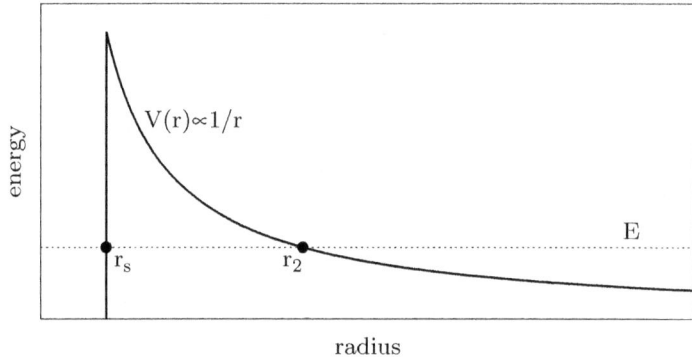

Fig. 15.1 A simple Coulomb barrier between two nuclei. At large separations, Coulomb repulsion leads to a $V(r) \propto 1/r$ potential. At small separations, the strong nuclear force creates a deep potential well. If a particle approaches from the right with an energy E that is lower than the peak, a classical analysis would say the particle can never reach the inner potential well. According to quantum mechanics, however, the particle can tunnel through the barrier. The tunneling probability is analyzed in Sect. 15.2.3

$$\text{Coulomb barrier} \sim \frac{Z_1 Z_2 e^2}{r_s}$$

where Z_1 and Z_2 are the atomic numbers of the two nuclei. In order for the kinetic energy to exceed the Coulomb barrier, the temperature must be higher than

$$T_{\text{fusion}} \sim \frac{2 Z_1 Z_2 e^2}{3 k r_s} \qquad \text{(classical)}$$

For hydrogen fusion, taking $r_s \sim 1$ fm gives

$$T_{\text{fusion}} \sim \frac{2 \times (1.52 \times 10^{-14}\,\text{kg}^{1/2}\,\text{m}^{3/2}\,\text{s}^{-1})^2}{3 \times (1.38 \times 10^{-23}\,\text{kg m}^2\,\text{s}^{-2}\,\text{K}^{-1}) \times (10^{-15}\,\text{m})} \sim 10^{10}\,\text{K}$$

The center of the Sun is hot (1.6×10^7 K), but not this hot. According to classical physics, fusion should not occur in the Sun.

Quantum Analysis

The picture changes when we consider quantum physics, thanks to **tunneling**. In the language of quantum mechanics, we discuss not particles themselves but rather their wavefunctions, which characterize the probability of finding particles in particular positions (or other quantum states). Approaching a barrier in the potential, a wavefunction need not vanish; it can penetrate the barrier and, if the barrier is not too thick, come out the other side with an amplitude that is smaller but still finite. This corresponds to a finite probability for the particle to "jump through" the barrier.

Tunneling means that particles do not necessarily have to have enough energy to go over the Coulomb barrier; they just need enough energy to tunnel through it. To make a toy model of this effect, recall wave/particle duality. We have already seen that wavelength and momentum are related for light through the relation $\lambda = h/p$ (see Sect. 13.1.4). Louis de Broglie suggested that the same idea applies to massive particles like electrons and nuclei. We imagine, then, that bringing two nuclei within a de Broglie wavelength of one another will allow them to tunnel through the Coulomb barrier and fuse (we will be more careful about the criterion below). Using $p = h/\lambda$, we can write the kinetic energy as

$$\text{kinetic energy} = \frac{1}{2}\mu v^2 = \frac{p^2}{2\mu} = \frac{h^2}{2\mu\lambda^2}$$

where $\mu = m_1 m_2/(m_1 + m_2)$ is the reduced mass of the system.[4] The Coulomb barrier has the same form as before, but now we take the separation to be λ:

$$\text{Coulomb barrier} \sim \frac{Z_1 Z_2 e^2}{\lambda}$$

The kinetic energy matches the Coulomb barrier for

$$\lambda \sim \frac{h^2}{2Z_1 Z_2 e^2 \mu}$$

Again equating the kinetic energy to $(3/2)kT$ gives

$$T_{\text{fusion}} \sim \frac{h^2}{3\mu k \lambda^2} \sim \frac{4Z_1^2 Z_2^2 e^4 \mu}{3k h^2} \qquad \text{(quantum)}$$

Plugging in numbers for hydrogen (with reduced mass $\mu = m_p/2$) yields

$$T_{\text{fusion}} \sim \frac{4 \times (1.52 \times 10^{-14}\,\text{kg}^{1/2}\,\text{m}^{3/2}\,\text{s}^{-1})^4 \times (8.36 \times 10^{-28}\,\text{kg})}{3 \times (1.38 \times 10^{-23}\,\text{kg m}^2\,\text{s}^{-2}\,\text{K}^{-1}) \times (6.626 \times 10^{-34}\,\text{kg m}^2\,\text{s}^{-1})^2} \sim 10^7\,\text{K}$$

This is quite close to the central temperature of the Sun. While our toy model should not be taken too literally, it does suggest that tunneling enables fusion in the Sun and therefore deserves a more careful treatment.

----

[4]We can work in the center of mass frame and convert the two-body Coulomb problem into an equivalent one-body problem, as we did with gravity in Sect. 4.1. The relevant mass is then the reduced mass.

15.2.3 Cross Section

The preceding analysis gives a general sense of the conditions required for fusion to be possible. Now let's get more specific. In Chap. 12 we saw that it is useful to discuss interactions between particles in terms of the **cross section**. Previously we pictured billiard ball collisions and took the cross section to be the physical size of the objects, but now we introduce a more general definition:

$$\sigma(E) = \frac{\text{number of reactions}/\text{nucleus}/\text{time}}{\text{number of incident particles}/\text{area}/\text{time}} = [L^2] \tag{15.1}$$

We write $\sigma(E)$ for nuclear reactions because the ability to cross the Coulomb barrier depends on energy. While the full energy dependence may be quite complicated, there are some key factors that we can identify.

First, the "size" of a nucleus is either its physical size or its de Broglie wavelength, whichever is larger. To estimate the de Broglie wavelength, we take the momentum to be $p \sim \mu v_{\text{rms}}$ where μ is the reduced mass and $v_{\text{rms}} = \sqrt{3kT/\mu}$ is the typical particle speed (see Eq. 12.6). At the center of the Sun, $T = 1.6 \times 10^7$ K (see Sect. 16.2.2) and hence $kT = 2.2 \times 10^{-16}$ J. Using $\mu = m_p/2$ for hydrogen then yields

$$\lambda \sim \frac{h}{(3\mu kT)^{1/2}}$$

$$\sim \frac{(6.626 \times 10^{-34}\,\text{kg m}^2\,\text{s}^{-1})}{[3 \times (8.36 \times 10^{-28}\,\text{kg}) \times (2.2 \times 10^{-16}\,\text{kg m}^2\,\text{s}^{-2})]^{1/2}}$$

$$\sim 10^{-12}\,\text{m}$$

This is much larger than the physical size of a nucleus ($\sim 10^{-15}$ m), so we can approximate the effective size of a nucleus as λ. Therefore we expect the cross section to scale as

$$\sigma(E) \propto \lambda^2 \propto \left(\frac{h}{\mu v}\right)^2 \propto \frac{1}{E} \tag{15.2}$$

Second, we need to account for the tunneling probability. Below we derive the probability for tunneling through a simple, fixed Coulomb barrier and show that it has the form

$$P = e^{-(E_c/E)^{1/2}} \tag{15.3}$$

where the energy scale E_c is defined by

$$E_c \equiv \frac{2\pi^2 Z_1^2 Z_2^2 e^4 \mu}{\hbar^2} \tag{15.4}$$

Even if the real interaction is more complicated than we have considered here, the two factors we have identified in Eqs. (15.2) and (15.3) should capture the strongest energy dependence. We therefore write

$$\sigma(E) = \frac{S(E)}{E} \, e^{-(E_c/E)^{1/2}} \tag{15.5}$$

and bundle any remaining effects into $S(E)$, which is called the **nuclear S-factor**. In principle, $S(E)$ needs to be determined for each reaction (mainly from experimental data; e.g., [3,4]). However, in the next section we will see that knowing the details of $S(E)$ is not essential for a general understanding of fusion (although it is vital for detailed investigations, of course).

Tunneling Probability

Quantum tunneling through a barrier that is wide compared with the de Broglie wavelength of a particle can be analyzed in the standard WKB (Wentzel-Kramers-Brillouin) approximation of quantum mechanics. (For more details, see Chap. 8 of *Introduction to Quantum Mechanics* by Griffiths [5] or a similar textbook.) Consider a barrier with potential $V(r)$. Conservation of energy gives the momentum:

$$E = \frac{p^2}{2\mu} + V(r) \quad \Rightarrow \quad p(r) = \{2\mu[E - V(r)]\}^{1/2}$$

where μ is the reduced mass of the system. If the energy E is below the peak of V, then $p(r)$ is imaginary within the barrier, but that is not a problem in a quantum analysis. The WKB approximation gives the probability that the particle can tunnel through the barrier as P such that

$$\ln P \approx -\frac{2}{\hbar} \int_{r_s}^{r_2} |p(r)| \, dr$$

where r_s and r_2 are indicated in Fig. 15.1. With a simple Coulomb barrier

$$V(r) = \frac{Z_1 Z_2 e^2}{r} \quad (r > r_s)$$

we can write the outer radius as

$$r_2 = \frac{Z_1 Z_2 e^2}{E} \tag{15.6}$$

The integral can then be evaluated as follows:

$$\ln P \approx -\frac{2}{\hbar}(2\mu)^{1/2} \int_{r_s}^{r_2} \left(\frac{Z_1 Z_2 e^2}{r} - E \right)^{1/2} dr$$

$$\approx -\frac{2}{\hbar} \left(2Z_1 Z_2 e^2 \mu\right)^{1/2} \int_{r_s}^{r_2} \left(\frac{1}{r} - \frac{1}{r_2}\right)^{1/2} dr$$

where we use Eq. (15.6) to replace E with r_2. In the limit $r_2 \gg r_s$, the integral evaluates to $(\pi/2) r_2^{1/2}$, yielding

$$\ln P \approx -\frac{\pi}{\hbar} \left(2Z_1 Z_2 e^2 \mu r_2\right)^{1/2} \approx -\frac{\pi Z_1 Z_2 e^2}{\hbar} \left(\frac{2\mu}{E}\right)^{1/2}$$

where we now use Eq. (15.6) to convert back to E. We can rewrite the result as

$$\ln P \approx -\left(\frac{E_c}{E}\right)^{1/2} \quad \text{where} \quad E_c = \frac{2\pi^2 Z_1^2 Z_2^2 e^4 \mu}{\hbar^2} \tag{15.7}$$

15.2.4 Reaction Rate

The cross section is one important factor in determining how many reactions occur. Another is the sheer number of nuclei that have the required energy. To figure out how to compute the reaction rate, let's go back to the definition of cross section (Eq. 15.1). The particles that can interact with a target nucleus in some time dt are those in a cylinder whose cross sectional area is σ and length is $v\,dt$. The number of reactions per nucleus in time dt can therefore be written as

$$\# \text{ reactions per nucleus} = \int n_2(v)\, \sigma\, v\, dt\, dv$$

where $n_2(v)\, dv$ is the number density of particles with speeds between v and $v + dv$.

Consider a volume V that contains N_1 target nuclei. To get the total number of reactions for all the nuclei, multiply the previous expression by N_1. To get the number of reactions per unit volume, divide by V. Finally, to get the number of reactions per unit volume per unit time, divide by dt. The result is the **reaction rate**,

$$r_{12} \equiv \text{number of reactions per unit volume, per unit time} = n_1 \int n_2(v)\, \sigma\, v\, dv$$

where $n_1 = N_1/V$ is the number density of nuclei. Using the Maxwell-Boltzmann distribution for $n_2(v)$ gives[5]

$$r_{12} = n_1 n_2 \left(\frac{\mu}{2\pi k T}\right)^{3/2} 4\pi \int_0^\infty v^3\, e^{-\mu v^2/2kT}\, \sigma\, dv$$

---

[5] Now n_2 indicates the total number density of reactants. We could write it as $n_{2,\text{tot}}$ following Eq. (12.3), but we omit "tot" to simplify the notation.

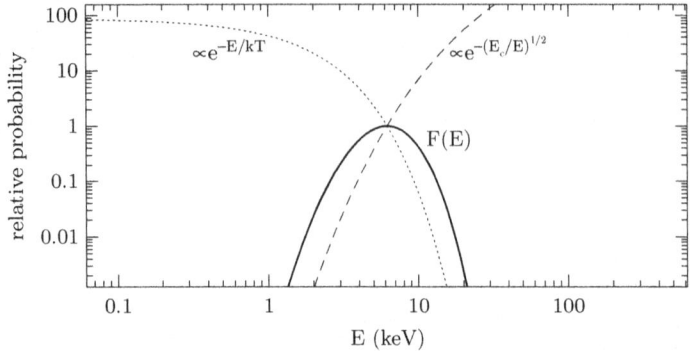

Fig. 15.2 The *dotted curve* shows the Boltzmann factor as a function of energy, while the *dashed curve* shows the tunneling probability factor for hydrogen in the Sun. The *solid curve* shows their product, which reveals the Gamow peak. For clarity, all curves are scaled to pass through 1 at E_0

Changing integration variables to $E = \mu v^2 / 2$ transforms the integral to

$$
r_{12} = n_1 n_2 \left(\frac{\mu}{2\pi kT} \right)^{3/2} 4\pi \int_0^\infty \left(\frac{2E}{\mu} \right)^{3/2} e^{-E/kT} \sigma \, \frac{dE}{(2\mu E)^{1/2}}
$$

$$
= \left(\frac{2}{kT} \right)^{3/2} \frac{n_1 n_2}{(\pi \mu)^{1/2}} \int_0^\infty E \, \sigma \, e^{-E/kT} \, dE
$$

$$
= \left(\frac{2}{kT} \right)^{3/2} \frac{n_1 n_2}{(\pi \mu)^{1/2}} \int_0^\infty S(E) \, e^{-(E_c/E)^{1/2}} \, e^{-E/kT} \, dE \qquad (15.8)
$$

In the last step we use Eq. (15.5) to substitute for $\sigma(E)$.

Notice the two exponential factors in the integrand. The Boltzmann factor $e^{-E/kT}$ decreases as energy increases, but the tunneling probability factor $e^{-(E_c/E)^{1/2}}$ increases with energy. The product of the two is a strongly peaked curve, as shown in Fig. 15.2. In Problem 15.3 you can show that the **Gamow peak** occurs at energy

$$
E_0 = \left[\frac{1}{4} E_c \, (kT)^2 \right]^{1/3} \qquad (15.9)
$$

Most of the contribution to the integral in Eq. (15.8) comes from energies near E_0. If the S-factor does not vary too rapidly (which is seen to be true for many reactions),[6] we can approximate $S(E)$ by its value at E_0 and pull $S(E_0)$ out of the integral. Then the integral can be evaluated analytically, using the method outlined in Problem 15.4.

---

[6] The exception is reactions in which "resonances" enhance the cross section at certain energies. Addressing nuclear resonances involves more detail than we want to get into here (see [3]).

Since fusion typically involves nuclei in a fairly narrow range of energies, the reaction rate can increase dramatically as the temperature increases and more particles are brought into the fusion energy range. It is convenient to characterize such a rapid temperature dependence as a power law relation,

$$r_{12} \propto T^{\alpha}$$

How do we determine the **power law index**, α? First, let's write the relation as

$$r_{12} = C \, T^{\alpha}$$

where C is a multiplicative factor that does not depend on temperature. Consider the derivative:

$$\frac{dr_{12}}{dT} = \alpha \, C \, T^{\alpha-1} = \frac{\alpha r_{12}}{T}$$

Therefore we can compute the power law index as

$$\alpha = \frac{T}{r_{12}} \frac{dr_{12}}{dT} \tag{15.10}$$

Alternatively, take the logarithm of the original relation:

$$\ln r_{12} = \ln(C T^{\alpha}) = \ln C + \alpha \ln T$$

Then we can find α using

$$\alpha = \frac{d(\ln r_{12})}{d(\ln T)}$$

which is mathematically equivalent to Eq. (15.10). Using either method, you can show in Problem 15.4 that the power law index for the fusion reaction rate has the form

$$\alpha = \left(\frac{E_c}{4kT} \right)^{1/3} - \frac{2}{3} \tag{15.11}$$

In Problem 15.5 you can quantify the temperature dependence of the reaction rates for two channels by which hydrogen fuses into helium.

Application to Hydrogen in the Sun

Using the reduced mass $\mu = m_p/2$, the energy factor E_c from Eq. (15.4) is

$$E_c = \frac{2\pi^2 \times (1.52 \times 10^{-14}\,\text{kg}^{1/2}\,\text{m}^{3/2}\,\text{s}^{-1})^4 \times (8.36 \times 10^{-28}\,\text{kg})}{(1.05 \times 10^{-34}\,\text{kg}\,\text{m}^2\,\text{s}^{-1})^2}$$

$$= 7.9 \times 10^{-14}\,\text{J}$$

$$= 490\,\text{keV}$$

where we convert to keV because (as we will see) this is a convenient unit given the energy scales in the Sun. For $T = 1.6 \times 10^7$ K, the thermal energy factors are

$$kT = (1.38 \times 10^{-23}\,\text{kg}\,\text{m}^2\,\text{s}^{-2}\,\text{K}^{-1}) \times (1.6 \times 10^7\,\text{K}) = 2.2 \times 10^{-16}\,\text{J} = 1.4\,\text{keV}$$

$$E_{\text{rms}} = \frac{3}{2}\,kT = 2.1\,\text{keV}$$

where E_{rms} is the typical kinetic energy in the Maxwell-Boltzmann distribution. The Gamow peak is then

$$E_0 = \left[\frac{1}{4} \times (490\,\text{keV}) \times (1.4\,\text{keV})^2\right]^{1/3} = 6.2\,\text{keV}$$

(This corresponds to a temperature of about 7×10^7 K.) In other words, the typical nuclei involved in fusion have energies about three times the RMS value. While such nuclei are not right at the peak of the Maxwell-Boltzmann distribution, they are still sufficiently abundant to make fusion work as an energy source.

15.3   Nuclear Reactions in Stars

Having discussed the general theory of fusion, we can now consider the specific reactions that occur in stars.

15.3.1   Cast of Characters

The first step is to list all the particles that might participate in nuclear reactions. The elements in the periodic table are composed of protons, neutrons, and electrons. All three particles have alternate versions called **antimatter** obtained by reversing the electric charge:

Matter		Antimatter
Electron, e^-	↔	Positron, e^+
Proton, p	↔	Antiproton, $\bar{p}$
Neutron, n	↔	Antineutron, $\bar{n}$

When matter and antimatter come together, they annihilate into energy in the form of photons of light (denoted by γ):

$$e^- + e^+ \rightarrow 2\gamma$$

Two photons are needed so the reaction can conserve momentum and energy simultaneously.

In the 1930s physicists noticed that certain nuclear reactions seemed to violate energy and momentum conservation. Wolfgang Pauli proposed that unseen particles carried the "missing" energy and momentum. They had to be neutral (there was no missing charge), so Enrico Fermi christened them **neutrinos** for "little neutral ones." As we will see in Sect. 15.4, we now understand that there are different types of neutrinos, but for the moment we focus on ones associated with electrons:

Matter		Antimatter
Neutrino, ν_e	$\leftrightarrow$	Antineutrino, $\bar{\nu}_e$

Whenever a nuclear reaction involves an electron or positron, a neutrino or antineutrino is produced as a result of the weak nuclear force.

15.3.2 Masses and Binding Energies

When we discuss atomic nuclei, we label them as

$$^A_Z X$$

where X is the abbreviation for the element in the periodic table, Z is the **atomic number** (the number of protons), and A is the **mass number** (the total number of protons and neutrons). For example, $^4_2$He indicates helium with two protons and two neutrons.

To quantify the energetics of fusion, we define the **binding energy** to be the difference between the actual mass of a nucleus (m_{nuc}) and the combined mass that all the constituent protons and neutrons would have if they were isolated:

$$E_b = \left[Z\, m_p + (A - Z)\, m_n - m_{\text{nuc}} \right] c^2 \tag{15.12}$$

This is the amount of energy that would be released if the nucleus were built from scratch in one step (although nuclei are not actually made that way, as we will see). To compare different nuclei, it can be valuable to consider the binding energy per nucleon, E_b/A. Here are the nuclear masses and binding energies of some low-mass isotopes, along with electrons and neutrons for comparison [6][7]:

---

[7]These are the masses of bare nuclei, not neutral atoms. Recall that we are considering reactions between nuclei in ionized gas.

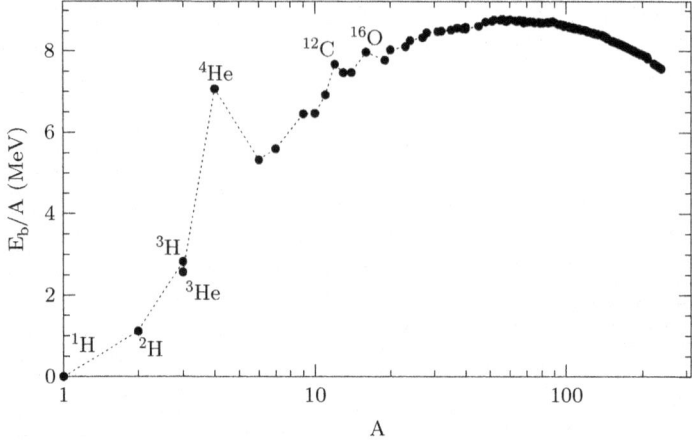

Fig. 15.3 Binding energy per nucleon, as a function of the mass number, A. Some important isotopes are identified (Data from [6])

	m (MeV/c^2)	E_b (MeV)	E_b/A (MeV)
e^-	0.511		
n	939.565		
p/${}_1^1$H	938.272		
${}_1^2$H	1,875.613	2.224	1.112
${}_2^3$He	2,808.391	7.718	2.573
${}_2^4$He	3,727.379	28.296	7.074
${}_3^7$Li	6,533.834	39.244	5.606
${}_4^8$Be	7,454.850	56.500	7.062
${}_5^8$B	7,472.319	37.737	4.717
${}_6^{12}$C	11,174.863	92.162	7.680

Notice that helium-4 has a higher binding energy per nucleon than other nearby nuclei. Moving up the periodic table, Fig. 15.3 shows that carbon-12 and oxygen-16 also have higher binding energies per nucleon than adjacent nuclei. That fact makes helium-4, carbon-12, and oxygen-16 unusually stable, which in turn makes them important players in the fusion reactions inside stars, and among the most abundant nuclei in the universe (after hydrogen).

15.3.3   Burning Hydrogen Into Helium

The reaction that powers stars for most of their lives[8] is the fusion of four hydrogen nuclei into helium-4. From theory and experiment, the overall reaction is

---

[8]In Chap. 16 we will examine reactions that occur in the late stages of stellar evolution.

$$4\,{}^1_1\text{H} \rightarrow {}^4_2\text{He} + 2e^+ + 2v_e + 2\gamma \tag{15.13}$$

This does not happen in one step, though. In fact, several sequences of reactions have this net effect.

Proton-Proton (PP) Chain

One sequence of reactions is called the proton-proton chain. There are actually several variants of this chain. The first, called **PP I**, begins with:

$$^1_1\text{H} + {}^1_1\text{H} \rightarrow {}^2_1\text{H} + e^+ + v_e \tag{15.14}$$

$$^2_1\text{H} + {}^1_1\text{H} \rightarrow {}^3_2\text{He} + \gamma \tag{15.15}$$

After this pair of reactions occurs twice, there are two helium-3 nuclei that can fuse via

$$^3_2\text{He} + {}^3_2\text{He} \rightarrow {}^4_2\text{He} + 2\,{}^1_1\text{H}$$

The first pair of steps uses three hydrogen nuclei; they must occur twice, consuming a total of six hydrogens. The final step produces two new hydrogens, though, so the net effect is that four hydrogens are consumed to create one helium-4 nucleus.

Note that the first reaction requires one proton to be converted into a neutron,[9]

$$p \rightarrow n + e^+ + v_e$$

There is something odd here: the neutron weighs *more* than the proton, so the right-hand side has more mass/energy (940.076 MeV, plus the neutrino energy) than the left-hand side (938.272 MeV). In other words, at least 1.8 MeV of energy is absorbed by this reaction. That is okay, though, because the fusion of a proton and neutron into deuterium (^2_1H) releases more than enough energy to compensate.

Once some helium-4 exists,[10] there is a chance that the helium-3 produced in the second step above will react with helium-4,

$$^3_2\text{He} + {}^4_2\text{He} \rightarrow {}^7_4\text{Be} + \gamma$$

After this, there are two possibilities. One branch called **PP II** involves:

$$^7_4\text{Be} + e^- \rightarrow {}^7_3\text{Li} + v_e \tag{15.16}$$

$$^7_3\text{Li} + {}^1_1\text{H} \rightarrow 2\,{}^4_2\text{He}$$

---

[9] This is a form of beta decay, which is different from hypothetical spontaneous proton decay.

[10] Helium-4 is produced in the big bang (see Chap. 20), so it can be present in stars even before PP I occurs.

The alternative called **PP III** has:

$$
\begin{aligned}
{}^{7}_{4}\text{Be} + {}^{1}_{1}\text{H} &\rightarrow {}^{8}_{5}\text{B} + \gamma \\
{}^{8}_{5}\text{B} &\rightarrow {}^{8}_{4}\text{Be} + e^{+} + \nu_e \\
{}^{8}_{4}\text{Be} &\rightarrow 2\,{}^{4}_{2}\text{He}
\end{aligned}
\tag{15.17}
$$

In principle, helium-3 can go straight to helium-4 through the reaction

$$
{}^{3}_{2}\text{He} + {}^{1}_{1}\text{H} \rightarrow {}^{4}_{2}\text{He} + e^{+} + \nu_e
\tag{15.18}
$$

This is known as **PP IV**, or **HeP** because it combines helium (He) with a proton (P). It has not actually been identified in the Sun because the predicted rate is too low.

In addition to Eq. (15.14), there is an alternate way to produce deuterium, which is known as **PEP** because it involves an electron in addition to two protons:

$$
{}^{1}_{1}\text{H} + e^{-} + {}^{1}_{1}\text{H} \rightarrow {}^{2}_{1}\text{H} + \nu_e
\tag{15.19}
$$

This step replaces Eq. (15.14) in about 1/400 of the reactions in the Sun.

Overall, 87.6 % of the reactions in the Sun follow the PP I branch, 10.7 % follow the PP II branch, and 0.9 % follow the PP III branch. (Only 0.8 % of reactions follow the alternate sequence known as the CNO cycle.) [7]

CNO Cycle

Another sequence of reactions for converting hydrogen into helium uses carbon, nitrogen, and oxygen as catalysts, and hence is called the **CNO cycle**. The main branch looks like this (**CNO I**):

$$
\begin{aligned}
{}^{12}_{6}\text{C} + {}^{1}_{1}\text{H} &\rightarrow {}^{13}_{7}\text{N} + \gamma \\
{}^{13}_{7}\text{N} &\rightarrow {}^{13}_{6}\text{C} + e^{+} + \nu_e \\
{}^{13}_{6}\text{C} + {}^{1}_{1}\text{H} &\rightarrow {}^{14}_{7}\text{N} + \gamma \\
{}^{14}_{7}\text{N} + {}^{1}_{1}\text{H} &\rightarrow {}^{15}_{8}\text{O} + \gamma \\
{}^{15}_{8}\text{O} &\rightarrow {}^{15}_{7}\text{N} + e^{+} + \nu_e \\
{}^{15}_{7}\text{N} + {}^{1}_{1}\text{H} &\rightarrow {}^{12}_{6}\text{C} + {}^{4}_{2}\text{He}
\end{aligned}
\tag{15.20}
\tag{15.21}
$$

This representation makes it look like carbon-12 is the start- and end-point, but the sequence is in fact a cycle that can be entered at any point. While various forms of carbon, nitrogen, and oxygen are created and consumed in the cycle, the net effect is the conversion of four hydrogens into one helium-4.

An alternative branch (**CNO II**) replaces the $^{15}_{7}N$ step above with a different series of reactions:

$$^{15}_{7}N + ^{1}_{1}H \rightarrow ^{16}_{8}O + \gamma$$
$$^{16}_{8}O + ^{1}_{1}H \rightarrow ^{17}_{9}F + \gamma$$
$$^{17}_{9}F \rightarrow ^{17}_{8}O + e^{+} + \nu_{e} \qquad (15.22)$$
$$^{17}_{8}O + ^{1}_{1}H \rightarrow ^{14}_{7}N + ^{4}_{2}He$$
$$^{14}_{7}N + ^{1}_{1}H \rightarrow ^{15}_{8}O + \gamma$$
$$^{15}_{8}O \rightarrow ^{15}_{7}N + e^{+} + \nu_{e} \qquad (15.23)$$

Notice that this branch does *not* replace the initial carbon. However, this branch is rare (0.04 %), so the overall carbon destruction rate is small. There are yet other branches of the CNO cycle that occur only in massive stars.

PP or CNO?

Which chain actually powers stars? As you can compute in Problem 15.5, the CNO cycle has a much stronger temperature dependence than the PP chain. The PP chain therefore dominates at low temperatures while the CNO cycle takes over at high temperatures (see Fig. 15.4). The transition temperature corresponds to the core of a main sequence star with a mass of about 1.2 $M_{\odot}$. Along the main sequence, in other words, stars with $M \lesssim 1.2\,M_{\odot}$ are mainly powered by the PP chain, while stars with $M \gtrsim 1.2\,M_{\odot}$ are mainly powered by the CNO cycle.

15.4 Solar Neutrinos

For the most part, we can test the theory of fusion only by building it into stellar models (see Sect. 16.2) and seeing how well the models reproduce properties of stars such as mass, size, and luminosity. With the Sun, however, we have access to the fusion reactions through the neutrinos they produce. These particles interact so weakly that they stream right out of the Sun. Most of them stream through Earth, too, but the few that are caught prove to be very informative.

15.4.1 Neutrino Production in the Sun

John Bahcall and collaborators [8] used a detailed model of the Sun (known as the Standard Solar Model; see Sect. 16.2.2) to predict the rate at which neutrinos

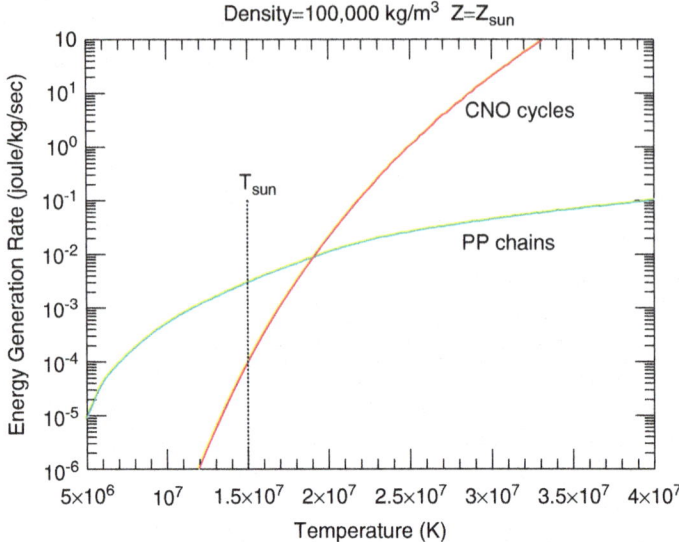

Fig. 15.4 Energy generation rate as a function of temperature for PP chains and CNO cycles, assuming the same chemical composition as the Sun and a density of $10^5\,\mathrm{kg\,m^{-3}}$ (Credit: F. Timmes, ASU)

are produced. Neutrinos hardly notice the matter in the Sun (see Problem 15.7), so the spectrum of neutrino energies at Earth should reflect the intrinsic distribution of energies in the core of the Sun. Figure 15.5 shows that the spectrum for each reaction has a cutoff set by the masses involved in the reaction (see Problem 15.8). This is important because, as we will see, it is easiest to detect neutrinos whose energies are above a few MeV. Most of the solar neutrinos that we detect therefore come from the boron decay reaction in the PP III branch. The boron reaction may be rare in the Sun, but it plays a major role in our understanding of stellar physics.

15.4.2 Neutrino Detection (I)

Figure 15.5 suggests that there are enormous numbers of neutrinos passing through Earth. How can we detect them? One reaction that can be used to "capture" a neutrino is

$$^{37}_{17}\mathrm{Cl} + \nu_e \ \rightarrow \ ^{37}_{18}\mathrm{Ar} + e^- \tag{15.24}$$

This reaction involves the isotope chlorine-37, which constitutes about 25 % of natural chlorine. In the 1960s, Ray Davis assembled about 600 tons of tetra-chlorethylene (cleaning fluid) to serve as the world's first neutrino detector. (See [9] for a contemporaneous article.) The idea was to let the tank sit for a few weeks,

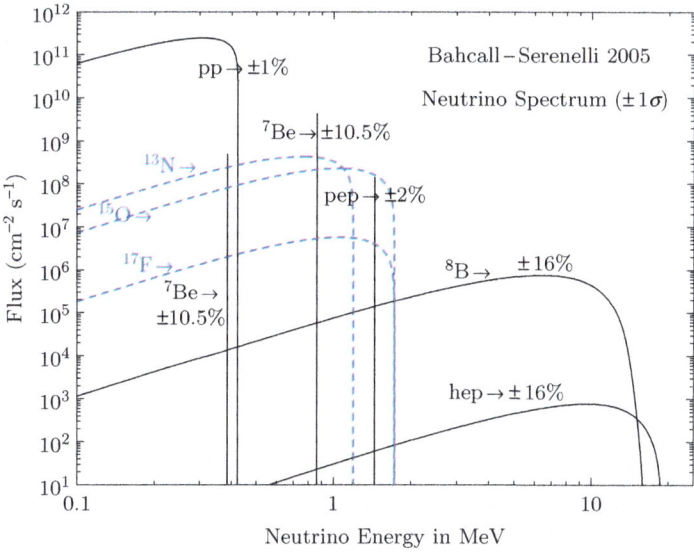

Fig. 15.5 Predicted fluxes (at Earth) of neutrinos from the Sun, as a function of neutrino energy. *Solid lines* correspond to different reactions in the PP chain, while *dashed lines* correspond to different steps in the CNO cycle. (Reactions that produce neutrinos are numbered in Sect. 15.3.3) (Credit: Bahcall et al. [8]. Reproduced by permission of the AAS)

then collect the argon atoms that were produced when chlorine-37 atoms captured neutrinos. It seems quite remarkable that Davis and his team were able to separate a few dozen argon atoms from $\sim 10^{30}$ chlorine atoms!

The experiment took place nearly 1.5 km underground, in the **Homestake Gold Mine** in South Dakota. It was located in a mine so the overlying rock could act as a natural shield against stray particles that might produce argon. Neutrinos pass through rock with ease (in fact, the Homestake experiment collected neutrinos during both day and night, because neutrinos could travel the long way through Earth and enter the bottom of the tank at night), but most other particles are stopped.

Homestake did detect neutrinos from the Sun, confirming our basic picture of fusion in the Sun's core. For this pioneering work, Davis shared the 2002 Nobel Prize in Physics. However, the measured flux of neutrinos was only about 30 % of the flux predicted by the Standard Solar Model. The discrepancy between theory and observations became known as the **solar neutrino problem**.

15.4.3   *Neutrino Oscillations*

The solar neutrino problem had two possible interpretations: either our model of the Sun was wrong, or neutrino physics was richer than anticipated. The second possibility was intriguing because the reactions that produce neutrinos in the Sun and the chlorine reactions used to capture neutrinos at Homestake all involve electrons or anti-electrons. According to particle physics, neutrinos associated with electrons are not the only kind that exist. The electron has two sister particles, known as the muon and the tau particle, and they have their own related neutrinos:

Electron	Muon	Tau
e^-	μ	τ
ν_e	ν_μ	ν_τ

(All six have associated antiparticles.) These particles are collectively known as **leptons**, and the three different classes are known as **flavors**.

The particle physics became relevant for astrophysics because theorists speculated that neutrinos might be able to **oscillate** between flavor states [10]. If so, then electron neutrinos produced in the Sun might transform to mu or tau neutrinos on the way to Earth, thereby becoming invisible to experiments like Homestake and causing the observed neutrino flux to be lower than expected. Such changes could occur only if the three types of neutrinos have different masses, but at the time it was not known whether they have any mass at all. The possibility that neutrino oscillations might solve the solar neutrino problem—and, conversely, that solar neutrinos might reveal new physics—inspired a new generation of experiments to detect other flavors of neutrinos and constrain their masses.

The oscillation process is modified slightly when neutrinos propagate through matter, because the presence of electrons changes the effective mass states. This process, known as the **MSW effect** after Mikheyev and Smirnov [11] and Wolfenstein [12], can affect neutrinos as they leave the Sun and also as they travel through Earth.

15.4.4 Neutrino Detection (II)

In order to test the hypothesis that neutrino oscillations solve the solar neutrino problem, new detectors needed to be sensitive not only to electron neutrinos but also to the mu and tau flavors. The two experiments that played the most significant roles were Super-Kamiokande in Japan and the Sudbury Neutrino Observatory in Canada.

Super-Kamiokande

Super-Kamiokande detects neutrinos when they scatter off electrons in 50,000 tons of water. (There is nothing particularly special about water; it is just a convenient medium that is translucent and feasible to obtain and purify in large quantities.) Formally, the interaction can be written as

$$\nu_x + e^- \rightarrow \nu_x + e^-$$

where ν_x refers to any flavor of neutrino. In detail, electron neutrinos have a higher electron scattering cross section than mu or tau neutrinos, but the difference is known and can be factored into data analysis. Super-Kamiokande focuses on neutrinos with energies above 5 MeV in order to discriminate them from background effects. When such an energetic neutrino scatters, it transfers some of its energy to an electron, causing the electron to reach a speed very close to c that can actually be faster than light moves in water.[11] A faster-than-light electron emits a flash of blue light called **Čerenkov radiation** that can be detected with photomultiplier tubes lining the tank.

Electron scattering does not directly distinguish neutrino flavors, but it still offers ways to look for evidence of oscillations. There may be a daily modulation in the solar neutrino flux: during daytime solar neutrinos come down from above, but at night they must travel through Earth before reaching the detector, which subjects them to the MSW effect. There may also be an annual modulation: as Earth's distance from the Sun varies, there is not only a $1/r^2$ change in flux (which can be accounted for) but also a small change in oscillations because of the varying propagation distance. Super-Kamiokande has analyzed these variations, in conjunction with the total flux and energy spectrum, to constrain neutrino oscillations and provide evidence that the neutrino flux at Earth is in fact consistent with predictions from the Standard Solar Model. (See [13] for a discussion of the methodology and initial results, and [14] for more recent results.)

Super-Kamiokande has also been used to study neutrinos that are produced when cosmic rays strike Earth's atmosphere [15], as well as neutrinos in artificial beams produced at accelerators [16]. The various studies are designed to understand different aspects of neutrino oscillations and implications for neutrino masses.

Sudbury Neutrino Observatory

Sudbury Neutrino Observatory (SNO) was designed not only to detect but also to distinguish between electron, mu, and tau neutrinos. It used 1,000 tons of **heavy water** (D_2O) in order to be sensitive to three different processes:

- **Electron scattering** (ES), as in Super-Kamiokande.
- **Charged current reaction** (CC). An electron neutrino can split a deuterium nucleus into a proton and neutron, and then convert the neutron into a proton and electron:

---

[11]The speed of light in a water is c/n where $n \approx 1.3$ is the index of refraction at visible wavelengths. Electrons can travel faster than light in water without violating the relativistic speed limit $v < c$.

$$\nu_e + {}^2_1\text{H} \rightarrow {}^1_1\text{H} + {}^1_1\text{H} + e^-$$

(This is the inverse of step (15.14) in the PP chain.) Electrons produced in this reaction can also emit Čerenkov radiation.

• **Neutral current reaction** (NC). Any flavor of neutrino can split a deuterium nucleus without transforming the neutron:

$$\nu_x + {}^2_1\text{H} \rightarrow {}^1_1\text{H} + n + \nu_x$$

The neutron can then be captured by another nucleus in a reaction that emits an energetic gamma ray. The gamma ray photon then scatters off an electron, accelerating the electron to the point that it can emit Čerenkov radiation.

Like Super-Kamiokande, SNO had a neutrino energy threshold of 5 MeV, so it was mainly sensitive to neutrinos from boron decay.

In the initial experiment, the three channels led to the following neutrino fluxes (in units of $10^6\,\text{cm}^{-2}\,\text{s}^{-1}$) [17]:

$$\text{CC:} \quad \phi_{\text{CC}} = 1.76^{+0.06}_{-0.05}(\text{stat.})^{+0.09}_{-0.09}(\text{syst.})$$

$$\text{ES:} \quad \phi_{\text{ES}} = 2.39^{+0.24}_{-0.23}(\text{stat.})^{+0.12}_{-0.12}(\text{syst.})$$

$$\text{NC:} \quad \phi_{\text{NC}} = 5.09^{+0.44}_{-0.43}(\text{stat.})^{+0.46}_{-0.43}(\text{syst.})$$

where the two sets of errorbars represent statistical and systematic uncertainties. The fact that the NC flux was higher than the CC and ES fluxes immediately revealed that some of the neutrinos from the Sun are being detected as mu and/or tau neutrinos, i.e., that neutrino oscillation does occur. Decomposing the three channel fluxes into contributions from electron neutrinos or mu/tau neutrinos (combined) yielded

$$\phi_e = 1.76^{+0.05}_{-0.05}(\text{stat.})^{+0.09}_{-0.09}(\text{syst.})$$

$$\phi_{\mu\tau} = 3.41^{+0.45}_{-0.45}(\text{stat.})^{+0.48}_{-0.45}(\text{syst.})$$

There were two key conclusions. First, the total flux of neutrinos agreed with predictions from the Standard Solar Model (5.05 in these units [18]). Second, only about 34 % of the detections involved electron neutrinos. Apparently the original solar neutrino problem arose not because the predictions were incorrect but because the Homestake experiment was unable to detect 2/3 of the solar neutrinos that pass through Earth.

In a second phase of the experiment, SNO added two tons of salt (NaCl) to the heavy water to enhance the sensitivity to NC reactions. (Chlorine is a good target for capturing neutrons released when deuterium is split; there were other technical gains as well.) This experiment yielded neutrino fluxes of [19]

$$\text{CC:} \quad \phi_{\text{CC}} = 1.68^{+0.06}_{-0.06}(\text{stat.})^{+0.08}_{-0.09}(\text{syst.})$$

$$\text{ES:} \quad \phi_{ES} = 2.35^{+0.22}_{-0.22}(\text{stat.})^{+0.15}_{-0.15}(\text{syst.})$$

$$\text{NC:} \quad \phi_{NC} = 4.94^{+0.21}_{-0.21}(\text{stat.})^{+0.38}_{-0.34}(\text{syst.})$$

In a third phase, SNO introduced an independent way to detect the NC neutrons by having them captured in helium-3 nuclei. This experiment yielded fluxes of [20]

$$\text{CC:} \quad \phi_{CC} = 1.67^{+0.05}_{-0.04}(\text{stat.})^{+0.07}_{-0.08}(\text{syst.})$$

$$\text{ES:} \quad \phi_{ES} = 1.77^{+0.24}_{-0.21}(\text{stat.})^{+0.09}_{-0.10}(\text{syst.})$$

$$\text{NC:} \quad \phi_{NC} = 5.54^{+0.33}_{-0.31}(\text{stat.})^{+036}_{-0.34}(\text{syst.})$$

Clearly the neutrino flux measurements are reproducible and robust to different methodologies.

The bottom line is that SNO solved the solar neutrino problem by demonstrating that neutrino oscillation is real and measuring a (total) neutrino flux that is consistent with predictions from the Standard Solar Model. The agreement between theory and experiment indicates that we understand quite a lot about what is happening deep inside the Sun. The little neutral ones have proven to be both harbingers of new particle physics and important messengers from the center of our star.

Problems

15.1. Does the fusion of heavy nuclei require higher or lower temperatures than the fusion of light nuclei? Why?

15.2. Suppose fusion in the Sun stopped with $^{3}_{2}\text{He}$ instead of proceeding to $^{4}_{2}\text{He}$. If everything else were the same, would the Sun's lifetime be longer or shorter? By how much?

15.3. Show that the function

$$F(E) = e^{-(E_c/E)^{1/2}} e^{-E/kT}$$

has a peak at energy

$$E_0 = \left[\frac{1}{4} E_c \, (kT)^2 \right]^{1/3}$$

This is the Gamow peak discussed in Sect. 15.2.4. Make sure to verify that it is a peak (i.e., a local maximum, not a local minimum).

15.4. The goal of this problem is to understand how the fusion reaction rate depends on temperature. In Eq. (15.8) we showed that the reaction rate can be written as

$$r_{12} = \left(\frac{2}{kT}\right)^{3/2} \frac{n_1 n_2}{(\pi \mu)^{1/2}} \int_0^\infty S(E) \, e^{-f(E)} \, dE \quad \text{where} \quad f(E) = \frac{E}{kT} + \left(\frac{E_c}{E}\right)^{1/2}$$

The function $e^{-f(E)}$ is sharply peaked near the Gamow peak E_0. Assuming that $S(E)$ is reasonably constant near E_0, we pull it out of the integral and write

$$r_{12} = \left(\frac{2}{kT}\right)^{3/2} \frac{n_1 n_2}{(\pi \mu)^{1/2}} S(E_0) \int_0^\infty e^{-f(E)} \, dE$$

Our task is to estimate the remaining integral. Here are the steps to do that.

(a) Expand the function $f(E)$ as a Taylor series around the Gamow peak, $E = E_0$. This means you can approximate f with the form

$$f(E) \approx b_0 + b_1(E - E_0) + b_2(E - E_0)^2 + \dots$$

What you need to do is determine the coefficients b_0, b_1, and b_2. Hint: you should find $b_1 = 0$ since E_0 is a local maximum of $f(E)$.

(b) Using part (a) and setting $x = E - E_0$ converts the integral into a Gaussian form that can be evaluated using expressions in Sect. A.7.[12] Use this with your values of b_0 and b_2 to write an approximation for r_{12}.

(c) Following Sect. 15.2.4, write the temperature dependence with a power law approximation $r_{12} \propto T^\alpha$ and show that α is given by Eq. (15.11).

15.5. Let's compute the temperature dependence of the reaction rate for two specific examples. Even if you have not worked through Problem 15.4, you can still use Eq. (15.11) here. Take the temperature to be $T = 1.6 \times 10^7$ K.

(a) PP chain: Consider the $_1^1 H + _1^1 H$ step and compute kT, E_c, E_0, and α.
(b) CNO cycle: Consider the $_7^{14} N + _1^1 H$ step and compute kT, E_c, E_0, and α.

15.6. A brown dwarf is a star that is not massive enough to have normal hydrogen fusion (see Problem 16.5).

(a) A brown dwarf can release energy by the Kelvin-Helmholtz mechanism. Suppose a brown dwarf of mass M and radius R is shrinking at a rate dR/dt. What is its luminosity? (As always, state any assumptions you make.)
(b) Consider a brown dwarf with $M = 0.05 \, M_\odot$ and $R = 0.1 \, R_\odot$, whose radius shrinks by 1 % over 10^9 yr. What is its luminosity (in units of $L_\odot$)? If it radiates like a blackbody, what is its effective temperature? In what portion of the electromagnetic spectrum does it emit most of its light?

---

[12] Section A.7 gives integrals of Gaussian functions over the range $-\infty < x < \infty$. The limits of integration in this problem are far enough from the peak that we can extend them to $\pm\infty$.

(c) Brown dwarfs can fuse deuterium[13] via the reaction $^2_1H + ^1_1H \rightarrow ^3_2He$. If a fraction $f = 4 \times 10^{-5}$ of a brown dwarf's mass is deuterium, how long could deuterium burning power the star at the same luminosity you estimated in part (b)?

15.7. The cross section for a typical solar neutrino to interact with an atomic nucleus is $\sigma \sim 10^{-47}\,m^2$ (it varies with the neutrino energy, but we will focus on order-of-magnitude estimates). Estimate the mean free path of neutrinos in the Sun. You may make reasonable assumptions and approximations to obtain an order-of-magnitude estimate.

15.8. In Fig. 15.5, there is an upper limit to the neutrino energy in each reaction. Consider the total mass at the start and end to find the maximum neutrino energy for each of the following reactions (as labeled in the figure):

(a) pp, Eq. (15.14)
(b) ^{8}B, Eq. (15.17)
(c) hep, Eq. (15.18)

15.9. Estimate the number of solar neutrinos passing through your body each second. Estimate the number of times a neutrino will hit a nucleus in your body during your lifetime (assuming a cross section of $\sigma \sim 10^{-47}\,m^2$). You will need to make a variety of assumptions and approximations; explain your reasoning.

References

1. B.W. Carroll, D.A. Ostlie, *An Introduction to Modern Astrophysics*, 2nd ed. (Addison-Wesley, San Francisco, 2007)
2. D. Maoz, *Astrophysics in a Nutshell* (Princeton University Press, Princeton, 2007)
3. W.A. Fowler, G.R. Caughlan, B.A. Zimmerman, Annu. Rev. Astron. Astrophys. **5**, 525 (1967)
4. C. Angulo et al., Nucl. Phys. A **656**, 3 (1999)
5. D. Griffiths, *Introduction to Quantum Mechanics* (Prentice Hall, Upper Saddle River, NJ 1995)
6. G. Audi, A.H. Wapstra, C. Thibault, Nucl. Phys. A **729**, 337 (2003)
7. J.N. Bahcall, A.M. Serenelli, S. Basu, Astrophys. J. Suppl. Ser. **165**, 400 (2006)
8. J.N. Bahcall, A.M. Serenelli, S. Basu, Astrophys. J. Lett. **621**, L85 (2005)
9. J.N. Bahcall, Sci. Am. **221**, 29 (1969)
10. V. Gribov, B. Pontecorvo, Phys. Lett. B **28**, 493 (1969)
11. S.P. Mikheyev, A.Y. Smirnov, Yad. Fizika **42**, 1441 (1985)
12. L. Wolfenstein, Phys. Rev. D **17**, 2369 (1978)
13. J. Hosaka et al., Phys. Rev. D **73**, 112001 (2006)
14. K. Abe et al., Phys. Rev. D **83**, 052010 (2011)
15. R. Wendell et al., Phys. Rev. D **81**, 092004 (2010)
16. K. Abe et al., Phys. Rev. Lett. **107**, 041801 (2011)

---

[13]They may contain deuterium left over from the big bang (see Sect. 20.2).

17. Q.R. Ahmad et al., Phys. Rev. Lett. **89**, 011301 (2002)
18. J.N. Bahcall, M.H. Pinsonneault, S. Basu, Astrophys. J. **555**, 990 (2001)
19. B. Aharmim et al., Phys. Rev. C **72**, 055502 (2005)
20. B. Aharmim et al., Phys. Rev. Lett. **101**, 111301 (2008)

Chapter 16
Stellar Structure and Evolution

In previous chapters we examined physical processes that occur near a star's surface (atomic excitation, ionization, absorption) and in the interior (nuclear fusion). Now we unite them in detailed models of stars. We use the models to analyze the structure of stars during the main stage of life when they burn hydrogen to generate the heat and pressure that balance gravity. We then consider what happens when the hydrogen fuel runs out. As we will see, old stars begin burning heavier nuclei and working their way up the periodic table of the elements. How far a star gets depends on its mass: stars with masses below about 8 $M_\odot$ reach carbon and oxygen before experiencing a relatively meek death; stars with masses above about 8 $M_\odot$, by contrast, create all the heavier elements and then literally go out with a bang.

16.1 Energy Transport

Before we can build stellar models, we need to think about how energy produced in the core can be transported to the surface, to be released as light. In Chap. 13 we saw one mechanism for moving energy: with **radiation**, energy is carried by light that can be absorbed by an atom or molecule and then reradiated in a different direction. Now we consider two other mechanisms that act in dense gas. With **conduction**, heat moves on microscopic scales by collisions between particles. With **convection**, heat travels across macroscopic scales by bulk motion of gas. Let's examine each in turn.

16.1.1 Conduction

Consider a box of gas (for now, one small enough that we can neglect the effects of gravity). In equilibrium, it has the same temperature throughout. Suppose we heat one end of the box just a little—not enough to substantially change the density

C. Keeton, *Principles of Astrophysics: Using Gravity and Stellar Physics to Explore the Cosmos*, Undergraduate Lecture Notes in Physics, DOI 10.1007/978-1-4614-9236-8_16, © Springer Science+Business Media New York 2014

and pressure, but enough to make the particles jiggle a little faster. As these faster particles fly around, they bump into other particles in the box and transfer some of their kinetic energy. In this way collisions can allow the extra energy (heat) to travel across the box. Can we find an equation to describe this process, specifically to describe how the temperature changes with time at different locations in the box?

The temperature can change only if T is not spatially uniform—heat can only "flow" from a hotter region to a colder region—so $\partial T / \partial t$ must be related to some spatial derivative of T. Can we have

$$\frac{\partial T}{\partial t} \propto \frac{\partial T}{\partial x} \quad ?$$

No, by symmetry. If this were true, the sign of $\partial T / \partial t$ would depend on whether we heat the left end or the right end of the box. That cannot be right! What about

$$\frac{\partial T}{\partial t} \propto \left| \frac{\partial T}{\partial x} \right| \quad ?$$

No again. Consider $T \propto x^2$. The temperature is non-uniform so there ought to be heat flow. But with this hypothesis we would have $\partial T / \partial t = 0$ at $x = 0$. That does not make sense.

If first derivatives do not work, what about

$$\frac{\partial T}{\partial t} \propto \frac{\partial^2 T}{\partial x^2} \quad ?$$

This makes sense physically: if we raise the temperature anywhere, $\partial^2 T / \partial x^2$ will be nonzero, and indeed it will be positive away from the source of heat. This is in fact the right dependence, so let's specify the proportionality constant, κ, and generalize to three dimensions:

$$\frac{\partial T}{\partial t} = \kappa \left(\frac{\partial^2 T}{\partial x^2} + \frac{\partial^2 T}{\partial y^2} + \frac{\partial^2 T}{\partial z^2} \right) = \kappa \nabla^2 T \tag{16.1}$$

This is the **heat equation**, and it has the general form of a **diffusion equation**. The coefficient κ is called the **thermal diffusivity**. By dimensional analysis, we must have

$$\kappa = [L^2 T^{-1}] \quad \Rightarrow \quad \kappa \sim \ell v \sim \frac{\ell^2}{\tau}$$

for some characteristic length ℓ and velocity v, which are presumably the mean free path and the typical particle velocity, respectively. Alternatively, we can write κ in terms of $\tau \sim \ell / v$, which is the typical time between collisions.

What is the time scale Δt_{diff} for heat to diffuse over some distance ΔL? If we suppose that some amount of heat ΔT moves, then we can approximate $\partial T / \partial t \sim \Delta T / \Delta t_{\mathrm{diff}}$, and we can approximate $\nabla^2 T \sim \Delta T / (\Delta L)^2$. Then the heat equation yields

Fig. 16.1 The *solid line* shows a random walk with $N = 50$ steps of equal length. The *dashed gray line* connects the start and end points (the walk can proceed in either direction)

$$\frac{\Delta T}{\Delta t_{\text{diff}}} \sim \kappa \frac{\Delta T}{(\Delta L)^2} \quad \Rightarrow \quad \Delta t_{\text{diff}} \sim \frac{(\Delta L)^2}{\kappa} \sim \frac{(\Delta L)^2}{\ell^2/\tau} \sim \tau \left(\frac{\Delta L}{\ell}\right)^2 \quad (16.2)$$

For comparison, how long does it take for a particle to move *freely* across a distance ΔL? This is called the "crossing time":

$$\Delta t_{\text{cross}} \sim \frac{\Delta L}{v} \sim \frac{\Delta L}{\ell/\tau} \sim \tau \frac{\Delta L}{\ell}$$

If the step size (ℓ) is small compared with the size of the box (ΔL), the diffusion time may be *much* longer than the crossing time. Why? Collisions have random directions, so particles do not travel straight across the box. Rather, they follow meandering trajectories known as random walks.

Random Walk

Suppose a particle starts at the origin and takes a series of steps that have fixed length ℓ but random directions. An example of a random walk with 50 steps is shown in Fig. 16.1. In general, after N steps the particle's position is

$$\mathbf{X} = \mathbf{x}_1 + \mathbf{x}_2 + \ldots + \mathbf{x}_N = \sum_{i=1}^{N} \mathbf{x}_i$$

The square of the net distance from the starting point (the origin) is

$$X^2 = \sum_{i,j} \mathbf{x}_i \cdot \mathbf{x}_j = \sum_{i=j} |\mathbf{x}_i|^2 + \sum_{i \neq j} \mathbf{x}_i \cdot \mathbf{x}_j = N \ell^2 + \sum_{i \neq j} \mathbf{x}_i \cdot \mathbf{x}_j$$

Here we separate the sum into a piece in which the indices match and a piece in which they differ, and then use the fact that the step size is fixed so $|\mathbf{x}_i| = \ell$ for all i.

We are interested in the typical distance traveled after N steps, which we define to be the root mean square distance $X_{rms} = \sqrt{\langle X^2 \rangle}$. The average has the form

$$\langle X^2 \rangle = N \, \ell^2 + \sum_{i \neq j} \langle \mathbf{x}_i \cdot \mathbf{x}_j \rangle$$

If the steps are independent of one another, then $\langle \mathbf{x}_i \cdot \mathbf{x}_j \rangle = \langle \mathbf{x}_i \rangle \cdot \langle \mathbf{x}_j \rangle$. If the directions are random, then $\langle \mathbf{x}_i \rangle = 0$ because the particle is equally likely to go right or left. The net result is

$$\langle X^2 \rangle = N \, \ell^2$$

Returning to the diffusion time, we can now estimate the number of steps needed to cross a distance ΔL. We set $X_{rms} = \Delta L$ to find

$$N = \left(\frac{\Delta L}{\ell} \right)^2$$

This is the typical number of steps; the actual number may be somewhat larger or smaller for individual random walks. If each step takes time τ, then the overall diffusion time is

$$\Delta t_{diff} \sim N\tau \sim \tau \left(\frac{\Delta L}{\ell} \right)^2$$

as in Eq. (16.2).

Example: What Is the Thermal Diffusivity of Earth's Atmosphere?

In Chap. 12 we estimated that nitrogen molecules in Earth's atmosphere have a mean free path of $\ell \sim 1.4 \times 10^{-7}$ m and a typical speed of $v \sim 5.1 \times 10^2$ m s^{-1}. Together these yield an estimate for the thermal diffusivity of

$$\kappa \sim \ell v \sim 7 \times 10^{-5} \, \mathrm{m^2 \, s^{-1}}$$

For comparison, the laboratory value is [1]

$$\kappa = 1.9 \times 10^{-5} \, \mathrm{m^2 \, s^{-1}}$$

What is the time scale for heat to diffuse across a room that is $\Delta L = 10$ m across?

$$\Delta t_{diff} \sim \frac{(\Delta L)^2}{\kappa} \sim \frac{(10 \, \mathrm{m})^2}{1.9 \times 10^{-5} \, \mathrm{m^2 \, s^{-1}}} \sim 5.3 \times 10^6 \, \mathrm{s} \sim 61 \, \mathrm{days}$$

Heat does not diffuse very quickly! Apparently we need to find another mechanism that can transport heat more effectively.

16.1.2 Convection

Adding a lot of heat to the box can induce mechanical forces that generate *bulk motion* in the gas. To analyze convection, we imagine taking a small bubble of gas from height z, moving it to height $z + \Delta z$, and asking what will happen next (also see [2,3]). If the bubble will fall back to its starting point, then small changes (such as the formation of a bubble) tend to damp out and the gas is stable. If the bubble will continue to rise, however, then small changes can grow into larger changes and convection can begin spontaneously.[1] Our goal in this section is to derive a condition under which gas is unstable to convection.

First, we need to recall some thermodynamics. **Specific heat** quantifies the amount of energy needed to raise the temperature of a substance. There are actually two specific heats: one if we increase the temperature while holding the pressure fixed, and another if we hold the volume fixed:

$$C_P = \frac{dQ}{dT}\bigg|_P \qquad \text{and} \qquad C_V = \frac{dQ}{dT}\bigg|_V$$

We define the ratio to be the **adiabatic index**,

$$\gamma = \frac{C_P}{C_V} \tag{16.3}$$

A non-relativistic ideal gas has $\gamma = 5/3$ for monatomic particles and $\gamma = 7/5$ for diatomic particles (such as N_2, O_2, etc.). A relativistic ideal gas has $\gamma = 4/3$. The adiabatic index characterizes a process in which no heat flows into or out of a system, which is a reasonable approximation for any "slow" thermodynamic process. Carroll and Ostlie [2] show that such a process is described by the **adiabatic equation of state**,

$$PV^\gamma = \text{constant} \quad \Rightarrow \quad P = K\rho^\gamma \tag{16.4}$$

where K is a constant.

Now consider an ideal gas whose density, temperature, and pressure vary with height. At height z the gas is described by T, P, and ρ, while at height $z + \Delta z$ it has $T + \Delta T$, $P + \Delta P$, and $\rho + \Delta \rho$. We can write

$$\Delta T = \frac{dT}{dz}\Delta z \qquad \Delta P = \frac{dP}{dz}\Delta z \qquad \Delta \rho = \frac{d\rho}{dz}\Delta z \tag{16.5}$$

From the ideal gas law, we know

---

[1]The situation is similar to water boiling, although that case is slightly more complicated because it involves the formation of *air* bubbles.

$$\rho = \bar{m} n = \frac{\bar{m} P}{k T}$$

where $\bar{m}$ is the average particle mass. Take the logarithm:

$$\ln \rho = \ln \bar{m} + \ln P - \ln k - \ln T$$

Take the derivative with respect to z, and multiply by Δz:

$$\frac{1}{\rho} \frac{d\rho}{dz} \Delta z = \frac{1}{P} \frac{dP}{dz} \Delta z - \frac{1}{T} \frac{dT}{dz} \Delta z$$

Using Eq. (16.5), we can write this as

$$\frac{\Delta \rho}{\rho} = \frac{\Delta P}{P} - \frac{\Delta T}{T} \tag{16.6}$$

Suppose we take a small bubble of gas from height z and move it to height $z + \Delta z$. It will adjust to some new density, pressure, and temperature $\rho + \delta \rho$, $P + \delta P$, and $T + \delta T$, which may or may not match the surrounding medium. The internal and external pressure forces must balance, so in fact $\delta P = \Delta P$. If we move the bubble quickly, there is no time for heat to flow into or out of the bubble and the enclosed gas must behave adiabatically:

$$\delta P = K \gamma \rho^{\gamma - 1} \delta \rho \quad \Rightarrow \quad \frac{\delta P}{P} = \gamma \frac{\delta \rho}{\rho} \tag{16.7}$$

If the density inside the bubble is lower than the density outside, the bubble will be buoyant and want to continue rising. The condition for buoyancy is therefore $\delta \rho < \Delta \rho$ or (using (16.6) and (16.7))

$$\frac{1}{\gamma} \frac{\delta P}{P} < \frac{\Delta P}{P} - \frac{\Delta T}{T}$$

Using $\delta P = \Delta P$ and rearranging yields

$$\Delta T < \left(1 - \frac{1}{\gamma} \right) \frac{T}{P} \Delta P \quad \Rightarrow \quad \frac{dT}{dz} < \frac{\gamma - 1}{\gamma} \frac{T}{P} \frac{dP}{dz}$$

(To obtain the last expression we divide through by Δz and then turn the differentials into derivatives.) In general, T and P both decrease with height, so the derivatives are negative. Switching to absolute values so we work with positive quantities, we have the following condition for buoyancy:

$$\left| \frac{dT}{dz} \right| > \frac{\gamma - 1}{\gamma} \frac{T}{P} \left| \frac{dP}{dz} \right| \tag{16.8}$$

An ideal gas in hydrostatic equilibrium has $P = nkT$ and $dP/dz = -g\rho = -g\bar{m}n$ (see Eq. 12.13). In this case, Eq. (16.8) becomes

$$\left|\frac{dT}{dz}\right| > \frac{\gamma-1}{\gamma}\frac{\bar{m}}{k}g \tag{16.9}$$

We learn that gas is unstable to convection if the temperature gradient exceeds a threshold determined by the pressure gradient, which in turn depends on the acceleration due to gravity.

Example: Earth

At Earth's surface, $g = 9.80\,\mathrm{m\,s^{-2}}$. For an ideal gas of molecular nitrogen (N_2), $\gamma = 7/5$ and $m = 28\,m_p$. With these numbers, the right-hand side of Eq. (16.9) becomes

$$\frac{\gamma-1}{\gamma}\frac{m}{k}g = \frac{2/5}{7/5}\frac{28\times 1.67\times 10^{-27}\,\mathrm{kg}}{1.38\times 10^{-23}\,\mathrm{kg\,m^2\,s^{-2}\,K^{-1}}}\times 9.80\,\mathrm{m\,s^{-2}} = 9.5\times 10^{-3}\,\mathrm{K\,m^{-1}}$$

In the lower part of Earth's atmosphere, the temperature varies roughly linearly with altitude, ranging from about 288 K at the ground to 210 K at an altitude of $15\,\mathrm{km} = 1.5\times 10^4$ m. This corresponds to a temperature gradient of

$$\left|\frac{dT}{dz}\right| = \frac{78\,\mathrm{K}}{1.5\times 10^4\,\mathrm{m}} = 5.2\times 10^{-3}\,\mathrm{K\,m^{-1}}$$

The temperature gradient is smaller than the value required for convective instability, so Earth's atmosphere is not convectively unstable (at sea level). This does not mean that convection cannot occur; it just means that convection will not begin spontaneously.

16.2 Stellar Models

Now we are ready to assemble the pieces and write down a set of equations that describe a model star. For simplicity, we assume the star is spherically symmetric and static (time-independent), so the density, pressure, temperature, etc. are functions only of r. Stars may not be perfectly spherical, especially if they rotate rapidly, and they are not truly static, but during the bulk of their lives they change slowly; so a spherical, static model is not a bad place to start.

16.2.1   Equations of Stellar Structure

The first set of equations describe how mass, luminosity, pressure, and temperature vary with radius. The equation relating mass to density is one we have seen before:

$$\text{enclosed mass} \qquad \frac{dM}{dr} = 4\pi r^2 \rho \qquad (16.10)$$

We can write an equation for luminosity that has a similar form if we define ϵ to be the energy generation rate per unit mass (in units of J s^{-1} kg^{-1}, for example). Then the energy output (or luminosity) from a spherical shell of radius r and thickness dr is $dL = 4\pi r^2 \rho \epsilon \, dr$, so the differential equation we need is

$$\text{energy generation} \qquad \frac{dL}{dr} = 4\pi r^2 \rho \epsilon \qquad (16.11)$$

The equation for pressure comes from hydrostatic equilibrium (12.13):

$$\text{hydrostatic equilibrium} \qquad \frac{dP}{dr} = -\frac{GM(r)\rho}{r^2} \qquad (16.12)$$

Finally, the equation for temperature depends on how energy is transported. In a convective regime, the temperature gradient will match the right-hand side of Eq. (16.9)[2]:

$$\text{convection} \qquad \frac{dT}{dr} = -\frac{\gamma - 1}{\gamma} \frac{\bar{m}}{k} \frac{GM(r)}{r^2} \qquad (16.13)$$

In a radiative regime, the temperature gradient depends on the **opacity** of the gas, $\bar{\kappa} = 1/\ell\rho$. If the opacity is high, light cannot transport energy very efficiently, so heat remains trapped and the temperature gradient is large. Conversely, if the opacity is low, light is able to move heat energy and the temperature gradient remains small. Carroll and Ostlic [2] and Maoz [3] derive the differential equation for temperature in a radiative regime:

$$\text{radiation} \qquad \frac{dT}{dr} = -\frac{3}{16\sigma} \frac{\bar{\kappa}\rho}{T^3} \frac{L(r)}{4\pi r^2} \qquad (16.14)$$

To supplement the four differential equations, we need equations of state that relate the pressure, energy generation rate, and opacity to the density, temperature, and composition of the gas. These are based on the gas physics, nuclear physics, thermodynamics, and other principles we have studied since Chap. 12. As an illustration, let's consider the pressure. The net gas pressure is the sum of partial pressures from all the constituents:

---

[2]Recall from Sect. 16.1.2 that this equation applies to an ideal gas in hydrostatic equilibrium.

$$P_{\text{gas}} = \sum_i n_i\, kT = n_{\text{tot}} kT$$

where $n_{\text{tot}} = \sum_i n_i$ and we are assuming an ideal gas. Pressure depends fundamentally on the *number* density of particles, but gravity depends on the *mass* density so our model is expressed in terms of ρ. Using the average particle mass, $\bar{m} \equiv \rho/n_{\text{tot}}$, we can write pressure in terms of mass density as

$$P_{\text{gas}} = \frac{\rho}{\bar{m}} kT \qquad (16.15)$$

As a star evolves, its mass density remains nearly unchanged,[3] but its composition varies as hydrogen gets converted to helium. Consider how this affects $\bar{m}$:

- Pure neutral hydrogen: $\bar{m} = m_p + m_e \approx m_p$
- Pure neutral helium: $\bar{m} = 2m_p + 2m_n + 2m_e \approx 4m_p$
- Pure ionized hydrogen: $\bar{m} = (m_p n_p + m_e n_e)/(n_p + n_e) \approx m_p/2$, where we assume $n_p = n_e$ for charge neutrality

(Since $m_e \ll m_p$, we can neglect the mass in electrons.) Clearly $\bar{m}$ and P_{gas} depend on the composition of the gas, which can vary throughout the star.

In addition to gas pressure, there is **radiation pressure** from the light. In Sect. 13.1.4 we derived the equation of state for photon pressure,

$$P_{\text{rad}} = 2.52 \times 10^{-16}\, \text{kg m}^{-1}\, \text{s}^{-2} \times \left(\frac{T}{\text{K}}\right)^4$$

Radiation pressure is actually negligible in the Sun. We will see below that the temperature at the center of the Sun is about $T = 1.567 \times 10^7$ K, so the radiation pressure is

$$P_{\text{rad}} = 2.52 \times 10^{-16}\, \text{kg m}^{-1}\, \text{s}^{-2} \times \left(1.567 \times 10^7\right)^4 = 1.5 \times 10^{13}\, \text{kg m}^{-1}\, \text{s}^{-2}$$

This is small compared with the gas pressure $P = 2.357 \times 10^{16}\, \text{kg m}^{-1}\, \text{s}^{-2}$. Radiation pressure is subdominant in most normal stars, but it becomes increasingly important as the mass increases and helps determine the upper limit on the mass of normal stars (see Problem 16.6).

The differential equations shown above only describe how quantities change with r. In order to obtain a complete solution we must specify starting or ending values. There are some natural **boundary conditions** for stars. At the center, as $r \to 0$ there is no mass or luminosity enclosed, so we must have

$$M(0) = 0 \qquad L(0) = 0$$

---

[3]Strictly speaking, some matter gets converted to energy, but that is a small fraction of the total.

At the surface, the mass, luminosity, and temperature should take on their overall values for the star, and at least for simplicity we might imagine that the density and pressure vanish:

$$\rho(R) = 0 \qquad P(R) = 0 \qquad M(R) = M_{\text{tot}} \qquad L(R) = L_{\text{tot}} \qquad T(R) = T_{\text{eff}}$$

In reality the boundary conditions can be more complicated: stars have diffuse but extended atmospheres, and they can lose mass (due to a "stellar wind"). But the simple boundary conditions given here provide a good starting point.

The equations of stellar structure represent a set of coupled differential equations where some of the components may not even be known analytically. (Nuclear reaction rates are often empirically calibrated.) In general they cannot be solved by hand, but they are suitable for numerical integration as discussed in Sect. A.6.

16.2.2 The Sun

John Bahcall and his collaborators have used the stellar structure equations to develop a detailed model of the Sun known as the **Standard Solar Model** [4]. It predicts the following values at the center of the Sun:

Density	$1.529 \times 10^5 \, \text{kg m}^{-3}$
Temperature	$1.567 \times 10^7 \, \text{K}$
Pressure	$2.357 \times 10^{16} \, \text{kg m}^{-1} \, \text{s}^{-2}$
Hydrogen mass fraction	34.61%
Helium mass fraction	63.37%

The density, temperature, pressure, and composition change with radius as shown in Fig. 16.2. The composition curves suggest that fusion occurs mainly in a core region whose radius is 15–20% of $R_\odot$. As we saw in Sect. 15.4, Bahcall et al. used the model to predict the abundance of neutrinos produced by the Sun, and subsequent detections have provided important evidence that the Standard Solar Model is accurate.

There are other ways to test the model as well. For example, some of the curves in Fig. 16.2 show a small "kink" around $0.7 R_\odot$. As you can show in Problem 16.3, this corresponds to the point at which the Sun becomes convectively unstable. The idea that the outer 30% of the Sun is convective is confirmed by **granulation** in the photosphere. Hot bubbles of gas rise to the surface, spread out, cool, and sink back down. Since the rising hot gas is brighter than the sinking cool gas (by the Stefan-Boltzmann law, Sect. 13.1.1), convection produces a patchwork of brighter and darker regions as shown in Fig. 16.3. Convection also causes the surface of the Sun effectively to oscillate. As with seismology on Earth, **helioseismology** uses

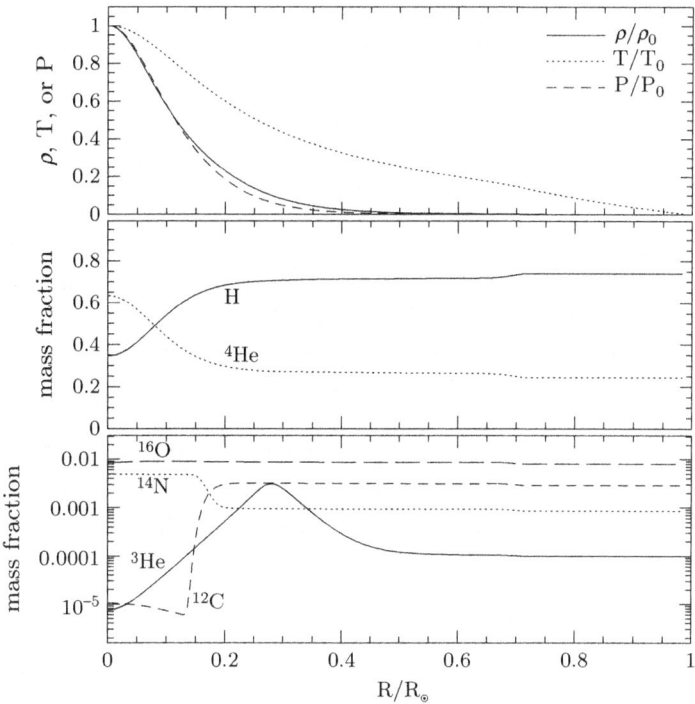

Fig. 16.2 Results from the Standard Solar Model. (*Top*) Density, temperature, and pressure as a function of radius, normalized by the values at the center: $\rho_0 = 1.529 \times 10^5 \, \text{kg m}^{-3}$, $T_0 = 1.567 \times 10^7 \, \text{K}$, and $P_0 = 2.357 \times 10^{16} \, \text{kg m}^{-1} \text{s}^{-2}$. (*Middle*) Mass fraction of hydrogen and helium-4 as a function of radius. (*Bottom*) Mass fraction of helium-3, carbon, nitrogen, and oxygen as a function of radius (Data from Bahcall et al. [4])

motions of the Sun's surface to probe the depths and test our models of the Sun's interior.

16.2.3 Other Stars

When the equations are applied to different kinds of stars, a pattern known as the **Vogt-Russell theorem** emerges: "The mass and the composition structure throughout a star uniquely determine its radius, luminosity, and internal structure, as well as its subsequent evolution."[4] This is not a rigorous mathematical theorem because quantities beyond mass and composition (such as magnetic fields and

---

[4]This phrasing comes from [2]. Also see [5].

Fig. 16.3 Granulation in the Sun's photosphere, as seen by the Solar Optical Telescope on the Hinode mission. The brighter regions correspond to hotter gas rising by convection, while the darker regions correspond to cooler gas sinking back down (Credit: Hinode JAXA/NASA/PPARC)

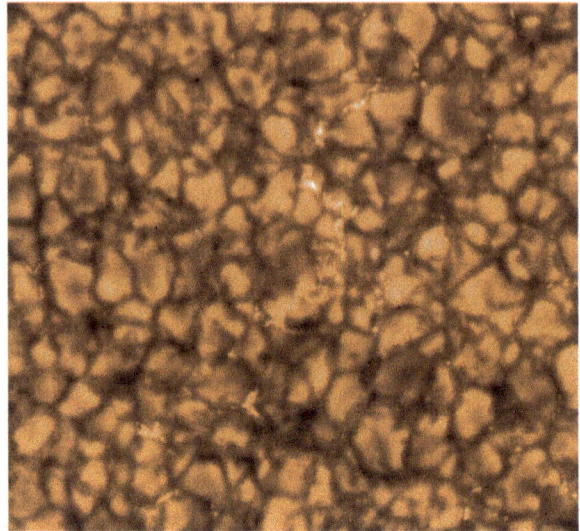

rotation) can in fact influence stellar structure. Nevertheless, the Vogt-Russell theorem is a good rule of thumb that seems to describe real stars quite well.

Mass is the main factor that determines star properties, because compositions are fairly uniform at birth—mostly hydrogen and helium in a ratio that reflects cosmic abundances (see Chap. 20), plus trace amounts of heavier elements—and they change slowly with time. Using the models to predict the luminosity and effective temperature as a function of mass yields a curve that follows the main sequence in the HR diagram (see Fig. 16.4). In other words, the observed main sequence is actually a sequence in mass, running from cool, faint, low-mass stars in the lower right of the HR diagram to hot, luminous, massive stars in the upper left. Physically, a more massive star needs a higher temperature and pressure to balance the stronger gravity. Those conditions yield more fusion and thus a higher energy output (luminosity).

What sets the endpoints of the main sequence? At lower masses the internal temperature is cooler, and at some point it is too low to support fusion. This is the bottom end of the main sequence—the lowest mass object we call a star. Models indicate that the fusion limit occurs around $0.08\,M_\odot$. At the top end, above $\sim 100\,M_\odot$ the fusion is so intense that radiation pressure makes the core unstable. (You can estimate these limits in Problems 16.5 and 16.6.)

There are a couple of points at low masses that are worth noting. In Sect. 15.3.3 we remarked that fusion in stars with $M \lesssim 1.2\,M_\odot$ mainly follows the PP chain, whereas fusion in more massive stars mainly follows the CNO cycle. For CNO-driven stars, the fusion rate depends strongly on temperature and hence radius. Radiation cannot transport energy rapidly enough, so a large temperature gradient builds up and induces convection. As a result, stars with $M \gtrsim 1.2\,M_\odot$ have convective cores. For PP-driven stars, by contrast, the fusion rate varies less strongly with temperature and hence radius, so radiation is sufficient to transport energy and

Fig. 16.4 A theory version of the HR diagram, based on stellar models. The *dotted line* shows luminosity and effective temperature as a function of mass (indicated in units of $M_\odot$); this corresponds to the main sequence in the observed HR diagram (see Fig. 14.6). *Solid lines* show evolutionary tracks once stars of different mass leave the main sequence. On the horizontal axis, temperature increases to the left to follow the convention in the observed HR diagram (Credit: Pols et al. [7], reproduced by permission of Oxford University Press on behalf of the Royal Astronomical Society)

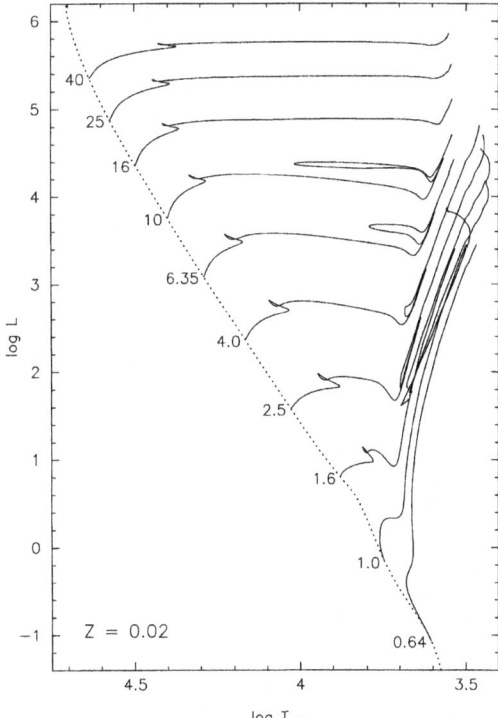

there is no convection in the core. Radiation becomes inefficient near the surface, though, because the lower temperature leads to less ionization and excitation and hence greater opacity. As a result, stars with $M \lesssim 1.2\,M_\odot$ have convection in the outer layers. As the mass decreases, the bottom of the convection zone moves progressively downwards until at $\sim 0.4\,M_\odot$ it reaches the core and the star becomes fully convective (e.g., [6]).

To summarize, we can make a table like this:

Mass ($M_\odot$)	Fusion	Convection
$\lesssim 0.08$	None	
0.08–0.4	PP	Fully convective
0.4–1.2	PP	Surface convection
1.2–100	CNO	Core convection
$\gtrsim 100$	Unstable	

16.3   Evolution of Low-Mass Stars ($M \lesssim 8\,M_{\odot}$)

Stars are not truly static; they evolve as fusion modifies the internal composition and hence the equations of state. Stellar models can be extended to track the evolution and predict how stars move through the HR diagram once they leave the main sequence. Figure 16.4 shows examples of evolutionary tracks for a few different masses. In this section we consider the main stages in the life of a star whose mass is less than $\sim\!8\,M_{\odot}$.[5]

16.3.1   Hydrogen, Helium, and Beyond

During most of a star's life it generates power by burning hydrogen to produce helium. Once a star exhausts its hydrogen fuel, it can burn helium to produce heavier elements. It turns out that the hydrogen and helium burning phases can each be divided into two sub-stages depending on where within the star the fusion occurs.

Hydrogen core burning. Stars that burn hydrogen in their cores lie on the main sequence in the HR diagram. Even during this stage there are some notable changes. As a star converts hydrogen into helium, its *internal* composition changes even if its *surface* composition (which determines its spectral type) does not. Recall from Eq. (16.15) that the pressure of a gas depends on the composition through the average particle mass, $\bar{m}$ (assuming the mass density is constant). As hydrogen changes to helium, $\bar{m}$ increases. In order to maintain the same pressure (to balance gravity), the temperature must likewise increase. That causes the fusion rate to increase, which in turn raises the star's luminosity. The net effect is that stars brighten a little as they age (by a factor of $\sim\!2$ for the Sun [8]), so they have a small vertical movement in the HR diagram. This contributes to the thickness of the main sequence.

Hydrogen shell burning. Once the hydrogen in the core is exhausted, the star has a helium core surrounded by an envelope of hydrogen. The temperature and pressure can be high enough to ignite hydrogen fusion in a shell around the helium core. The temperature in the shell is actually higher than it was during the earlier stage of hydrogen core burning, so the energy production rate and luminosity are higher. Some of the energy goes into making the outer envelope expand, which causes the surface to cool and become redder. The star becomes luminous, large, and cool—a **red giant**. As the star continues to age, it moves up the red giant branch in the HR diagram.

Helium core burning. Once enough helium accumulates to cross a threshold in mass, the helium core begins to collapse due to its gravity. At some point the central

---

[5]Carroll and Ostlie [2] discuss stellar evolution at a similar technical level but in more detail.

temperature ($T \sim 10^8\,$K) and density ($\rho \sim 10^7\,\mathrm{kg\,m^{-3}}$) become high enough to initiate fusion that converts three helium nuclei into carbon:

$$^4_2\mathrm{He} + ^4_2\mathrm{He} \;\rightleftharpoons\; ^8_4\mathrm{Be}$$

$$^8_4\mathrm{Be} + ^4_2\mathrm{He} \;\rightarrow\; ^{12}_6\mathrm{C}$$

Helium-4 nuclei are known as "alpha particles," so this set of reactions is called the **triple alpha process**. Beryllium-8 can spontaneously decay (hence the reverse arrow above), so it needs to react quickly with another helium-4 nucleus to form carbon-12. Thus, the triple alpha process can proceed only if the temperature and density are high enough to achieve a sufficient reaction rate. Once some carbon exists, it can combine with more helium to form oxygen:

$$^{12}_6\mathrm{C} + ^4_2\mathrm{He} \;\rightarrow\; ^{16}_8\mathrm{O}$$

With these new energy sources, the core is able to expand a little, which lowers the core temperature and reduces the luminosity. That, in turn, allows the envelope to shrink, causing the *surface* temperature to rise. Once helium ignites, in other words, the star shifts down and to the left in the HR diagram, moving onto the **horizontal branch**.

Helium shell burning. Once helium in the core is used up, the process above repeats itself, only at higher temperatures. Now the star burns hydrogen in an outer shell and helium in an inner shell, all surrounding a carbon/oxygen core. Such a star lies on the **asymptotic giant branch** (AGB) in the HR diagram.

AGB stars are large and have cool envelopes ($T \sim 3{,}000\,$K) that contain carbon and oxygen dredged up from the core by convection. The conditions are right to generate dust that is rich in silicates and/or graphite. Such dust will eventually return to the interstellar medium, where it can be detected by the way it scatters light.

The energy production rate in AGB stars is high enough to drive a **stellar wind** that carries away some of the mass. The mass loss can be $\mathrm{d}M/\mathrm{d}t \sim -10^{-4}\,M_\odot\,\mathrm{yr}^{-1}$, which may not seem like a lot but is enough that a star could lose most of its mass in as little as 10,000 years.

Planetary nebula. Eventually the entire outer envelope is expelled to form an expanding gas cloud known as a planetary nebula. (These have nothing to do with planets; they got the name because in early telescopes they appeared as circular disks and thus resembled images of nearby planets.) The planetary nebula expands at a rate of 10–30 km/s, so after 10,000–50,000 years it will be so large and diffuse that it will have dispersed into the interstellar medium.

White dwarf. The remnants of the carbon/oxygen core settle into a hot, dense object known as a white dwarf. The gas has changed from an ideal gas in which the pressure is produced by motion of the gas particles to a "degenerate" gas in which the pressure arises from a quantum mechanical effect that prevents particles from being in the same quantum state. We will study a gas supported by electron degeneracy pressure in Chap. 17. White dwarfs cool slowly, and they can be around for a long time.

To recap, here are the stages in the life of a low-mass star, the corresponding parts of the HR diagram, and the approximate duration of each stage (for a solar mass star [7,9]):

Hydrogen core burning	Main sequence	10^{10} yr
Hydrogen shell burning	Red giant branch	10^9 yr
Helium core burning	Horizontal branch	10^8 yr
Helium shell burning	Asymptotic giant branch	10^7 yr
Mass loss	Planetary nebula	10^4 yr
Electron degeneracy	White dwarf	

Again, this evolutionary process creates elements up through carbon and oxygen.

16.3.2 Observations

We cannot observe all the stages of evolution for any single star; even the "short" stages are longer than a human lifetime. Nevertheless, it is possible to see the whole evolutionary pathway. The key idea is that all stars with $M \lesssim 8\,M_\odot$ follow the same set of steps, *but at different rates*. More massive stars pass through the sequence more quickly than less massive stars (they have more fuel but burn it faster).

Consider a set of stars with different masses that all formed at the same time. After a few billion years, the more massive stars will have progressed through the evolutionary sequence to reach, say, the asymptotic giant branch. Stars that are a little less massive will have reached the horizontal branch. Still smaller stars will be on the red giant branch. The lowest mass stars will remain on the main sequence. At a snapshot in time, the stars will trace out the full evolutionary track in the HR diagram (with position along the track determined by mass).

The universe kindly provides exactly what we need to see this. **Star clusters** are collections of stars that formed at approximately the same time (when a gas cloud collapsed and fragmented; see Chap. 19). Figure 16.5 shows HR diagrams for two observed star clusters. Stellar evolution theory can be used to predict curves showing the positions of stars that have different masses but the same age, known as **isochrones** (from *iso* = sam + *chrone* = time). Matching an isochrone to the observed HR diagram makes it possible to determine the age of a star cluster.

A particularly important point in the HR diagram of a cluster is the **main sequence turn-off (MSTO)**. Less massive stars have longer main sequence life-times than more massive stars, so as time passes the turn-off point moves down the main sequence. Matching observed MSTO points with theoretical predictions

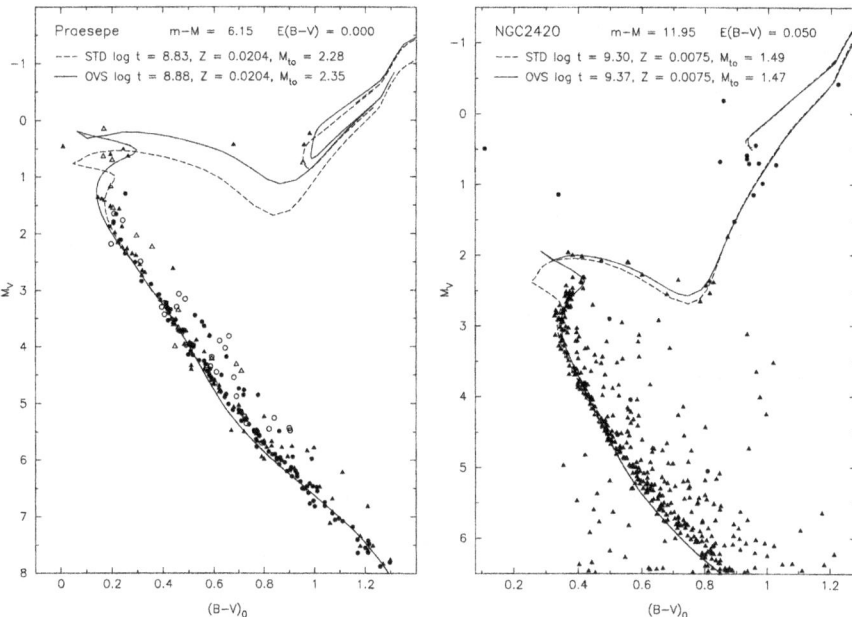

Fig. 16.5 HR diagrams for two star clusters. The horizontal axis is an index astronomers use to quantify color (see Fig. 14.6 for the corresponding spectral types). The vertical axis is $M_V = -2.5\log_{10} L + $ constant, with the minus sign causing brighter stars to have smaller values of M_V. In each panel, the points indicate individual stars while the curves represent isochrones from different evolution models. Praesepe (*left*) is a relatively young cluster with an age of about 0.7 Gyr, while NGC 2420 (*right*) is an older cluster with an age of about 2 Gyr (Credit: Pols et al. [7], reproduced by permission of Oxford University Press on behalf of the Royal Astronomical Society)

reveals how much time has passed since the cluster formed—the age of the cluster. Such "age dating" has revealed that the oldest known star clusters are around 11.5 Gyr old [10], whereas the universe is about 13.8 Gyr old [11]. (The two clusters shown in Fig. 16.5 are relatively young.)

16.4   Evolution of High-Mass Stars ($M \gtrsim 8\,M_\odot$)

High-mass stars also go through stages of core H burning, shell H burning, core He burning, and shell He burning. What makes them different from low-mass stars is that they do not stop at helium. The higher mass leads to higher temperatures and densities that can drive fusion further up the periodic table of the elements.

16.4.1   Beyond Carbon and Oxygen

After helium burning produces carbon and oxygen, the temperature in a high-mass star remains high enough ($T \gtrsim 10^8$ K) that carbon and oxygen can burn by reacting with helium:

$$^{12}_{6}C + ^{4}_{2}He \rightarrow ^{16}_{8}O + \gamma$$

$$^{16}_{8}O + ^{4}_{2}He \rightarrow ^{20}_{10}Ne + \gamma$$

As the temperature increases still further, carbon and oxygen can burn in new ways, producing a whole range of byproducts. Above 6×10^8 K, carbon/carbon reactions become possible:

$$^{12}_{6}C + ^{12}_{6}C \rightarrow ^{16}_{8}O + 2\,^{4}_{2}He$$

$$\rightarrow ^{20}_{10}Ne + ^{4}_{2}He$$

$$\rightarrow ^{23}_{11}Na + p$$

$$\rightarrow ^{23}_{12}Mg + n$$

$$\rightarrow ^{24}_{12}Mg + \gamma$$

Above 10^9 K, oxygen/oxygen reactions begin:

$$^{16}_{8}O + ^{16}_{8}O \rightarrow ^{24}_{12}Mg + 2\,^{4}_{2}He$$

$$\rightarrow ^{28}_{14}Si + ^{4}_{2}He$$

$$\rightarrow ^{31}_{15}P + p$$

$$\rightarrow ^{31}_{16}S + n$$

$$\rightarrow ^{32}_{16}S + \gamma$$

And above 3×10^9 K, even heavier elements can burn:

$$^{28}_{14}Si + ^{4}_{2}He \rightleftharpoons ^{32}_{16}S + \gamma$$

$$^{32}_{16}S + ^{4}_{2}He \rightleftharpoons ^{36}_{18}Ar + \gamma$$

$$\vdots$$

$$^{48}_{24}Cr + ^{4}_{2}He \rightleftharpoons ^{52}_{26}Fe + \gamma$$

$$^{52}_{26}Fe + ^{4}_{2}He \rightleftharpoons ^{56}_{28}Ni + \gamma$$

These reactions are known collectively as **silicon burning**.

Where does the process stop? As fusion creates heavier and heavier nuclei, it releases less and less energy because the available binding energy decreases (see Fig. 16.6). The burning must go faster and faster in order to supply the energy the

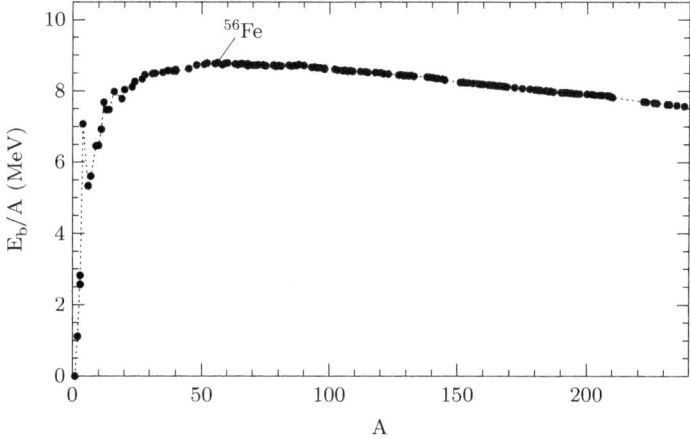

Fig. 16.6 Binding energy per nucleon, as a function of the atomic mass number (similar to Fig. 15.3, but better showing the high-mass end) (Data from [12])

star needs to avoid collapsing. Here are the durations of the various stages for a 25 $M_\odot$ star [3]:

Core hydrogen burning	5×10^6 yr
Core helium burning	5×10^5 yr
Core carbon burning	500 yr
Core silicon burning	1 day

Once silicon burning produces nickel-56, the core has crossed the peak in the binding energy curve. At that point there is no more energy to be released by fusion.[6] When the star loses its ability to create energy, things get really wild.

16.4.2 Explosion: Supernova

By the time the star is burning silicon, the temperature is so high that photons possess enough energy to destroy heavy nuclei; this process is called **photodisintegration** (and it is why the silicon burning reactions above have reverse arrows). Two reactions that absorb energy from the star are:

$$^{56}_{26}\text{Fe} + \gamma \;\to\; 13\, ^4_2\text{He} + 4\, n$$

---

[6] A little energy is available from radioactive decay of nickel-56 into cobalt-56 and then into iron-56, but it is not enough to support the star.

$$^4_2\text{He} + \gamma \rightarrow 2p + 2n$$

Another significant reaction is **inverse beta decay**,

$$p + e^- \rightarrow n + \nu_e$$

Essentially, the pressure is high enough to squeeze protons and electrons together to form neutrons and neutrinos. This process is important for two reasons. First, the neutrinos carry away huge amounts of energy. Second, electrons that had been helping to support the core (via electron degeneracy pressure) are removed. When they disappear, the core quickly collapses. How quickly? In Problem 16.9 you can derive the **freefall** time for a sphere of density ρ to collapse under its own gravity:

$$t_{\text{ff}} = \left(\frac{3\pi}{32G\rho} \right)^{1/2} \tag{16.16}$$

The degenerate core of a massive star can be comparable to the mass of our Sun. In Sect. 1.3.2 we used dimensional analysis to estimate that an object of mass $M \sim M_\odot$ supported by electron degeneracy pressure has a radius of $R \sim 6 \times 10^6$ m. Thus the mean density is

$$\rho \sim \frac{3M}{4\pi R^3} \sim \frac{3 \times (1.99 \times 10^{30}\,\text{kg})}{4\pi \times (6 \times 10^6\,\text{m})^3} \sim 2 \times 10^9\,\text{kg m}^{-3}$$

The freefall time scale is then

$$t_{\text{ff}} \sim \left[\frac{3\pi}{32 \times (6.67 \times 10^{-11}\,\text{m}^3\,\text{kg}^{-1}\,\text{s}^{-2}) \times (2 \times 10^9\,\text{kg m}^{-3})} \right]^{1/2} \sim 1.4\,\text{s}$$

With no pressure support, it takes a core that is the mass of the Sun and the size of Earth only about a second to collapse.

Once the core shrinks to a size of $\sim$10 km, the density is comparable to that of an atomic nucleus. The gas develops **neutron degeneracy pressure** that prevents the core from collapsing further. (This is an example of a relativistic degenerate gas, which we will study in Sect. 17.1.) The collapse releases a tremendous amount of gravitational potential energy, going from

$$U_{\text{before}} \sim -\frac{(6.67 \times 10^{-11}\,\text{m}^3\,\text{kg}^{-1}\,\text{s}^{-2}) \times (1.99 \times 10^{30}\,\text{kg})^2}{6 \times 10^6\,\text{m}} \sim -4 \times 10^{43}\,\text{J}$$

to

$$U_{\text{after}} \sim -\frac{(6.67 \times 10^{-11}\,\text{m}^3\,\text{kg}^{-1}\,\text{s}^{-2}) \times (1.99 \times 10^{30}\,\text{kg})^2}{10^4\,\text{m}} \sim -3 \times 10^{46}\,\text{J}$$

Fig. 16.7 Images of M51, the Whirlpool Galaxy. The *left panel* shows a typical view of the galaxy. In the *middle* and *right panels*, the crosshairs mark different supernovae that were observed in 2005 and 2011, respectively (Image © 2011 R. Jay GaBany, Cosmotography.com, reproduced by permission)

Something in the ballpark of 10^{46} J of gravitational energy is released. Most of it is carried by the neutrinos produced when the protons and electrons combined to form neutrons. Only a fraction of the energy goes into kinetic energy, but it is enough to create a shock wave that blows apart the gaseous envelope of the star. The result is an enormous stellar explosion called a **type II (core collapse) supernova**.[7] An even smaller fraction of the energy goes into photons—but that is still a lot of light energy. At peak brightness a supernova can have a luminosity of around $10^9\,L_\odot$, so it stands out even against the background galaxy (see Fig. 16.7). Overall, the rough energy budget for the explosion is as follows [3]:

Neutrinos	$\sim 10^{46}$ J
Kinetic energy	$\sim 10^{44}$ J
Photons	$\sim 10^{42}$ J

As a rule of thumb, a typical massive star-forming galaxy has about one supernova per century. In the Milky Way, a type II supernova observed in 1054 left a remnant known as the Crab Nebula (Fig. 16.8).[8] Another one seen in 1987 in the Large Magellanic Cloud (a satellite galaxy of the Milky Way) has provided a wealth of information about core collapse supernovae. In particular, Supernova

---

[7]There is another class of supernova, called type Ia, that occurs when a white dwarf crosses a threshold in mass known as the Chandrasekhar limit (see Sect. 17.2.2). These are the supernovae that are used to study the expansion of the universe (see Chap. 18).

[8]In addition, there were supernovae that were probably type Ia seen in the years 1006, 1572 ("Tycho's supernova"), and 1604 ("Kepler's supernova").

Fig. 16.8 The Crab Nebula is the remnant of a Type II supernova that was observed in the year 1054 (Credit: NASA, ESA, J. Hester and A. Loll) (Arizona State University)

1987A was close enough (cosmically speaking) that three neutrino detectors on Earth were able to record a total of 24 neutrinos from the explosion. While that may not seem like a large number, it was consistent with predictions given the distance to the supernova and the small cross section for neutrino interaction. This marked the first direct confirmation that neutrinos are produced in copious amounts as part of a core collapse event.

16.4.3   Beyond Iron

Where do elements heavier than iron come from? There are two sets of reactions that do not produce energy but can occur when neutrons are abundant. The first is **neutron capture**:

$$_Z^A X + n \;\rightarrow\; _Z^{A+1} X + \gamma$$

The second is **beta decay**, which turns a neutron within a nucleus into a proton and a free electron:

$$_Z^{A+1} X \;\rightarrow\; _{Z+1}^{A+1} X + e^- + \bar{\nu}_e + \gamma$$

There are two cases:

- If beta decay is more rapid than neutron capture, then heavy elements build up "slowly." This *s*-**process** tends to create stable nuclei.

- If neutrons are captured more rapidly than beta decay can eliminate them, then heavy, neutron-rich elements build up "rapidly." This is the r-**process**.

With all of this as background, we can now understand where all of the elements in the universe come from:

Hydrogen, helium, lithium	Big bang (see Chap. 20)
Beryllium through carbon and oxygen	Low- and high-mass stars
All heavier elements	High-mass stars

Nuclear processes in dying stars is responsible for all elements in the universe heavier than hydrogen, helium, and lithium.

Problems

16.1. Throughout most of the Sun the gas is fully ionized. In this problem, you may ignore elements heavier than helium, but remember to account for the free electrons.

(a) In intermediate layers of the Sun ($r \approx 0.5 R_\odot$), there are about 86 helium nuclei for every 1,000 hydrogen nuclei. What is the average particle mass $\bar{m}$ in this region, in units of the proton mass m_p?

(b) In the core of the Sun, the average particle mass approaches $0.84 m_p$. For every 1,000 hydrogen nuclei in the Sun's core, how many helium nuclei are there?

16.2. How long does it take a photon to escape from the center of the Sun? A photon has to random walk its way out because it scatters off free electrons with cross section $\sigma_T = (8\pi e^4)/(3 m_e^2 c^4) = 6.65 \times 10^{-29} \, \text{m}^2$ (the Thomson cross section). Make a rough estimate of the travel time using the average density of the Sun. (See [13] and references therein for a more detailed treatment.)

16.3. The data file for Bahcall's Standard Solar Model (Sect. 16.2.2, [4]) is available online. You may use any appropriate software to analyze the data and make plots. Note that in electronic files, we abbreviate scientific notation, so that 6.02×10^{23} becomes 6.02E+23 (or just 6.02E23) and 1.38×10^{-16} becomes 1.38E−16.

(a) Use Eq. (16.15) to compute and plot the average particle mass (in units of m_p) versus radius (in units of $R_\odot$). Make sure to label the axes with appropriate units. You should see three fairly distinct zones: (i) the inner 20 % of the Sun; (ii) the region between 20 and 95 % of the Sun's radius; (iii) the outer 5 % of the Sun. What is happening physically that distinguishes these three zones? Do your results agree with Problem 16.1?

(b) On a new graph, plot both of these quantities

$$\left|\frac{dT}{dr}\right| \quad \text{and} \quad \frac{\gamma-1}{\gamma}\frac{\bar{m}}{k}g$$

as a function of radius. Show that the outer layer of the Sun is convective. Hint: to take a derivative numerically, use

$$\frac{dT}{dr} \approx \frac{T(r_{i+1})-T(r_i)}{r_{i+1}-r_i}$$

Use the appropriate adiabatic index γ for a non-relativistic ideal monatomic gas, and recall that the acceleration due to gravity is $g(r) = GM(r)/r^2$.

16.4. Problems 12.6 and 14.3 involve a model star with a uniform density of hydrogen gas in hydrostatic equilibrium. Would such a star be stable or unstable to convection?

16.5. The lower limit to the main sequence occurs when the core temperature of a star is not sufficient to fuse hydrogen into helium. In Sect. 1.3.2 we estimated the central temperature of a star of mass M and radius R. For main sequence stars, radius and mass are correlated: $R \propto M^\alpha$ with $\alpha \approx 0.7$ [14, 15]. In Sect. 15.2.2 we estimated the central temperature required to support fusion. Put the pieces together to estimate the mass (in $M_\odot$) of the smallest star whose core is hot enough to ignite fusion.

16.6. The upper end of the main sequence occurs where radiation pressure is strong enough to make a star unstable. In Sect. 1.3.2 we used dimensional analysis to derive scaling relations for the central pressure and temperature of a star with mass M and radius R. Filling in the constants of proportionality by working in reference to the Sun, we can write

$$P_c = P_\odot \frac{(M/M_\odot)^2}{(R/R_\odot)^4} \quad \text{and} \quad T_c = T_\odot \frac{(M/M_\odot)}{(R/R_\odot)}$$

This is the pressure required to counteract the pull of gravity. If a star is hot enough that the pressure from photons (see Sect. 13.1.4),

$$P_{rad} = \frac{\pi^2(kT)^4}{45(\hbar c)^3}$$

is strong enough to counteract gravity, the star will be unstable. Find an expression (in terms of symbols) for the mass at which this occurs. Then plug in numbers to estimate the mass (in $M_\odot$) at the upper end of the main sequence.

16.7. Estimate the duration of the core hydrogen and helium burning phases in the life of a 4 $M_\odot$ star. In both phases, the star's luminosity is roughly 500 $L_\odot$ (see Fig. 16.4). Why is the helium burning phase much shorter than the hydrogen burning phase?

16.8. Supernova 1987A occurred in the Large Magellanic Cloud about 50 kpc from Earth. Models indicate that about 1.5×10^{44} J went into the kinetic energy of the explosion, with an ejected mass of about 20 $M_\odot$. Estimate the typical speed of the ejecta. About how long would it take for the ejecta to expand to subtend a radius of 0.1 arcsec on the sky such that we could resolve the debris as a supernova remnant? You may assume for simplicity that the shell expands at a constant speed, but please comment on how this assumption affects your answer.

16.9. Here is how you can derive the freefall time for gravitational collapse.

(a) Consider dropping a test particle from rest at a height r_0 above a mass M. Use conservation of energy to determine the speed v of the particle at any height r.
(b) With $v(r) = dr/dt$ we have a differential equation for r, which can be solved by writing

$$\frac{dr}{v(r)} = dt$$

and integrating both sides from the initial state ($t = 0$, $r = r_0$) to the final state ($t = t_{ff}$, $r = 0$). Evaluate the integral to find an expression for t_{ff} in terms of M, r_0, and constants. Hint: to evaluate the r integral, you may find it helpful to change variables using $r = r_0 \cos^2 \theta$.
(c) Suppose the mass M was initially spread out into a sphere of radius r_0 and density ρ_0. Rewrite t_{ff} in terms of ρ_0. (This is reasonable because the sphere collapses as the particle falls, so the preceding analysis remains valid.)

References

1. T. Glickman, American Meteorological Society, *Glossary of Meteorology* (American Meteorological Society, Boston, 2000)
2. B.W. Carroll, D.A. Ostlie, *An Introduction to Modern Astrophysics*, 2nd edn. (Addison-Wesley, San Francisco, 2007)
3. D. Maoz, *Astrophysics in a Nutshell* (Princeton University Press, Princeton, 2007)
4. J.N. Bahcall, A.M. Serenelli, S. Basu, Astrophys. J. Lett. **621**, L85 (2005)
5. J. Cox, R. Giuli, *Principles of Stellar Structure: Applications to Stars*. Principles of Stellar Structure (Gordon and Breach, New York, 1968)
6. J.L. van Saders, M.H. Pinsonneault, Astrophys. J. **746**, 16 (2012)
7. O.R. Pols, K.P. Schröder, J.R. Hurley, C.A. Tout, P.P. Eggleton, Mon. Not. R. Astron. Soc. **298**, 525 (1998)
8. J.N. Bahcall, M.H. Pinsonneault, S. Basu, Astrophys. J. **555**, 990 (2001)
9. J.R. Hurley, O.R. Pols, C.A. Tout, Mon. Not. R. Astron. Soc. **315**, 543 (2000)
10. B. Chaboyer, Phys. Rep. **307**, 23 (1998)

11. Planck Collaboration, ArXiv e-prints arXiv:1303.5076 (2013)
12. G. Audi, A.H. Wapstra, C. Thibault, Nucl. Phys. A **729**, 337 (2003)
13. R. Mitalas, K.R. Sills, Astrophys. J. **401**, 759 (1992)
14. O. Demircan, G. Kahraman, Astrophys. Space Sci. **181**, 313 (1991)
15. G. Torres, J. Andersen, A. Giménez, Astron. Astrophys. Rev. **18**, 67 (2010)

Chapter 17
Stellar Remnants

White dwarfs and neutron stars are dense objects left behind when low- and high-mass stars die (respectively). These objects have no ongoing fusion to generate the heat and pressure that normally counteract gravity, so the gas gets crushed to incredibly dense states that are quite unfamiliar to us on Earth. Essentially, gravity squeezes the gas until quantum mechanics pushes back. In this chapter we study white dwarfs in some detail and discuss neutron stars briefly.[1]

17.1 Cold, Degenerate Gas

In physics, the term **degeneracy** describes a situation in which multiple states have the same energy. If the lowest energy states in a gas are completely filled, we say the gas is degenerate. Our first task is to determine the equation of state for such a system. In Sect. 12.1.3 we derived a general expression for the pressure,

$$P = \frac{1}{3} \int p\, v\, n(\mathbf{p})\, d\mathbf{p}$$

where $n(\mathbf{p})\, d\mathbf{p}$ is the number density of particles with momentum between $\mathbf{p}$ and $\mathbf{p} + d\mathbf{p}$ (now in vector form). Let's introduce the concept of **phase space** as an abstract space in which each possible state of a system is represented as a unique point. For a particle, phase space has six dimensions; in Cartesian coordinates the six dimensions are (x, y, z, p_x, p_y, p_z), but it is possible to use other coordinate systems as well. Then n is the number of particles per unit phase space volume. This quantity is more generally known as the **phase space distribution function**,

---

[1]This presentation follows part of the book *Black Holes, White Dwarfs, and Neutron Stars: The Physics of Compact Objects* by Shapiro and Teukolsky [1], which gives considerably more detail.

C. Keeton, *Principles of Astrophysics: Using Gravity and Stellar Physics to Explore the Cosmos*, Undergraduate Lecture Notes in Physics, DOI 10.1007/978-1-4614-9236-8_17, © Springer Science+Business Media New York 2014

$$\mathscr{F} = \text{number of particles per unit phase space volume}$$

Thus, we can write a generalized version of the pressure integral as

$$P = \frac{1}{3} \int p\, v\, \mathscr{F} \, \mathrm{d}\mathbf{p} \tag{17.1}$$

In classical mechanics we treat phase space as continuous, but in quantum mechanics we think of it being discretized into cells whose 6-d volume is h^3, where h is Planck's constant.

Let's consider gas that is "cold." We will specify what this means in a moment; for now it lets us say the system will arrange itself to minimize the total energy. In an ideal gas the energy is all kinetic (there is no interaction potential), so the lowest energy state has the particles packed as close as possible to the origin of momentum space. The fundamental limitation comes from the Pauli exclusion principle: for spin-1/2 particles like protons, neutrons, or electrons, the maximum number of particles in each phase space cell is two (spin-up and spin-down). The configuration that minimizes the total energy is a sphere centered on $\mathbf{p} = 0$ that has two particles in every cell out to some momentum threshold p_F, which is referred to as the **Fermi momentum**. In other words, the distribution function for the **Fermi sphere** is

$$\mathscr{F} = \begin{cases} 2h^{-3} & p < p_F \\ 0 & p > p_F \end{cases} \tag{17.2}$$

Since white dwarfs and neutron stars emerge from the cores of dying stars, it may not be clear whether they qualify as "cold." The key is how the thermal energy (E_T, Sect. 12.1.2) compares with the kinetic energy at the surface of the Fermi sphere (E_F). If $E_T \gg E_F$, then random thermal motions prevent particles from settling into the Fermi sphere, and the gas is not degenerate. If $E_T \ll E_F$, by contrast, then particles can settle into the Fermi sphere with little or no thermal fluctuations above E_F. In scenarios where E_F is large, particles can have what we might consider to be a high temperature—they can even be relativistic—yet still qualify as "cold" in terms of the criterion for degeneracy. (See Problem 17.2 for quantitative examples.)

Given the distribution function, we can obtain the number density by integrating over all momenta:

$$n \equiv \int \mathscr{F} \, \mathrm{d}\mathbf{p} = \int_0^{p_F} \frac{2}{h^3} 4\pi p^2 \, \mathrm{d}p = \frac{8\pi p_F^3}{3h^3} \tag{17.3}$$

Turning this around, we can express the Fermi momentum in terms of the density:

$$p_F = \left(\frac{3h^3 n}{8\pi} \right)^{1/3} = \left(3\pi^2 \hbar^3 n \right)^{1/3} \tag{17.4}$$

Finally, combining the phase space distribution function (17.2) with the relativistic expression (10.27) for v lets us write the pressure integral as

$$P = \frac{8\pi}{3h^3} \int_0^{p_F} \frac{p^4 c^2}{(m^2 c^4 + p^2 c^2)^{1/2}} \, dp \qquad (17.5)$$

This is the general expression for the pressure of a cold, degenerate gas.

Non-relativistic Case

If the gas is non-relativistic, having $v \ll c$ and hence $p \ll mc$ lets us simplify the integrand to p^4/m. Then we can evaluate the integral:

$$P = \frac{8\pi}{3h^3 m} \int_0^{p_F} p^4 \, dp = \frac{8\pi}{15h^3 m} p_F^5 = \frac{8\pi}{15h^3 m} \left(\frac{3h^3 n}{8\pi} \right)^{5/3} \qquad (17.6)$$

where we use Eq. (17.3) for p_F. Collecting constants yields

$$P = \frac{(3\pi^2)^{2/3}}{5} \frac{\hbar^2}{m} n^{5/3} \qquad (17.7)$$

This is the *exact* equation of state for a non-relativistic, cold, degenerate gas. You may recall that we obtained this expression—up to the numerical factors—in Chap. 1 using dimensional analysis. Now we see where it comes from in detail.

Ultra-relativistic Case

If the gas is ultra-relativistic, having $p \gg mc$ modifies the analysis. In Problem 17.1 you can work through this case to derive the equation of state

$$P = \frac{(3\pi^2)^{1/3}}{4} \hbar c \, n^{4/3} \qquad (17.8)$$

Again, we found this expression using dimensional analysis in Chap. 1, but now we see the details (and the numerical factors).

17.2   White Dwarfs

We have seen that the pressure of a degenerate gas depends on the number density of particles, and in the non-relativistic case it depends inversely on the particle mass. White dwarfs are typically composed of ionized carbon and/or oxygen (the byproducts of the helium burning that is the last stage of fusion in a low-mass star; see Sect. 16.3). Electrons outnumber nuclei and have smaller masses, so the pressure comes mainly from electrons even though the mass is mostly in nuclei. In this section we develop a model for a star composed of a cold, degenerate electron gas and compare the model predictions with the properties of observed white dwarfs.

17.2.1 Equation of State

The analysis in Sect. 17.1 gives the pressure in terms of the number density of particles, but in astrophysics we find it more convenient to work in terms of mass density. To relate the electron number density to the mass density (which is dominated by protons and neutrons, collectively known as nucleons), we use:

$$n_e = \left(\frac{\# \text{ electrons}}{\text{nucleon}}\right)\left(\frac{\# \text{ nucleons}}{\text{volume}}\right) = \frac{Z}{A}\frac{\rho}{m_p} \tag{17.9}$$

where Z and A are the atomic number and atomic mass, respectively, and for our purposes here it is adequate to say that all nucleons have mass m_p. In the non-relativistic case we can combine Eqs. (17.7) and (17.9) to write the pressure in terms of ρ as

$$P = \frac{(3\pi^2)^{2/3}}{5}\frac{\hbar^2}{m_e}\left(\frac{Z}{A}\frac{\rho}{m_p}\right)^{5/3} \tag{17.10}$$

while in the relativistic case we instead use (17.8) to obtain

$$P = \frac{(3\pi^2)^{1/3}}{4}\hbar c\left(\frac{Z}{A}\frac{\rho}{m_p}\right)^{4/3} \tag{17.11}$$

In both cases, the pressure has the form of a **polytropic equation of state**,

$$P = K\rho^{1+1/n} \tag{17.12}$$

where the constant K depends on the gas composition through Z and A, and the **polytropic index** is

$$n = \begin{cases} 3/2 & \text{non-relativistic} \\ 3 & \text{relativistic} \end{cases}$$

(Please do not confuse this n with number density. The notation is unfortunate, but it is so common that we will stick with it. It should be clear from context whether n represents number density or polytropic index.)

17.2.2 Polytropic Stars

Let's return to the stellar structure equations in Sect. 16.2 and consider a star that has gravity and polytropic pressure, but no energy production or transport. This is a simplified but useful model for a white dwarf. (You can test the assumptions in Problem 17.2.) In this case, the key equation of stellar structure is the equation of hydrostatic equilibrium,

$$\frac{dP}{dr} = -\rho \frac{GM(r)}{r^2} \quad \Rightarrow \quad \frac{r^2}{\rho} \frac{dP}{dr} = -GM(r)$$

Take the derivative of both sides, and use the mass equation $dM/dr = 4\pi r^2 \rho$:

$$\frac{1}{r^2} \frac{d}{dr} \left(\frac{r^2}{\rho} \frac{dP}{dr} \right) = -4\pi G\rho \tag{17.13}$$

Now use the polytropic equation of state. It is convenient to introduce some new variables. Let θ be a dimensionless density variable defined by

$$\rho = \rho_c \theta^n \tag{17.14}$$

where ρ_c is the central density. Then the polytropic equation of state is

$$P = K \rho_c^{1+1/n} \theta^{n+1}$$

Also, let ξ be a dimensionless radial coordinate defined by

$$r = a\xi \quad \text{where} \quad a = \left[\frac{n+1}{4\pi G} K \rho_c^{1/n-1} \right]^{1/2} \tag{17.15}$$

It is useful to invert the last equation and solve for ρ_c:

$$\rho_c = \left[\frac{(n+1)K}{4\pi G} \right]^{n/(n-1)} a^{-2n/(n-1)} \tag{17.16}$$

The motivation for these choices becomes clear when we use the new variables in Eq. (17.13):

$$\frac{1}{a^2 \xi^2} \frac{1}{a} \frac{d}{d\xi} \left[\frac{a^2 \xi^2}{\rho_c \theta^n} K \rho_c^{1+1/n} (n+1) \theta^n \frac{1}{a} \frac{d\theta}{d\xi} \right] = -4\pi G \rho_c \theta^n$$

Simplifying, and substituting for a using Eq. (17.15), yields

$$\frac{1}{\xi^2} \frac{d}{d\xi} \left(\xi^2 \frac{d\theta}{d\xi} \right) = -\theta^n \tag{17.17}$$

This is a well-known differential equation known as the **Lane-Emden equation**. For most values of n the solutions must be found numerically, but they are well studied. We can use simple boundary conditions: $\theta(0) = 1$ by construction from

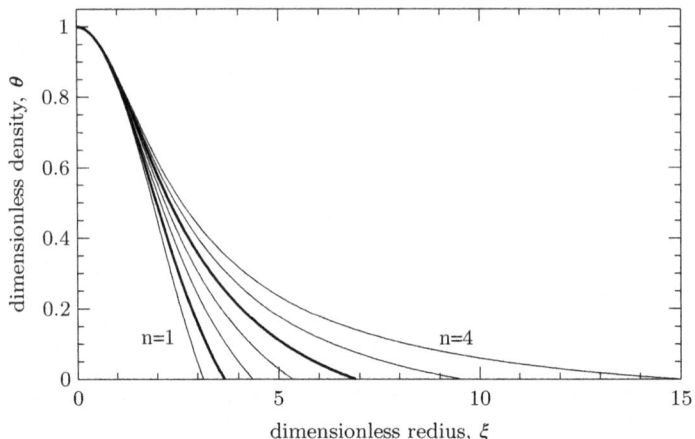

Fig. 17.1 Solutions of the Lane-Emden equation. From left to right, the curves have $n = 1$–4 in steps of $1/2$. The solutions with $n = 3/2$ and 3 are highlighted

Eq. (17.14), and $\theta'(0) = 0$ because we want the density to be smooth at the center. Then there is a unique solution for each value of n, some of which are shown in Fig. 17.1. For all cases $n < 5$ the density goes to zero at some finite value of radius, which corresponds to the surface:

$$\text{surface:} \quad \xi = \xi_1 \quad R = a\xi_1 \quad (17.18)$$

We can tabulate some important properties of the solutions that are relevant for non-relativistic and relativistic white dwarfs:

n	ξ_1	$\xi_1^2 \,|\theta'(\xi_1)|$
3/2	3.65	2.71
3	6.90	2.02

Let's use the Lane-Emden equation to understand some physical properties of the star. The total mass is

$$M = \int_0^R 4\pi r^2 \rho \, dr$$

$$= 4\pi a^3 \rho_c \int_0^{\xi_1} \xi^2 \, \theta^n \, d\xi$$

$$= 4\pi a^3 \rho_c \int_0^{\xi_1} \xi^2 \left[-\frac{1}{\xi^2} \frac{d}{d\xi} \left(\xi^2 \frac{d\theta}{d\xi} \right) \right] d\xi$$

$$= 4\pi a^3 \rho_c \, \xi_1^2 \, |\theta'(\xi_1)|$$

In the third step we use Eq. (17.17) to replace θ''; then we evaluate the integral in terms of ξ_1 and $\theta'(\xi_1)$, which are tabulated above. Now replace ρ_c using Eq. (17.16):

$$M = 4\pi a^{3-2n/(n-1)} \left[\frac{(n+1)K}{4\pi G} \right]^{n/(n-1)} \xi_1^2 \, |\theta'(\xi_1)|$$

Finally, use $a = R/\xi_1$ from Eq. (17.18):

$$M = 4\pi \, R^{(n-3)/(n-1)} \left[\frac{(n+1)K}{4\pi G} \right]^{n/(n-1)} \xi_1^{(3-n)/(n-1)} \, \xi_1^2 \, |\theta'(\xi_1)| \qquad (17.19)$$

What was the point of all of this? We have obtained a relation between mass and radius for a polytropic star. The collection of constants looks a little messy, but the important scaling is

$$M \propto R^{(n-3)/(n-1)}$$

For the **non-relativistic** case we have $n = 3/2$ and hence

$$M \propto R^{-3} \qquad \Leftrightarrow \qquad R \propto M^{-1/3}$$

We found this scaling using dimensional analysis in Chap. 1, but now we have shown it rigorously. Physically, a star with more mass needs more pressure to balance gravity. If the star is supported by degeneracy pressure, electrons need to move closer together in order for P to increase. Consequently, more massive stars must be smaller.

We can go further and fill in the constants. For the non-relativistic case we found

$$K = \frac{(3\pi^2)^{2/3}}{5} \frac{\hbar^2}{m_e} \left(\frac{Z}{Am_p} \right)^{5/3} = 9.915 \times 10^{16} \, \mathrm{kg^{-2/3} \, m^4 \, s^{-2}} \times \left(\frac{Z}{A} \right)^{5/3}$$

We listed properties of the Lane-Emden solutions in the table above. Putting everything together, we obtain for the non-relativistic case

$$M = 21.5 \, M_\odot \times \left(\frac{R}{10^4 \, \mathrm{km}} \right)^{-3} \times \left(\frac{Z}{A} \right)^5 \qquad \text{(non-relativistic)} \qquad (17.20)$$

A carbon/oxygen white dwarf has $Z/A = 0.5$, yielding

$$M = 0.67 \, M_\odot \times \left(\frac{R}{10^4 \, \mathrm{km}} \right)^{-3}$$

For comparison, the **ultra-relativistic** case has $n = 3$ and hence

$$M \propto R^0 = \text{constant}$$

Filling in the numerical factors yields

$$K = \frac{(3\pi^2)^{1/3}}{4} \, \hbar c \left(\frac{Z}{Am_p}\right)^{4/3} = 1.232 \times 10^{10} \, \text{kg}^{-1/3} \, \text{m}^3 \, \text{s}^{-2} \times \left(\frac{Z}{A}\right)^{4/3}$$

so the mass is

$$M = 5.75 \, M_\odot \times \left(\frac{Z}{A}\right)^2 \quad \text{(ultra-relativistic)} \tag{17.21}$$

This analysis reveals that *all relativistic polytropic stars have the same mass*, up to the composition-dependent factor Z/A. For $Z/A = 0.5$ as appropriate for a carbon/oxygen white dwarf, the mass is

$$M = 1.44 \, M_\odot \tag{17.22}$$

We can connect the non-relativistic and ultra-relativistic cases with the following physical picture. As a white dwarf becomes more massive, it shrinks according to $R \propto M^{-1/3}$. The rising density increases the Fermi momentum (Eq. 17.4) and makes the system increasingly relativistic. Once the star reaches $1.44 \, M_\odot$ (for $Z/A = 0.5$) it is ultra-relativistic, and it cannot get any more massive and still be supported by electron degeneracy pressure. The upper limit on the mass of a white dwarf is called the **Chandrasekhar limit** after Subramanyan Chandrasekhar, who made the theoretical prediction in 1930.

If a white dwarf exceeds the Chandrasekhar limit, it will explode as a type Ia supernova. These are objects that cosmologists have used as standard candles to chart the expanding universe (see Chap. 18).

17.2.3   Testing the Theory

To recap, our detailed analysis of white dwarfs has yielded three key conceptual points:

1. A white dwarf the mass of the Sun is about the size of Earth.
2. More massive white dwarfs are smaller, with $M \propto R^{-3}$.
3. There is a maximum allowed mass for an object supported by electron degeneracy pressure.

How can we test these predictions?

Fig. 17.2 Hubble Space Telescope image of Sirius A (the bright one) and its white dwarf companion Sirius B (the faint spot in the lower left). Sirius A appears big because it is overexposed; its size is not actually resolved. The diagonal spikes are caused by diffraction within the telescope (Courtesy: NASA, ESA, H. Bond and E. Nelan (STScI), M. Barstow and M. Burleigh (Univ. of Leicester), and J. Holberg (Univ. of Arizona))

White dwarfs were seen long ago but not recognized as particularly unusual until the early twentieth century. One is a dim star discovered by William Herschel in 1783 in the triple star system 40 Eridani [2].[2] Another is in a binary system with Sirius, the brightest star in our night sky. Between 1834 and 1844, Friedrich Bessel observed that Sirius moves as if it has a companion [3]. Christian Peters used the motion to infer the orbit in 1851, and Alvan Clark identified the companion itself in 1862 [4]. Today the Hubble Space Telescope can easily resolve the two stars (see Fig. 17.2). The orbital motion reveals that the bright star has mass $M_A = 2.0\,M_\odot$ while the faint star has mass $M_B = 1.0\,M_\odot$ [5]. The small difference in mass is surprising given the large difference in luminosity (a factor of nearly 1,000).

Even more striking are the spectral properties of these stars. In 1910, Henry Norris Russell, Edward Pickering, and Williamina Fleming used spectral classification to realize that 40 Eridani B lies far below the main sequence in the HR diagram [6, 7]. In 1915, Walter Adams discovered that the spectrum of Sirius B is very similar to that of Sirius A despite the large difference in luminosity [8]. Now we know that the faint star is actually hotter than the bright star ($T_B \approx 25{,}000$ K vs. $T_A \approx 10{,}000$ K [5]). According to the Stefan-Boltzmann law (Eq. 13.1), an object with a high temperature but low luminosity must be very small (see Problem 13.4).

Today many more white dwarfs are known, although it is still challenging to measure masses and radii precisely enough to test theoretical predictions. (For a long time, one obstacle was knowing distances well enough to convert flux to luminosity.

---

[2]The brightest stars in our sky have individual names, but most stars are labeled by the name of the constellation in which they appear on the sky, and a letter or number that indicates how they rank among stars in that constellation.

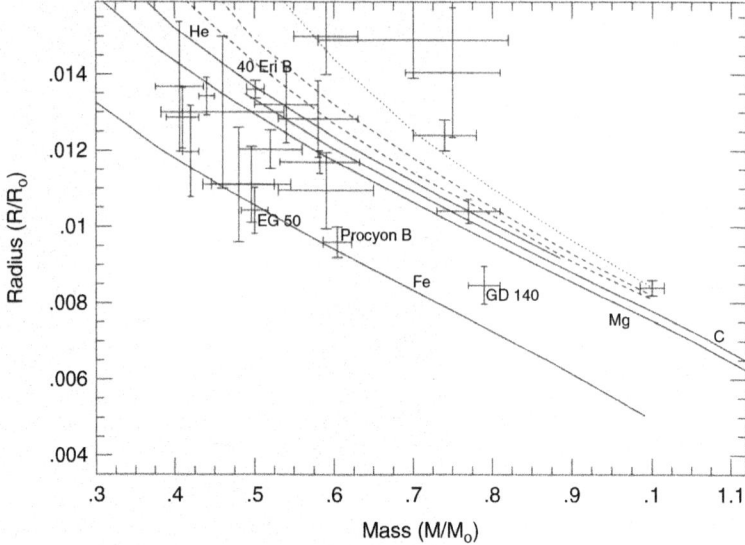

Fig. 17.3 White dwarf mass/radius relation. The curves show theoretical relations for different compositions and assumptions about the star's atmosphere. The points with errorbars show measurements of real white darfs (Credit: Provencal et al. [9]. Reproduced by permission of the AAS)

In the 1990s the Hipparcos satellite measured precise distances for a large sample of nearby stars, including some white dwarfs.) Here is a brief summary of the analysis.

Radius. This is mainly determined from the star's luminosity (inferred from its flux and distance) and effective temperature (from a spectrum). Using the Stefan-Boltzmann law (Eq. 13.1), we can write

$$L = 4\pi R^2 \sigma T^4 \quad \Rightarrow \quad R = \frac{1}{T^2}\left(\frac{L}{4\pi\sigma}\right)^{1/2} \tag{17.23}$$

Mass. Three approaches have been used, depending on what information is available.

1. *Binary star.* If the white dwarf is in a binary system, the orbital motion yields reliable masses (see Sect. 4.2).
2. *Surface gravity.* The width of spectral lines depends on the strength of gravity at the star's surface, $g = GM/R^2$. Increasing g raises the gas pressure and density, leading to more frequent collisions that perturb atomic and molecular energy levels. Measuring g from the star's spectrum and R from (17.23) makes it possible to infer the mass.
3. *Gravitational redshift.* Photons emitted by a star lose some energy as they climb out of the gravitational potential well, and thus shift to slightly longer

wavelengths (see Sect. 10.3.4). This gravitational redshift offers another way to determine the strength of gravity at the star's surface, which can be combined with the radius to find the mass.

Observational data are compared with theoretical predictions in Fig. 17.3. Many of the errorbars are fairly large, indicating the challenge in making the measurements. Nevertheless, the agreement with theory is quite good, especially for the predictions that white dwarfs are comparable in size to Earth and more massive white dwarfs are smaller. There is some scatter among the observed white dwarfs, which can be interpreted as evidence that the stars have different internal compositions and/or different atmospheres (see [9]). The bottom line is that objects made of the novel state of matter known as degenerate gas do exist, and we understand their properties.

17.3 Neutron Stars and Pulsars

Neutrons stars are also supported by degeneracy pressure, but from neutrons rather than electrons. The difference is important because degeneracy pressure depends on the *number* density of particles. Since they are so much more massive, neutrons must have a higher *mass* density than electrons to create a comparable number density and hence pressure. Neutron stars are therefore much smaller than white dwarfs—typically about 10 km in radius (see Problem 17.6).

Being so compact, neutron stars have very strong gravity, which means general relativity is needed for any detailed analysis. Furthermore, the neutrons are so close together that interactions between them cannot be ignored, which means the equation of state for dense nuclear matter plays a role as well. These two facts make neutron stars more complicated in detail than white dwarfs, but many of the key conceptual ideas are similar. (See [1] for details.)

Observationally, we study neutrons stars primarily as **pulsars**. Neutron stars tend to rotate very rapidly. (If a spinning star shrinks, it must spin faster to conserve angular momentum.) They also have strong magnetic fields (another consequence of having shrunk), which causes them to emit strong beams of radio waves from their magnetic poles. If the magnetic poles are not aligned with the spin axis, the radio beams sweep through space like beams from a lighthouse, and if one reaches Earth we detect periodic pulses of radio waves. Hence the name pulsar.

Pulsars are observed to have spin periods in the range of seconds down to milliseconds.[3] Moreover, they are extremely regular, and the periods can be measured incredibly precisely; for example, the binary pulsar system PSR J0737−3039 has an orbital period of 0.10225156248 day, with an uncertainty of ±5 in the last digit [10]. Such precision is rare in astronomy, and it makes pulsars important tools for testing

---

[3] A whole star spinning in a few milliseconds—wow!

general relativity. Pulsar discoveries have played a role in two Nobel Prizes: for Antony Hewish in 1974, and Russell Hulse and Joseph Taylor Jr. in 1993. Using pulsar timing to test relativity continues to be the focus of exciting research (e.g., [10, 11]).

Problems

17.1. Evaluate the pressure integral equation (17.5) in the ultra-relativistic case to derive the equation of state (17.8).

17.2. We have assumed the gas in a white dwarf is cold and degenerate, while the gas inside the Sun is not degenerate. Measurements suggest the assumptions are reasonable (see Fig. 17.3), but we should check the numbers. As discussed in Sect. 17.1, the key is how the thermal and Fermi energies compare. For each of the following cases, calculate E_T and E_F and determine whether the gas is degenerate.

	T_c (K)	ρ_c (kg m^{-3})	Core composition
(a) Sun today	1.6×10^7	1.5×10^5	~50/50 mix of H/He
(b) Sun on giant branch	2.7×10^7	5.1×10^7	He
(c) 5 $M_\odot$ star on giant branch	1.1×10^8	7.7×10^6	He
(d) 0.6 $M_\odot$ white dwarf	1.1×10^7	1.1×10^9	C/O

17.3. It is possible to analyze an electron gas with a finite temperature. In this case the distribution function has the form

$$\mathscr{F} = \frac{2}{h^3} \frac{1}{e^{(E-E_F)/kT} + 1} \qquad (17.24)$$

where E_F is the Fermi energy. If the gas is non-relativistic, the number density and pressure are given by

$$n = \int \mathscr{F} \, d\mathbf{p} \quad \text{and} \quad P = \frac{1}{3m} \int p^2 \, \mathscr{F} \, d\mathbf{p} \qquad (17.25)$$

(a) Plot $\mathscr{F}$ as a function of E/E_F for $kT/E_F = 0.01, 0.1$, and 0.2.
(b) Explain qualitatively whether a gas with $T > 0$ will have *higher* or *lower* pressure than a gas with $T = 0$ and the same density.
(c) Change integration variables in Eq. (17.25) to show that

$$n = \frac{4\pi}{h^3} (2m)^{3/2} E_F^{3/2} \int \frac{x^{1/2} \, dx}{e^{(x-1)E_F/kT} + 1}$$

Find an analogous expression for P. Then show that the equation of state can be written as

$$P = \frac{(3\pi^2)^{2/3}}{5} \frac{\hbar^2}{m} n^{5/3} \times \frac{I_3}{I_1^{5/3}} \tag{17.26}$$

where

$$I_1 = \frac{3}{2} \int \frac{x^{1/2}\, dx}{e^{(x-1)E_F/kT} + 1} \quad \text{and} \quad I_3 = \frac{5}{2} \int \frac{x^{3/2}\, dx}{e^{(x-1)E_F/kT} + 1}$$

(d) Equation (17.26) differs from the zero temperature case (Eq. 17.7) by the factor $I_3/I_1^{5/3}$. Use numerical integration to compute this factor for $kT/E_F = 0.01$, 0.1, and 0.2. Is the result consistent with your answer to part (b)? Would you expect a small but finite temperature to significantly change the results from this chapter?

17.4. In Problem 12.6 we studied a uniform density star (also see Problems 14.3 and 16.4). The same model can be treated using the framework of Sect. 17.2.2.

(a) What is the appropriate value of the polytropic index n for this model?
(b) Express $P(r)$ from Problem 12.6 in terms of the scaled variables as $\theta(\xi)$.
(c) Verify that your expression solves the Lane-Emden equation (17.17).

17.5. In the text we studied non-relativistic white dwarfs under the assumption $p \ll mc$. Now let's see whether the derived star properties are consistent with that assumption.

(a) Following our non-relativistic analysis in Sect. 17.2.2, find the Fermi momentum at the center of the star (i.e., using the central density), and compute the ratio $p_F/(m_e c)$. Work symbolically; express your answer in terms of the star's mass M and composition factor Z/A, along with constants.
(b) Evaluate the ratio for a carbon/oxygen white dwarf with $M = 0.7\, M_\odot$, and again for $M = M_\odot$.

17.6. In the text we computed the mass of an ultra-relativistic degenerate star. Here is how to estimate the size.

(a) Following our relativistic analysis in Sect. 17.2.2, derive an expression for the star's radius in terms of the Fermi momentum p_F, the composition factor Z/A, and constants.
(b) For an ultra-relativistic white dwarf, we expect $p_F = \eta m_e c$ where $\eta \gg 1$. Compute the star's radius in terms of η, and evaluate the result using $\eta \sim 10$. Assume a carbon/oxygen composition.
(c) To analyze a neutron star we should really use general relativity, but let's forge ahead with our Newtonian approach. Repeat part (b) assuming $p_F = \eta m_n c$ and $\eta \sim 10$. (What should you use for Z/A?)

References

1. S.L. Shapiro, S.A. Teukolsky, *Black Holes, White Dwarfs and Neutron Stars: The Physics of Compact Objects* (Wiley, New York, 1986)
2. W. Herschel, R. Soc. Lond. Philos. Trans. Ser. I **75**, 40 (1785)
3. F.W. Bessel, Mon. Not. R. Astron. Soc. **6**, 136 (1844)
4. C. Flammarion, Astron. Regist. **15**, 186 (1877)
5. J. Liebert, P.A. Young, D. Arnett, J.B. Holberg, K.A. Williams, Astrophys. J. Lett. **630**, L69 (2005)
6. E.L. Schatzman, *White Dwarfs*. Series in Astrophysics (North-Holland, Amsterdam, 1958)
7. W.S. Adams, Publ. Astron. Soc. Pac. **26**, 198 (1914)
8. W.S. Adams, Publ. Astron. Soc. Pac. **27**, 236 (1915)
9. J.L. Provencal, H.L. Shipman, E. Høg, P. Thejll, Astrophys. J. Lett. **494**, 759 (1998)
10. M. Kramer et al., Science **314**, 97 (2006)
11. J.M. Weisberg, D.J. Nice, J.H. Taylor, Astrophys. J. **722**, 1030 (2010)

Chapter 18
Charting the Universe with Stars

The physics of stars has turned out to be surprisingly important for cosmology. If we understand how stars work, we can use their observed properties to infer their intrinsic luminosities, and combine those with measured fluxes to determine distances. Finding stars at different distances then allows us to map the geometry of the universe. Two types of stars have come to play vital roles in cosmology: pulsating stars called Cepheids, and exploding white dwarfs called type Ia supernovae.

18.1 Stellar Pulsations

As we will see, we do not actually need to understand the physics of Cepheid variable stars in detail to use them as distance indicators. Nevertheless, the general ideas (if not the full details) are interesting and within our reach, so it is worthwhile to take a brief look at stellar pulsations.

18.1.1 Observations

For at least 400 years (and probably longer) people have noted that certain stars vary in brightness. In 1595, David Fabricius saw that o Ceti faded to the point that it became invisible to the naked eye, then returned to visibility, with a period of 11 months (see Fig. 18.1). He named the star Mira, meaning "wonderful." In 1784, John Goodricke discovered that δ Cephei varies with a period of a little over 5 days. This is the prototype for a class of stars now called **Cepheids**.

In the early twentieth century, Henrietta Swan Leavitt was one of the "computers" working at Harvard (like Williamina Fleming and Annie Jump Cannon, whom we encountered in Chap. 14). Leavitt discovered some 2,400 Cepheids by painstakingly comparing photographs of star fields taken at different times and identifying stars

C. Keeton, *Principles of Astrophysics: Using Gravity and Stellar Physics to Explore the Cosmos*, Undergraduate Lecture Notes in Physics, DOI 10.1007/978-1-4614-9236-8_18, © Springer Science+Business Media New York 2014

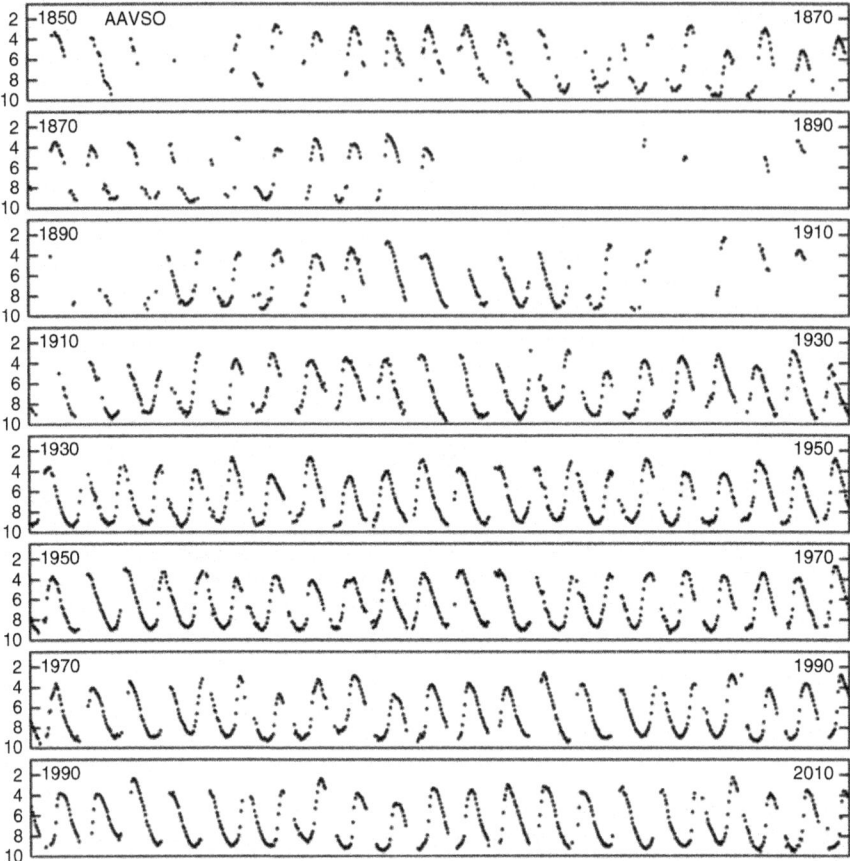

Fig. 18.1 Light curve (brightness versus time) of the variable star Mira, from 1850 to the present. Each point is the average brightness over 10 days. The vertical axis is $-2.5 \log L +$ constant; because of the minus sign, the number is smaller when the star is brighter. Courtesy of the American Association of Variable Star Observers (AAVSO)

that varied. Many of them were in the Small Magellanic Cloud, a dwarf galaxy orbiting the Milky Way. Leavitt noticed that the brighter Cepheids seemed to have longer periods. Since all the stars were in the SMC, she knew the stars that appeared to be brighter were intrinsically more luminous. To check for a connection between period and luminosity, Leavitt plotted the two quantities as shown in Fig. 18.2. Her discovery of the Cepheid **period/luminosity relation**—now known as the **Leavitt law** [1]—was a breakthrough in our ability to chart the universe (as we will see in Sect. 18.2).

Fig. 18.2 Henrietta Swan Leavitt's measurement of the relation between period and brightness for variable stars in the Small Magellanic Cloud. The horizontal axis is the logarithm of the period in days, while the vertical axis is $-2.5 \log L +$ constant. The two sets of points represent the maximum and minimum brightness for each star (Credit: Leavitt and Pickering [2])

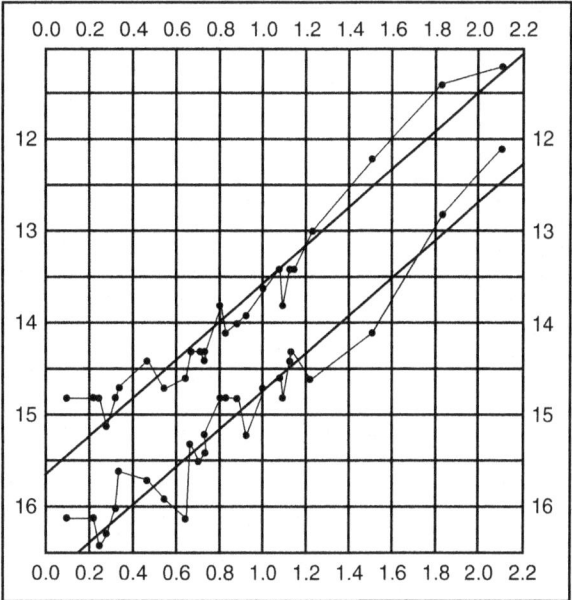

Now many different kinds of variable stars are known, with periods as long as hundreds of days (as for Mira) or as short as hours or even minutes. Several classes of variable stars, including Cepheids, lie in a particular region of the HR diagram. This "instability strip" intersects the horizontal branch of the stellar evolution tracks, implying that the stars are in the helium core burning stage of their lives. We now understand that these stars are pulsating—rhythmically expanding and contracting or undergoing even more complex oscillations.

18.1.2 Theory

Pulsations in Cepheids are thought to be driven by a mechanism known as the "Eddington valve" [3, 4]. Suppose there is region of the star where compressing the gas causes the opacity to rise. If the star contracts, the opacity increases and acts as a closed valve, trapping light and causing heat and pressure to build. Conversely, if the star expands, the opacity decreases and the valve opens. A star has to have a region with a significant amount of partially ionized helium in order for opacity to behave this way, which is why opacity-driven pulsations occur in a specific region of the HR diagram.

Treating the valve mechanism in detail is beyond the scope of our analysis, but we can make a model that captures the basic physics. The key ingredient is having pressure vary inversely with size so that squeezing a star causes it to recoil. Let's postulate an equation of state of the form[1]

$$P \propto R^{-K} \quad \Rightarrow \quad P = P_0 \left(\frac{R}{R_0} \right)^{-K} \tag{18.1}$$

Furthermore, let's assume the star expands and contracts as a whole, maintaining spherical symmetry. Then we can obtain an equation of motion by considering a spherical shell of mass m at the surface of a star of mass M and radius R. Applying Newton's second law to the shell yields

$$m \frac{d^2 R}{dt^2} = -\frac{GMm}{R^2} + 4\pi R^2 P \tag{18.2}$$

where the first term on the right-hand side represents gravity pulling inward, while the second term represents pressure pushing outward. If the star were in equilibrium it would have some radius R_0 and pressure P_0. Setting $d^2 R/dt^2 = 0$ in equilibrium, we can solve the equation of motion to find

$$P_0 = \frac{GMm}{4\pi R_0^4} \tag{18.3}$$

To analyze departures from equilibrium, let's introduce scaled variables:

$$\xi = \frac{R}{R_0} \quad \text{and} \quad \tau = \frac{t}{t_0}$$

where

$$t_0 = \left(\frac{R_0^3}{GM} \right)^{1/2} \tag{18.4}$$

is a time scale. Rewriting the equation of motion in terms of the scaled variables, and using Eqs. (18.3) and (18.1), yields

$$\frac{mR_0}{t_0^2} \frac{d^2\xi}{d\tau^2} = -\frac{GMm}{R_0^2 \xi^2} + \frac{GMm}{R_0^2} \xi^{2-K}$$

which simplifies to

$$\frac{d^2\xi}{d\tau^2} = -\xi^{-2} + \xi^{2-K} \tag{18.5}$$

---

[1] An adiabatic equation of state would fit the bill: from Eq. (16.4), $P \propto R^{-3\gamma}$ for an adiabatic process. We do not necessarily assume that pulsations are adiabatic, but we do consider the possibility below.

This is the scaled equation of motion for pulsations if the equation of state has the form of Eq. (18.1). As with polytropic stars in Sect. 17.2.2, we have made some progress with the differential equation by identifying the key physical scalings so that the remaining piece is dimensionless.

In Problem 18.1 you can work with numerical solutions to Eq. (18.5). To make further analytic process here, let's imagine that the oscillations have a small amplitude such that

$$\xi = 1 + \delta\xi \quad \text{with} \quad \delta\xi \ll 1$$

Then we can make a Taylor series expansion of Eq. (18.5) to first order in ξ and obtain:

$$\frac{d^2(\delta\xi)}{d\tau^2} \approx -(K - 4)\,\delta\xi \tag{18.6}$$

If $K > 4$, this is the equation of motion for a simple harmonic oscillator. The angular frequency in scaled units is $(K - 4)^{1/2}$, so when we put in the physical scaling from Eq. (18.4) we obtain the angular frequency and period in physical units as follows:

$$\omega = \left[\frac{GM}{R_0^3}(K - 4)\right]^{1/2} \quad \text{and} \quad \Pi = \frac{2\pi}{\omega} = 2\pi\left[\frac{R_0^3}{GM(K - 4)}\right]^{1/2} \tag{18.7}$$

We see that the period depends on the star properties through the combination M/R_0^3, which is proportional to the mean density. From the Vogt-Russell theorem of stellar structure (see Sect. 16.2.3), the radius and mean density are determined mainly by the mass, and to a lesser extent by the composition; thus, there should be a reasonably tight relation between period and mass. Since pulsating stars lie on the horizontal branch in the HR diagram, and thus are in a similar phase of evolution, there is a close relation between mass and luminosity. Putting the pieces together, we imagine there to be a reasonably tight relation between period and luminosity—which is exactly what Henrietta Swan Leavitt discovered.

What can we say quantitatively about the scaling of period with mass? If we postulate that stars on the horizontal branch have a mass–radius relation of the form $R \propto M^\alpha$, then the period scales as $\Pi \propto M^{(3\alpha-1)/2}$. If $\alpha > 1/3$ then the period will increase as the mass increases. In the examples below we see that this is indeed the case.

The period clearly depends on K. Without getting into details of the Eddington valve, we might consider a simple case in which the fluctuations are adiabatic. From Eq. (16.4), the adiabatic equation of state is $P \propto R^{-3\gamma}$ where the adiabatic index γ is the ratio of specific heats. An ideal, non-relativistic, monatomic gas has $\gamma = 5/3$ and hence $K = 5$. We use this in Eq. (18.7) and plug in numbers to work in reference to the Sun:

$$\Pi = 2\pi \left(\frac{R_\odot^3}{G\,M_\odot} \right)^{1/2} \times \frac{(R/R_\odot)^{3/2}}{(M/M_\odot)^{1/2}}$$

$$= 2\pi \left[\frac{(6.96 \times 10^8\,\text{m})^3}{(6.67 \times 10^{-11}\,\text{m}^3\,\text{kg}^{-1}\,\text{s}^{-2}) \times (1.99 \times 10^{30}\,\text{kg})} \right]^{1/2} \times \frac{(R/R_\odot)^{3/2}}{(M/M_\odot)^{1/2}}$$

$$= (1.0 \times 10^4\,\text{s}) \times \frac{(R/R_\odot)^{3/2}}{(M/M_\odot)^{1/2}}$$

$$= 0.12\,\text{day} \times \frac{(R/R_\odot)^{3/2}}{(M/M_\odot)^{1/2}}$$

This leads to the following quantitative examples:

- If the Sun were to pulsate in the way we have described, it would have a period of about 0.12 day. (It would not actually be a Cepheid, though, because it is not in the helium burning phase.)
- The star δ Cephei is the original Cepheid variable. It has $M \approx 5\,M_\odot$ and $R \approx 45\,R_\odot$ [5,6], so our model predicts $\Pi = 15.6$ day. For comparison, the observed period is 5.4 days.
- Well-studied Cepheids in our galaxy have masses in the range 4–11 $M_\odot$ and radii in the range 30–120 $R_\odot$ [6,7]. The corresponding range of periods, according to our model, is then:

$$\text{low mass}: \quad \Pi_{\text{lo}} = 0.12\,\text{day} \times \frac{30^{3/2}}{4^{1/2}} = 9.5\,\text{day}$$

$$\text{high mass}: \quad \Pi_{\text{hi}} = 0.12\,\text{day} \times \frac{120^{3/2}}{11^{1/2}} = 46\,\text{day}$$

For comparison, the observed range of periods is about 3–40 day.

Our model is not highly accurate because it is based on a simplified treatment of the pulsation physics. Nevertheless, it gives useful estimates, and it reveals in a general way how the pulsation period depends on the properties (mass and radius) of a star.

18.2   Standard Candles

The practical value of Cepheids comes from their use as distance indicators. Since we cannot lay down a ruler to another star, let alone another galaxy, we need to find indirect ways to measure distances. Distance appears, of course, in the inverse square law relating flux and luminosity,[2]

---

[2]In an expanding universe, the d in (18.8) is the "luminosity distance" (see Sect. 11.3.2).

$$F = \frac{L}{4\pi d^2} \qquad (18.8)$$

We measure F, so if we somehow knew L then we could use Eq. (18.8) to infer d. Put another way, if we had a set of **standard candles** with known luminosities, we could use their fluxes to determine their distances.

The Leavitt law made that possible by letting astronomers determine a Cepheid's luminosity from a direct measurement of its pulsation period. In fact, it was by combining Cepheid-based distance measurements with Doppler-based velocity measurements that Edwin Hubble discovered the expanding universe (see Sect. 11.1). Astronomers have continued to refine and extend the use of Cepheids as standard candles, even making it a Key Project for the Hubble Space Telescope [8].

There are some challenges in using standard candles to chart the universe. As a matter of principle, we might be concerned that we lack a complete theoretical interpretation of the Leavitt law. While we can understand the general physics of stellar pulsations (as in Sect. 18.1), that is not the same as predicting Cepheid properties from first principles. The period/luminosity relation that we use is primarily empirical, and we cannot be certain that it captures all of the important physics. For example, luminosity may depend on variables beyond period, such as color and metallicity (see [1]). If additional parameters are important, neglecting them could create scatter in the period/luminosity relation; that would be inconvenient but not terrible. More troublesome is the possibility that there could be a systematic shift in the period/luminosity relation between different galaxies that have, say, different distributions of metallicity. Fortunately, such a possibility can be tested empirically by comparing the period/luminosity relation in many different galaxies.

As a matter of practice, we need to calibrate Cepheids as standard candles before we can use them to measure distances. Strictly speaking, what Henrietta Swan Leavitt discovered was a relation between period and flux for a set of Cepheids in the Small Magellanic Cloud (SMC). Since the stars were all at (essentially) the same distance from Earth, there was a direct proportionality between flux and luminosity, but in order to determine the proportionality constant—and thereby calibrate the period/luminosity relation—Leavitt needed an independent measurement of the distance to the SMC. Today, astronomers use two main techniques to establish the calibration. One is to measure distances to Cepheids in our own Milky Way galaxy using parallax.[3] The other is to observe Cepheids in the Large Magellanic Cloud (LMC), and then measure the distance to the LMC independently (which can be done several different ways; see [1] for a review). Figure 18.3 shows that the two techniques yield consistent results for the relation between period and luminosity.

Last but not least, Cepheids can be observed (even with the Hubble Space Telescope) only in galaxies that are relatively nearby in cosmic terms. If we want to chart galaxies that are more distant, we need to find a different standard candle.

---

[3]Parallax is a kind of triangulation that uses Earth's motion around the Sun to provide a different perspective on nearby stars relative to background objects (see Sect. 2.1).

Fig. 18.3 Modern version of Leavitt law for Cepheids in the Milky Way (*filled symbols*) and the Large Magellanic Cloud (LMC, *open symbols*). The vertical axis is $-2.5 \log L$ + constant. The different shaded bands indicate brightnesses measured in different filters, from *blue* (*bottom*) to *infrared* (*top*) wavelengths (Credit: Freedman and Madore [1], reproduced by permission)

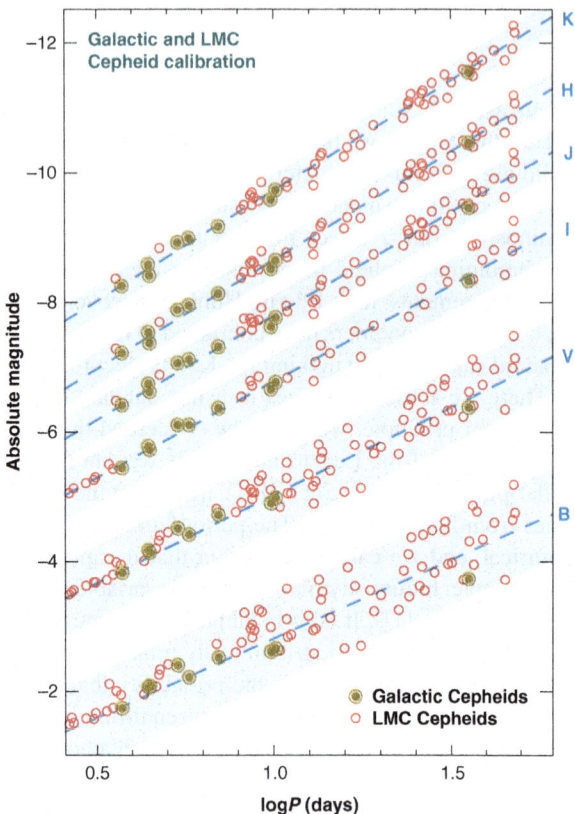

This is where type Ia supernovae became important tools for cosmology. The light curves of these exploding stars exhibit a characteristic rise and fall, and the duration and shape of the curve turn out to be related to the peak luminosity (as shown in the top panel of Fig. 18.4). Correcting for the "stretch" makes it possible to place supernovae on a common curve so they serve as standard candles (bottom panel of Fig. 18.4).

Type Ia supernovae are bright enough to be seen at great distances (out to cosmological redshifts of $z \sim 1$ and beyond, corresponding to billions of light years). As with Cepheids, though, we lack a theoretical understanding of the all-important relation—in this case, between light curve shape and luminosity. We know that type Ia supernovae occur when a white dwarf accretes enough mass from a binary companion to cross the Chandrasekhar limit and explode (see Sect. 17.2.2), but we do not know whether the companion is another white dwarf or a non-degenerate star. While we do not necessarily need to understand all of the details in order to use type Ia supernovae to measure distances, we do need to consider the possibility that the explosions might not always be the same. This, again, is an issue that can be tested empirically by observing a large sample of supernovae in different galaxies.

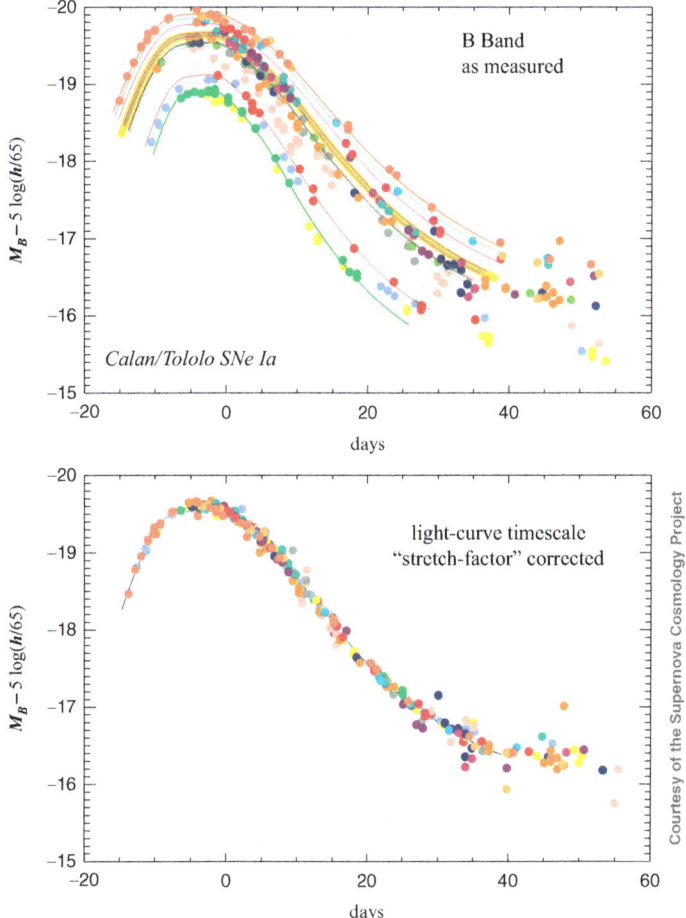

Fig. 18.4 Light curves for type Ia supernovae; different colors denote different observed explosions. For the vertical axis, $M_B = -2.5 \log L +$ constant. (The term $5 \log(h/65)$ is a way astronomers have dealt with imperfect knowledge of the Hubble constant.) For the horizontal axis, time is measured from the peak brightness; the explosion occurs a few weeks before the peak. The *top panel* shows that brighter supernovae tend to have wider light curves. In the *bottom panel*, the observed relation has been used to derive a "stretch" factor that places all of the supernovae on a common curve (Reprinted with permission from Perlmutter [9]. Copyright 2003, American Institute of Physics)

To calibrate type Ia supernovae as standard candles, we can piggyback off of Cepheids: if a supernova is observed in a galaxy whose distance has been measured with Cepheids, we can use that supernova and its distance as an anchor for measuring other supernova distances. The idea of using different but overlapping distance indicators is known as building a **distance ladder**. While modern distance ladders have a number of parallel and interlocking rungs, stellar pulsations and explosions still play a central role in charting the universe. In fact, it was Cepheid-calibrated type Ia supernovae that led to the discovery of the accelerated expansion of the universe (see Fig. 11.4).

Problems

18.1. We can study pulsations numerically to complement the analysis in Sect. 18.1.2.[4] Section A.6 discusses a computational algorithm to solve a second-order differential equation such as (18.5). Implement the algorithm in a spreadsheet or computer program. To obtain a specific solution you need to give initial conditions ξ_{init} and v_{init}; one reasonable approach is to set $v_{init} = 0$.

(a) Briefly explain how you solve the differential equation.
(b) Assume an adiabatic equation of state with $\gamma = 5/3$. Plot the size of the star and the velocity of the surface (in scaled units) assuming $\xi_{init} = 1.05$. Determine the oscillation period of your numerical solution, and explain how you did it. Calculate the expected period of small-amplitude oscillations from Eq. (18.7). How well do the period values agree?
(c) Pick a different value of K, find the numerical solution, and compare the oscillation period with the prediction from Eq. (18.7).
(d) Now go back to the case with $\gamma = 5/3$ and try $\xi_{init} = 0.65$. What is the oscillation period? Why don't these plots look like your previous ones? Why doesn't the analytic prediction do a good job here?

18.2. In this problem you can get a sense of how to use Cepheids as standard candles to measure distances.

(a) Imagine you observe 5 Cepheids in a galaxy 1 Mpc away, with the following periods and fluxes:

Star	P (day)	$F/\hat{F}$
A	10.75	1.17
B	46.53	6.80
C	6.19	0.48
D	36.07	4.15
E	7.89	0.86

Here $\hat{F}$ is a reference flux value, taken to be the flux of a Cepheid with a period of $\hat{P} = 10$ day when seen from a distance of 1 Mpc. Make a plot of $\log P$

---

[4]This problem is inspired by Problem 14.13 in the book by Carroll and Ostlie [10], but we work with the dimensionless version of the equation of motion (18.5).

and $\log F$ (similar to Fig. 18.3) and use it to calibrate a relation of the form $F/\hat{F} = (P/\hat{P})^\alpha$ by determining α. Hints: a power law relation appears as a straight line in a log-log plot; there is some "noise" in the data, so the plot will not be a perfect line.

(b) Now suppose you observe a second galaxy and discover a Cepheid with period $P_2 = 18.34$ day and flux $F_2/\hat{F} = 0.22$. How far away is galaxy #2?

18.3. In this problem you can see how to construct a distance ladder. Suppose you observe 10 galaxies: in #1 you measure a well-calibrated Cepheid; in #2–3 you see both Cepheids and type Ia supernovae; and in #4–10 you find SN Ia. Here are the data, including distance and Doppler velocity where known. (The flux units are arbitrary.) For this problem, assume Cepheids have a simple period/luminosity

	Galaxy data		Cepheid data		SN Ia data
	v (km s^{-1})	D (Mpc)	P (day)	Flux	Flux
#1		10	6.85	5.45×10^{-4}	
#2			15.57	2.07×10^{-4}	1.33×10^2
#3			18.97	9.98×10^{-5}	5.26×10^1
#4	4,675				1.99×10^1
#5	5,734				1.33×10^1
#6	7,056				8.78
#7	7,081				8.69
#8	12,361				2.85
#9	13,224				2.49
#10	13,472				2.40

relation of the form $L \propto P$, and assume the SN Ia fluxes have been corrected for light curve shape.

(a) Use the Cepheids to determine the distances to galaxies #2–3.
(b) Use the supernovae in galaxies #2–3 to calibrate the SN flux/distance relation. This amounts to finding the intrinsic luminosity of a type Ia supernova. Verify that both galaxies give consistent results.
(c) Use the supernovae to determine distances to at least two of the remaining galaxies. Then determine the Hubble constant H_0 in the relation $v = H_0 D$.

References

1. W.L. Freedman, B.F. Madore, Ann. Rev. Astron. Astrophys. **48**, 673 (2010)
2. H.S. Leavitt, E.C. Pickering, Har. Coll. Obs. Circ. **173**, 1 (1912)
3. A.S. Eddington, Obs. **40**, 290 (1917)
4. S. Webb, *Measuring the Universe: The Cosmological Distance Ladder*. Springer Praxis Books/Space Exploration (Springer, Chichester, UK, 1999)

5. J.T. Armstrong, T.E. Nordgren, M.E. Germain, A.R. Hajian, R.B. Hindsley, C.A. Hummel, D. Mozurkewich, R.N. Thessin, Astron. J. **121**, 476 (2001)
6. F. Caputo, G. Bono, G. Fiorentino, M. Marconi, I. Musella, Astrophys. J. **629**, 1021 (2005)
7. G. Bono, W.P. Gieren, M. Marconi, P. Fouqué, F. Caputo, Astrophys. J. **563**, 319 (2001)
8. W.L. Freedman et al., Astrophys. J. **553**, 47 (2001)
9. S. Perlmutter, Phys. Today **56**(4), 53 (2003)
10. B.W. Carroll, D.A. Ostlie, *An Introduction to Modern Astrophysics*, 2nd edn. (Addison-Wesley, San Francisco, 2007)

Chapter 19
Star and Planet Formation

Now that we have discussed the life and death of stars, we should say a few words about their birth as well. Stars form out of gas clouds when gravity conquers pressure, causing the clouds to collapse and fragment into pieces that can become individual solar systems. The process is complicated, and we will not try to capture all of the details, but with some judicious simplifications we can elucidate the key physics.[1]

19.1 Gravitational Collapse

Many gas clouds achieve a state of (near) equilibrium in which the inward force of gravity is balanced by the outward force from pressure. In order to form stars, the gas must be able to collapse to the high densities (and temperatures) that enable fusion. To understand how that happens, we need to understand both the state of equilibrium and the conditions for collapse.

19.1.1 Equilibrium: Virial Temperature

Consider a gas cloud in equilibrium with pressure balancing gravity. Suppose the gas is cold, so there is no fusion, and the only force is gravity. (The pressure is created by particles flying around; it is not a different force.) In Sect. 8.1.3 we proved the virial theorem, which describes the relation between the (average) kinetic and potential energies for an equilibrium system of particles interacting via gravity:

$$2K + U = 0$$

---

[1] See the books by Carroll and Ostlie [1] and Maoz [2] for additional discussions of star formation.

C. Keeton, *Principles of Astrophysics: Using Gravity and Stellar Physics to Explore the Cosmos*, Undergraduate Lecture Notes in Physics, DOI 10.1007/978-1-4614-9236-8_19, © Springer Science+Business Media New York 2014

Strictly speaking, K and U are energies averaged over time, although for a system in equilibrium we do not have to worry too much about the difference between instantaneous and time-averaged quantities.

Consider a spherical cloud of mass M and radius R, whose average particle mass is $\bar{m}$. The average mass density is $\rho = (3M)/(4\pi R^3)$, and the number density is then $n = \rho/\bar{m}$. It is common to specify the mass and number density, so let's express R in terms of them:

$$R = \left(\frac{3M}{4\pi\bar{m}n}\right)^{1/3}$$

$$= \left[\frac{3 \times (1.99 \times 10^{30}\,\text{kg})}{4\pi \times (1.67 \times 10^{-27}\,\text{kg}) \times (10^6\,\text{m}^{-3})}\right]^{1/3}$$

$$\times \left(\frac{M}{M_\odot}\right)^{1/3} \left(\frac{\bar{m}}{m_p}\right)^{-1/3} \left(\frac{n}{10^6\,\text{m}^{-3}}\right)^{-1/3}$$

$$= 2\,\text{pc} \times \left(\frac{M}{M_\odot}\right)^{1/3} \left(\frac{\bar{m}}{m_p}\right)^{-1/3} \left(\frac{n}{10^6\,\text{m}^{-3}}\right)^{-1/3} \tag{19.1}$$

From the Maxwell-Boltzmann distribution (Sect. 12.1.2), the total kinetic energy is

$$K = \frac{3}{2}NkT = \frac{3}{2}\frac{M}{\bar{m}}kT$$

The gravitational potential energy is

$$U \sim -\frac{3}{5}\frac{GM^2}{R} \sim -\frac{3}{5}\left(\frac{4\pi}{3}\right)^{1/3} G\,M^{5/3}\,\bar{m}^{1/3}\,n^{1/3}$$

The factor of $3/5$ is appropriate for a uniform density sphere. Other geometries would lead to a different dimensionless factor of order unity, but we will use the uniform sphere as a simple example. If the system is in equilibrium, we can use the virial theorem to set $2K = |U|$ and then solve for temperature:

$$T_{\text{vir}} \sim \frac{GM\bar{m}}{5kR}$$

$$\sim \frac{1}{5}\left(\frac{4\pi}{3}\right)^{1/3}\frac{G}{k}\,M^{2/3}\,\bar{m}^{4/3}\,n^{1/3}$$

$$\sim 0.049\,\text{K} \times \left(\frac{M}{M_\odot}\right)^{2/3}\left(\frac{\bar{m}}{m_p}\right)^{4/3}\left(\frac{n}{10^6\,\text{m}^{-3}}\right)^{1/3} \tag{19.2}$$

This is known as the **virial temperature**, and it indicates how hot a gas cloud must be in order to support itself against gravity.

Example: Giant Molecular Cloud

Our galaxy contains many gas clouds that are cool enough to have molecules whose rotational modes allow them to emit light at radio wavelengths (see Sect. 13.4.4). Despite the presence of the complex molecules, the gas clouds are still predominantly hydrogen. Some typical numbers are $M \sim 1,000\,M_\odot$ and $n \sim 10^8$–$10^{10}\,\mathrm{m}^{-3}$. Since the hydrogen is mostly in molecular form (H_2), the average particle mass is $\bar{m} \approx 2m_p$. For the quoted range of densities, the virial temperature is in the range $T_{\mathrm{vir}} \sim 60$–$270\,\mathrm{K}$. For comparison, the clouds are actually at temperatures more like 20 K. We will soon see what this means for such clouds.

Example: Protogalaxy

While our focus in this chapter is on stars, the physical processes that underlie the virial temperature also apply to gas clouds out of which galaxies form, so it is worthwhile to take a brief sidetrack. A protogalaxy might have something like $M \sim 10^{12}\,M_\odot$ and $R \sim 300\,\mathrm{kpc}$. This yields a mean mass density of $\rho \sim 6 \times 10^{-25}\,\mathrm{kg\,m}^{-3}$, or a mean number density of $n \sim 7 \times 10^2\,\mathrm{m}^{-3}$ for $\bar{m} = m_p/2$ as appropriate for ionized hydrogen. The corresponding virial temperature is $T \sim 2 \times 10^5\,\mathrm{K}$. Protogalaxies are quite hot (albeit diffuse) because a lot of energy is liberated by gravitational collapse.

19.1.2 Conditions for Collapse

What happens if the actual temperature is below the virial temperature? The particles move slowly, so the kinetic energy is small, and there is no way to have $2K + U = 0$. This means the virial theorem does not apply and the cloud cannot be in equilibrium. Rather, it must collapse.

James Jeans first formulated the conditions under which a cloud will collapse. He studied a general form of the problem [3, 4], but we can get at the essential physics through a toy model. Consider a cloud of mass M and radius R. Suppose we squeeze it so the radius decreases by an amount dR. (For clarity, we assume $dR > 0$ and keep track of signs explicitly.) Originally, the gravitational potential energy is

$$U \sim -\frac{3}{5}\frac{GM^2}{R}$$

while once we squeeze the cloud the potential energy decreases by

$$dU \sim -\frac{dU}{dR}dR \sim -\frac{3}{5}\frac{GM^2}{R^2}dR$$

Since the gas has pressure $P = nkT$, when we squeeze we do work on it, which increases the thermal energy by

$$\mathrm{d}K \sim P\,|\mathrm{d}V| \sim nkT\,4\pi R^2\,\mathrm{d}R \sim \frac{3M}{4\pi R^3 \bar{m}}\,kT\,4\pi R^2\,\mathrm{d}R \sim \frac{3MkT}{\bar{m}R}\,\mathrm{d}R$$

The total change in energy is then

$$\mathrm{d}E = \mathrm{d}U + \mathrm{d}K = \left(-\frac{3}{5}\frac{GM^2}{R^2} + \frac{3MkT}{\bar{m}R} \right)\mathrm{d}R \qquad (19.3)$$

If the term in parentheses is negative, it is energetically favorable for the cloud to shrink, so the cloud will spontaneously begin to collapse. In order for this to happen, the mass must exceed the following threshold:

$$M > \frac{5kTR}{G\bar{m}}$$

We can express the condition for collapse in a different way if we go back to Eq. (19.3) and write R in terms of M and the mean density ρ. This yields a limit on the mass (at fixed density):

$$M > \left(\frac{5kT}{G\bar{m}} \right)^{3/2} \left(\frac{3}{4\pi\rho} \right)^{1/2} \equiv M_J \qquad (19.4)$$

This threshold is called the **Jeans mass**. We can rewrite the threshold in terms of the density at fixed mass:

$$\rho > \frac{3}{4\pi M^2} \left(\frac{5kT}{G\bar{m}} \right)^3 \equiv \rho_J \qquad (19.5)$$

This is the **Jeans density**. The threshold values are not highly precise because they were derived under the assumptions of spherical symmetry and uniform density, but they are still useful as estimates of the conditions required for a cloud to be able to collapse under its own gravity.

19.1.3 Fragmentation

Let's go back to the example of giant molecular clouds in Sect. 19.1.1. Recall the typical numbers: mass $M \sim 1{,}000\,M_\odot$, average particle mass $\bar{m} \approx 2m_p$, and temperature $T \sim 20$ K. Then the Jeans density corresponds to a number density of

$$n_J = \frac{\rho_J}{\bar{m}} = \frac{3}{4\pi M^2 \bar{m}^4} \left(\frac{5kT}{G}\right)^3$$

$$\sim \frac{3}{4\pi (10^3 \times 1.99 \times 10^{30} \, \text{kg})^2 \times (2 \times 1.67 \times 10^{-27} \, \text{kg})^4}$$

$$\times \left[\frac{5 \times (1.38 \times 10^{-23} \, \text{kg m}^2 \, \text{s}^{-2} \, \text{K}^{-1}) \times (20 \, \text{K})}{6.67 \times 10^{-11} \, \text{m}^3 \, \text{kg}^{-1} \, \text{s}^{-2}}\right]^3$$

$$\sim 4 \times 10^6 \, \text{m}^{-3}$$

The actual density of the cloud ($n \sim 10^8$–$10^{10} \, \text{m}^{-3}$) exceeds the Jeans threshold by several orders of magnitude, so the cloud ought to be able to collapse. The fact that molecular clouds still exist suggests that having a density above the Jeans density may be a necessary but not sufficient condition for collapse. In Sect. 19.3.3 we will discuss other processes that might support a cloud against collapse.

For now, though, let's carry on and think about what happens when the cloud does start to collapse. Suppose the cloud is not perfectly uniform but has some lumps where the density is a little higher. Consider a lump of mass $\sim M_\odot$. Since $n_J \propto M^{-2}$, the Jeans density for the lump is more like $n_J \sim 10^{12} \, \text{m}^{-3}$. This is much higher than the density of the lump itself, so the lump cannot collapse any further within the cloud.

However, as the whole cloud collapses, everything becomes more dense. Eventually the density gets high enough that the $\sim M_\odot$ lump crosses its Jeans density. At that point the lump essentially separates from the global cloud and collapses on its own within the cloud.

This process can repeat itself throughout the cloud, wherever the density was a little higher than average. The cloud **fragments** into a bunch of smaller, collapsing lumps. It is those lumps that will eventually form individual stars.

The fragmentation can happen over a variety of scales, meaning there are lumps of different masses. That, in turn, leads to a population of stars with different masses. We describe the resulting distribution of masses with the **initial mass function** (**IMF**), which we write as dN/dM such that

$$\frac{dN}{dM} \, dM = \text{number of stars with mass between } M \text{ and } M + dM$$

In practice, there are a few high-mass stars and many low-mass stars. Salpeter [5] analyzed the population of stars near the Sun and inferred that the IMF could be approximated by the function

$$\frac{dN}{dM} \propto M^{-2.35} \tag{19.6}$$

at least over the range 0.4–10 $M_\odot$. Since the exponent is negative, the number of stars decreases as the mass increases. There is still a lot of work underway to understand the IMF in detail (see [6]), but we will use the Salpeter IMF as a simple model for quantitative estimates.

19.1.4   Collapse Time Scale

What happens if the Jeans criterion is satisfied and the cloud can collapse? If the temperature is low enough that the pressure is negligible, the cloud will go into freefall and collapse on the freefall time scale (see Sect. 16.4.2),

$$
t_{\mathrm{ff}} = \left(\frac{3\pi}{32G\rho} \right)^{1/2} \sim 5 \times 10^7 \, \mathrm{yr} \times \left(\frac{\bar{m}}{m_p} \right)^{-1/2} \left(\frac{n}{10^6 \, \mathrm{m}^{-3}} \right)^{-1/2} \qquad (19.7)
$$

For the giant molecular cloud we have been considering, with $\bar{m} \approx 2m_p$ and $n \sim 10^8$–$10^{10} \, \mathrm{m}^{-3}$, the freefall time scale is in the range of hundreds of thousands to millions of years. That is relatively short on astronomical time scales.

19.2   Gas Cooling

The Jeans criteria merely indicate whether it is possible for a cloud to begin to collapse; they do not say anything about what happens next. Consider: as the cloud shrinks, the gravitational potential energy decreases. By conservation of energy, the kinetic energy must increase. This corresponds to a higher temperature, which by the ideal gas law translates into more pressure. In other words, even if a cloud is initially able to collapse, at some point the pressure will rise to the point that it halts the collapse. Stars are much, much smaller than the clouds out of which they form, so star forming clouds must have some way to get rid of excess energy and prevent pressure from building up. How do they do that?

When we studied ideal gases (Sect. 12.1), we treated the gas particles as billiard balls with no internal energy states, so all collisions between particles were elastic. Real gas particles do have internal energy states, though. This makes it possible to have inelastic collisions that transfer some of the kinetic energy into particles' internal energy. Depending on the physical state of the gas, various things can happen[2]:

- The collision can break molecules apart. The **dissociation** of molecular hydrogen (H_2) absorbs 4.5 eV per molecule.
- The collision can ionize an atom. Ionization of atomic hydrogen absorbs 13.6 eV per atom.
- The collision can excite an electron inside an atom (without ionizing it). The atom will subsequently decay by emitting light. If the gas is optically thin, the light can escape, carrying away the energy.
- The collision can excite a vibrational or rotational mode of a molecule. The molecule will subsequently decay and emit light.

---

[2]We discussed collisional excitation and ionization in Sect. 14.1, and molecular vibration and rotation in Sect. 13.4.

In the first two cases the gas itself absorbs energy, while in the last two cases the gas converts kinetic energy into light. Either way, the bottom line is that kinetic energy decreases—the gas cools.

When we quantify gas cooling, it is convenient to write the energy loss rate in a form that explicitly identifies the dependence on global quantities (such as volume and density) so the piece that depends on atomic physics is scale-free. The energy loss rate has dimensions of energy/time, so it is equivalent to a luminosity. It surely scales with the overall volume of the gas cloud. It must depend on density through n^2 because collisions involve two particles and the collision rate scales with n^2. Putting the pieces together, we write

$$L_{\text{cool}} = V\, n^2\, \Lambda \tag{19.8}$$

where the factor Λ is called the **cooling function**, which has dimensions

$$[\Lambda] = \left[\frac{\text{energy}}{\text{time}} \times L^3 \right]$$

We will say more about the cooling function as it applies to star-forming clouds below. For now, we assume Λ is known and ask how the cooling rate compares with the rate at which gas is heated by collapse. If we neglect dimensionless factors, we can take $V \sim R^3 \sim M/(\bar{m}n)$ and write

$$L_{\text{cool}} \sim \frac{M}{\bar{m}}\, n\, \Lambda \tag{19.9}$$

The kinetic energy from collapse is

$$K \sim \frac{1}{2} M \left(\frac{\mathrm{d}R}{\mathrm{d}t} \right)^2$$

The heating rate is the derivative of this,

$$
\begin{aligned}
\frac{\mathrm{d}K}{\mathrm{d}t} &\sim M\, \frac{\mathrm{d}R}{\mathrm{d}t}\, \frac{\mathrm{d}^2 R}{\mathrm{d}t^2} \\
&\sim M \left(\frac{GM}{R} \right)^{1/2} \frac{GM}{R^2} \\
&\sim G^{3/2} \left(\frac{M}{R} \right)^{5/2} \\
&\sim G^{3/2}\, M^{5/3}\, (\bar{m}n)^{5/6} \tag{19.10}
\end{aligned}
$$

In the second line we use $\mathrm{d}R/\mathrm{d}t \sim (GM/R)^{1/2}$ and $\mathrm{d}^2 R/\mathrm{d}t^2 \sim GM/R^2$, which come from dimensional analysis and are valid for freefall collapse. Now let's define the **cooling efficiency** to be the ratio of the cooling and heating rates:

$$\varepsilon_{\text{cool}} \equiv \frac{L_{\text{cool}}}{dK/dt} \sim \frac{n^{1/6} \Lambda}{G^{3/2} \bar{m}^{11/6} M^{2/3}} \qquad (19.11)$$

using Eqs. (19.9) and (19.10). The quantitative scaling is

$$\varepsilon_{\text{cool}} \sim 1.4 \times 10^3 \times \left(\frac{M}{M_\odot}\right)^{-2/3} \left(\frac{\bar{m}}{m_p}\right)^{-11/6} \left(\frac{n}{10^6 \, \text{m}^{-3}}\right)^{1/6} \left(\frac{\Lambda}{10^{-42} \, \text{J m}^3 \, \text{s}^{-1}}\right)$$

The idea now is that if $\varepsilon_{\text{cool}} \gtrsim 1$ then cooling is efficient and the collapse will occur on the freefall time scale. By contrast, if $\varepsilon_{\text{cool}} \ll 1$ then cooling is inefficient so the collapse will be limited by the time it takes the gas cloud to get rid of its excess kinetic energy.

Application to Star-Forming Clouds

Let's return to the example of a giant molecular cloud with $M \sim 1,000 \, M_\odot$, $n \sim 10^8$–$10^{10} \, \text{m}^{-3}$, $\bar{m} \approx 2m_p$, and $T \sim 20 \, \text{K}$. At this temperature, molecular rotation is the main coolant because the kinetic energy is too low for collisions to dissociate molecules, ionize atoms, or even excite electrons. Neufeld et al. [7] study rotational cooling of dense molecular gas in detail. At a density of $n \sim 10^9 \, \text{m}^{-3}$ they find $\Lambda \sim 10^{-42} \, \text{J m}^3 \, \text{s}^{-1}$, which translates into a cooling efficiency of $\varepsilon_{\text{cool}} \sim 13$. Therefore the gas should be able to cool quickly enough to collapse on a time scale close to the freefall time, which is $t_{\text{ff}} \sim 10^6 \, \text{yr}$ for this density. For comparison, at a density of $n \sim 10^{10} \, \text{m}^{-3}$ Neufeld et al. find $\Lambda \sim 10^{-43} \, \text{J m}^3 \, \text{s}^{-1}$, which translates into $\varepsilon_{\text{cool}} \sim 2$. While the efficiency is lower, it should still be adequate to allow the collapse to proceed at a rate not too much slower than freefall, which has a time scale of $t_{\text{ff}} \sim 4 \times 10^5 \, \text{yr}$ for this density.

In other words, our estimates suggest that cooling should allow molecular clouds to collapse, and to do so on time scales of millions of years. Why, then, do molecular clouds still exist in a galaxy like the Milky Way, which is billions of years old? There must be other physical processes that come into play; we will mention some possibilities in Sect. 19.3.3 below.

19.3   Halting the Collapse

If freefall and cooling continued forever, gas clouds would collapse all the way to black holes. That does not happen, so we must ask what processes could slow or halt the collapse.

19.3.1 Cessation of Cooling

As the gas collapses and becomes more dense, eventually the mean free path for light becomes smaller than the size of the cloud. Once that happens, light emitted by atoms and molecules will be scattered and/or reabsorbed before it can escape. The photons begin to bounce around inside the gas, so the gas becomes opaque, and the system begins to behave like a blackbody (see Sect. 13.1). A **protostar** has formed.

The protostar will continue to contract, but more slowly than before because all it can do is convert gravitational energy into blackbody radiation. This is the Kelvin-Helmholtz mechanism that we discussed in Sect. 15.1 (when we ruled it out as the power source for the Sun). As we saw there, the time scale for Kelvin-Helmholtz collapse is $\sim 10^7$ yr for a star like the Sun. The star goes through a series of steps before igniting regular hydrogen fusion and settling onto the main sequence in the HR diagram. (Other books, such as *An Introduction to Modern Astrophysics* by Carroll and Ostlie [1], discuss pre-main-sequence evolution in more detail.)

19.3.2 Radiation Pressure

Once the protostar begins to glow as a blackbody, the light itself can exert an outward force that opposes gravity. (Recall our discussion of radiation pressure in Sect. 13.1.4.) If the luminosity is high enough, the outward force from light can overpower the inward force from gravity and halt the collapse. What is the condition for this to occur?

First consider the force from radiation pressure. If the luminosity is L, the energy flux (energy per unit area per unit time) at distance r is $L/(4\pi r^2)$. Each photon carries energy $h\nu$, so the number of photons per unit area per unit time is $L/(4\pi r^2 h\nu)$. Light interacts mainly with electrons, and the cross section for a photon to interact with an electron is given by the **Thomson cross section**,

$$\sigma_T = \frac{8\pi e^4}{3m_e^2 c^4} = 6.65 \times 10^{-29}\,\text{m}^2 \qquad (19.12)$$

Therefore the number of photons per unit time that interact with a given electron is $L\sigma_T/(4\pi r^2 h\nu)$. Each photon carries momentum $h\nu/c$, so the change in the momentum of the electron is

$$F_{\text{light}} = \frac{h\nu}{c} \times \frac{L\sigma_T}{4\pi r^2 h\nu} = \frac{2\,e^4\,L}{3\,m_e^2\,c^5\,r^2} \qquad (19.13)$$

Now consider gravity. This acts mainly on protons, because they are so much more massive than electrons; but the gas is mainly hydrogen, so there are equal numbers of protons and electrons and it is fine to work with electrons for the light force and protons for the gravity force. The force on a proton at distance r is

$$F_{\text{grav}} = \frac{G M m_p}{r^2} \tag{19.14}$$

Equating (19.13) and (19.14) lets us find the luminosity for which the outward light force exactly balances the inward gravity force:

$$L_E = \frac{3 G M m_p m_e^2 c^5}{2 e^4}$$

$$= 1.3 \times 10^{31} \, \text{kg} \, \text{m}^2 \, \text{s}^{-3} \times \left(\frac{M}{M_\odot} \right)$$

$$= 3.3 \times 10^4 L_\odot \times \left(\frac{M}{M_\odot} \right) \tag{19.15}$$

This is known as the **Eddington luminosity**. Here are some examples, which involve mature stars rather than protostars but are illustrative:

• The Sun's luminosity is well below the Eddington value for its mass, so radiation pressure has a negligible effect.
• O stars are much more luminous than the Sun. Consider an O star with $M = 60 \, M_\odot$ and $L = 8 \times 10^5 \, L_\odot$. How does the actual luminosity compare with the Eddington value for this mass?

$$\frac{L}{L_E} = \frac{8 \times 10^5 \, L_\odot}{3.3 \times 10^4 \, L_\odot \times 60} = 0.4$$

Radiation pressure does not dominate gravity, but it is hardly negligible.

Aside: Quasars

While our focus in this chapter is on stars, the effects of radiation pressure are interesting in a different context. As discussed in Sect. 3.2.3, quasars and other active galactic nuclei are small but bright objects in which the light is emitted by matter that is heated as it falls into a supermassive black hole. Infall can occur only if the luminosity does not exceed the Eddington limit:

$$L < \frac{3 G M m_p m_e^2 c^5}{2 e^4}$$

For a given luminosity, we can turn this into a lower limit on the mass of the black hole:

$$M > \frac{2\,e^4\,L}{3\,G\,m_p\,m_e^2\,c^5} \quad \text{or} \quad \frac{M}{M_\odot} > \frac{L}{3.3 \times 10^4\,L_\odot}$$

Some of the most luminous quasars have $L \sim 10^{13}\,L_\odot$, yielding

$$\frac{M}{M_\odot} \gtrsim 3.0 \times 10^8$$

Before direct dynamical measurements were available (see Sect. 3.2.2), this argument provided indirect evidence that black holes could be very massive.

19.3.3 Other Effects

In our discussion of giant molecular clouds we have found that the density typically exceeds the Jeans threshold for collapse, and cooling ought to be quite efficient, so we might wonder why the clouds still exist. The answer, presumably, is that some physical effects we have not considered play a role. One possibility is **external heating**. Light from bright stars near a gas cloud may ionize gas atoms and inject heat through the kinetic energy of the ejected electrons. Also, cosmic rays are charged particles that can carry enormous amounts of kinetic energy (up to 10^{14} MeV), some of which may be transferred to the gas through collisions.

In addition, there may be **turbulence** in the gas, and it is not easy for gas to get rid of kinetic energy associated with turbulent motion. Finally, if there is a **magnetic field** in the cloud it can provide additional support against collapse. Magnetic fields act a little bit like rubber bands threading the gas. The tension in the rubber bands provides resistance when the gas tries to collapse.

Clearly our discussion of the physics of star formation is far from complete. Nevertheless, with gravitational collapse and gas cooling we have understood some of the most important concepts.

Recap

Even our simplified discussion of star/galaxy formation has covered a lot of ground, so let's review the key elements:

- Stars form by the gravitational collapse and fragmentation of gas clouds.
- A cloud must exceed the Jeans threshold (on mass or density) in order to collapse.
- Gravitational collapse heats the gas, so a cloud needs to cool if it is to shrink enough to form stars. It does so by dissociating molecules, ionizing atoms, and/or radiating light.
- A cloud will fragment as it collapses, creating stars with a wide range of masses.

- The collapse halts with the formation of a protostar that is dense enough to trap light inside.

19.4   Protoplanetary Disks

Now that we have developed a basic picture of how stars are born, let's think about how planets form around those stars. The key idea is that angular momentum keeps some of the matter from falling into the star itself; this matter settles into a disk in which planets can grow. Our goal here is to understand gross features of our Solar System like the difference between the inner, terrestrial planets and the outer, Jovian planets. As we will see, long-standing ideas about planet formation have been challenged by the properties of planets orbiting other stars.

19.4.1   Temperature Structure

The formation of planets within a protoplanetary disk will be influenced by the temperature of the material out of which the planets form. We can predict how temperature varies with position in any sort of accretion disk if we make the following assumptions [1, 2]:

1. The disk is roughly in equilibrium, so the amount of mass and the temperature at a given position are nearly independent of time.
2. Mass is flowing *through* the disk onto the star. Let $\dot{m} \equiv \mathrm{d}m/\mathrm{d}t$ be the **mass accretion rate**.
3. The disk radiates as a blackbody.
4. The total mass in the disk is small compared with the mass of the star, so we can neglect the self-gravity of the disk.

This model is clearly simplistic. It fails to describe the processes of planet formation that are taking place within the disk. The disk must eventually disappear, so it cannot truly be in equilibrium. The model does not specify how much mass flows through the disk. Nevertheless, as we have often done in this chapter and throughout the book, we will use a simplified model to uncover some important concepts and scalings.

Consider a narrow annulus of the disk extending from radius r to $r + \mathrm{d}r$. Imagine that a small packet of mass $\mathrm{d}m$ moves through the annulus, thus changing its gravitational potential energy. For circular orbits (and, more generally, from the virial theorem), half of the potential energy goes into the kinetic energy of motion while the other half is released into the gas to be radiated away. The energy released is thus

$$dE = \frac{1}{2} dU \quad \Rightarrow \quad \frac{dE}{dr} = \frac{1}{2} \frac{dU}{dr} = \frac{1}{2} \frac{G M \, dm}{r^2} \quad \Rightarrow \quad dE = \frac{G M \, dm}{2r^2} dr$$

where we use assumption #4 to write the potential energy as $U = -G M \, dm/r$. If the energy release happens in time dt, with mass accretion rate $\dot{m} = dm/dt$ the energy release rate is

$$\frac{dE}{dt} = \frac{G M \dot{m}}{2r^2} dr$$

By assumption #1, this amount of energy must be radiated away to keep the disk in equilibrium. By assumption #3, the energy is radiated as blackbody radiation. The Stefan-Boltzmann law then says the luminosity from the annulus is the area times σT^4. The area is $2\pi r \, dr$, and there is another factor of 2 because radiation can leave both the "top" and "bottom" of the disk. Thus, the luminosity from the annulus is

$$L = 4\pi r \, \sigma \, T^4 \, dr$$

Setting $L = dE/dt$ lets us solve for the temperature:

$$T = \left(\frac{G M \dot{m}}{8\pi \, \sigma \, r^3} \right)^{1/4} \tag{19.16}$$

In our model, temperature decreases with radius in the disk with the scaling

$$T \propto r^{-3/4}$$

Recall that this scaling describes heat generated gravitationally as material falls through the disk toward the star. For comparison, in Sect. 13.2 we saw that heat generated radiatively (i.e., direct heating by starlight) yields a shallower radial dependence $T \propto r^{-1/2}$.

19.4.2 Picture of Planet Formation

The temperature structure of a protoplanetary disk becomes important when we consider what happens *within* the disk. Any small grains of ice or dust that happen to exist will occasionally bump into each other and stick together. These seeds will begin to pull in nearby material by gravity.[3] As they grow larger, they will collect more and more material, becoming **planetesimals**. Those planetesimals will themselves collide and aggregate into planets.

---

[3]We neglected the self gravity of the disk when modeling disk temperature in Sect. 19.4.1, but we cannot neglect it when discussing how planets form.

The temperature structure determines what materials can participate in this process. Based on our specific model of disk temperature, and on general physical reasoning, we expect the inner region of a protoplanetary disk to be warm. Heavy atoms and molecules (like rocky and metallic compounds) can condense into solid grains, but light atoms and molecules (like hydrogen and water) cannot. As as result, the main material that can participate in planet building here is rocks and metals—the principal constituents of terrestrial planets.[4]

Beyond some threshold in radius, which is known as the **snow line** or **frost line**, water is able to condense into ice and join in the planet formation process. In general there is more ice than rocks and metals in protoplanetary disks, so including ice allows planetesimals to become much more massive. That, in turn, lets young planets in the outer disk develop enough gravity to capture light gases like hydrogen and helium. Since those two elements are so much more abundant than everything else, any planet that can incorporate hydrogen and helium can become *much* more massive than planets that only have rocks and metals.

From Eq. (19.16) we see that the temperature scaling, and by extension the location of the frost line, depends on both the mass of the central star and the accretion rate in the protoplanetary disk. Even without knowing $\dot{m}$, we can infer that the frost line in our protoplanetary disk must have been somewhere between the orbits of Mars and Jupiter. The story we have told then explains why the inner planets are small and rocky while the outer planets are large and gaseous. But is there any actual observational support for this story? Several pieces of evidence are circumstantial but compelling:

- *Impacts.* Bodies such as the Moon and Mercury are covered with impact craters, suggesting that collisions were common at some point in the past.
- *Asteroids.* These seem to be planetesimals that were not able to aggregate into a planet.
- *Moons.* Some small moons could be planetesimals that got captured by a planet. Some large moons (like our own) could have been created by a violent collision between a planet and a big planetesimal.
- *Rotation.* Most of the planets rotate in the same direction as they orbit the Sun, which makes sense if they grew in a spinning disk. However, Venus rotates "backward," and Uranus is tipped on its side. These unusual rotations could have been produced by particularly violent collisions.

In other words, the picture we have described seems to do a good job explaining the general features of our Solar System. That is certainly reassuring, but perhaps not surprising: theories of planet formation have presumably been tuned to agree with

---

[4]Earth does have light molecules in the form of oceans and atmosphere, but they do not invalidate the general argument because they constitute a tiny fraction of Earth's mass. Also, evidence from isotopic ratios suggests that Earth received at least some of its water from planetesimals and comets that formed farther out in the protoplanetary disk (see [8] and references therein).

what we observe. To test the theories we really need to consider other planetary systems.

That has been possible since about 1995, as the discovery of multitudes of planets orbiting other stars has opened the "era of comparative exoplanetology" [9]. We discussed exoplanets in Sect. 4.3. For our purposes here, a momentous discovery was the existence of planets like Jupiter in orbits *very* close to their stars. For example, the planet orbiting HD 209458 has a mass of 0.68 M_J and a semimajor axis of just 0.047 AU. Not all exoplanets are such "hot Jupiters," but enough of them are that we need to understand how such objects came to be.

It is hard to find fault with the idea that Jovian planets can only form in the outer parts of a protoplanetary disk, beyond the frost line. So the hypothesis has emerged that hot Jupiters formed in large orbits and then **migrated** into the small orbits where they are found today. Gravitational interactions between a young planet and the remnants of the protoplanetary disk could pull the planet toward the star [10]. While the general idea makes a lot of sense, there are still a number of issues to understand in detail:

- Why did the planet stop at some specific, small radius? Why didn't it migrate all the way into the star?
- Why did some Jovian planets move a long way while others did not? In particular, why didn't Jupiter migrate very far?
- Were there other (Earth-like) planets in hot Jupiter systems? What happened to them when the big planet migrated?
- What does all of this imply for the existence of habitable planets and the possibility of extraterrestrial life?

There are ideas about how to answer these questions, but so far no complete theory that explains the full diversity of planetary systems. This is the subject of active research, so stay tuned (e.g., [11]).

Problems

19.1. In Problem 12.6 you analyzed a uniform density star and found that the pressure and temperature are not uniform. Compute the average temperature inside the star and compare it with our expression for the virial temperature.

19.2. Suppose a galaxy forms stars at an average rate of 1 $M_\odot$ yr^{-1} with a Salpeter IMF (Eq. 19.6) in the range 0.1–100 $M_\odot$. After the first $\sim$ 100 Myr, the galaxy will have a steady rate of type II supernovae from the deaths of stars more massive than 8 $M_\odot$ (see Sect. 16.4.2). What is the average time between supernovae? Hint: think about the average *number* of new stars that form each year, and the fraction of those that will explode; stellar lifetimes do not affect the answer as long as the star formation and supernova rates are steady.

19.3. The distribution of masses discussed in Sect. 19.1.3 is the *initial* mass function. The distribution changes with time because stars with different masses evolve at different rates (see Sect. 16.3). Let's make a simple model for main sequence (MS) stars. Suppose the IMF is a power law $dN/dM \propto M^{-\alpha}$ with $\alpha \approx 2.35$. Also suppose main sequence stars have a power law relation between mass and luminosity, $L \propto M^{\beta}$ with $\beta \approx 4$ [12]. The luminosity reflects the rate at which stars consume fuel, so the main sequence lifetime is roughly $\tau_{MS} \propto M/L \propto M^{1-\beta}$. The proportionalities can be calibrated with respect to the Sun, which has $\tau_{MS,\odot} \sim 10^{10}$ yr. Finally, suppose a galaxy has been forming stars at a constant rate $\dot{N}$ for time τ_{gal}.

(a) At low masses, all the stars that ever formed are still on the main sequence. How does the number of MS stars scale with mass in this regime?

(b) At high masses, only stars that formed "recently" are still on the main sequence. How does the number of MS stars scale with mass in this regime?

(c) What mass marks the transition between the two regimes? How does the transition mass vary with time?

19.4. Detailed studies of gas cooling are quite involved [7], but we can make a simple model to understand the basic physics of the cooling function Λ. Consider a gas cloud of mass M containing particles of mass m at temperature T. Low-temperature gas cools when collisions excite rotational modes that decay and emit light. Suppose the photon energy is E and the time between collisions is τ. (We assume the time for a rotational mode to decay is shorter than τ.) Then the energy loss rate per molecule is E/τ, and the energy loss rate for the whole cloud is E/τ times the number of molecules.

(a) Estimate the energy loss rate for the cloud. Hint: recall Sect. 12.1.4 and 13.4.4; where necessary, assume the molecule has some characteristic size r_0.

(b) Write your answer from (a) in the same form as Eq. (19.9) and extract an expression for Λ in our model. Apply the model to the example in Sect. 19.2. Hint: the expression involves m, T, and constants.

(c) Our model so far assumes all particles contribute equally and all collisions yield photons. In reality, cooling in low-temperature gas is dominated by carbon monoxide even though CO is rare (an abundance of about 1 part in 10^4). How does that affect our estimate of Λ? How does our final estimate compare with the values quoted in Sect. 19.2?

The model developed here is admittedly simplistic; it does not capture, for example, how Λ depends on density in detailed calculations. Nevertheless, it serves the purpose of a toy model, which is to identify the key physical principles.

19.5. We said in Sect. 19.2 that ionization can absorb energy and allow a gas cloud to shrink. Let's see how this affects a spherical cloud of hydrogen with total mass M.[5]

---

[5]This question is inspired in part by an analysis in the book by Maoz [2].

(a) Recall our analysis of hydrogen ionization in Sect. 14.1.3. Use the Saha equation (14.5) to estimate the temperature at which hydrogen can be ionized; specifically, plot the ionization fraction X as a function of temperature and find T such that $X \approx 0.5$. Assume the cloud is in equilibrium and use the analysis in Sect. 19.1.1 to express the number density n in terms of M, $\bar{m}$, and T.

(b) How big is the cloud when it has the temperature required for ionization? What is its gravitational potential energy?

(c) Ionization absorbs 13.6 eV of kinetic energy per atom. What is the new gravitational potential energy after the gas is fully ionized? How big is the cloud after ionization? Hint: recall our use of the virial theorem in Sect. 8.3.1.

19.6. The ideas we have developed for star formation can also be applied to galaxy formation. One key difference is that the total mass is larger than the normal (or "baryonic") mass because of dark matter; we need to keep track of when to use the total mass (M) and when to use the baryonic mass (M_b). The two are related by the cosmic baryon fraction $f_b = M_b/M$.

(a) Use the virial theorem to estimate the total thermal energy of a gas cloud with mass M and radius R. What is the thermal energy in baryons (K_b)?

(b) Gas cooling only involves the baryons. Drawing on Sect. 19.2, write the cooling rate L_{cool} in terms of the baryonic mass (M_b), density (ρ_b), and particle mass (m).

(c) Combine your results from (a) and (b) to estimate the cooling timescale, $t_{cool} \sim K_b/L_{cool}$.

(d) Numerical simulations of structure formation suggest that all virialized cosmic halos have the same total density, $\bar{\rho} \approx 2 \times 10^{-24} \, \mathrm{kg \, m^{-3}}$. Express R in terms of M and $\bar{\rho}$, and then show that

$$t_{cool} \sim \frac{3}{10} \left(\frac{4\pi}{3} \right)^{1/3} \frac{GM^{2/3}m^2}{f_b \bar{\rho}^{2/3} \Lambda}$$

(e) Estimate the mass of the largest object that could cool within the age of the universe (13.8 Gyr). Observations indicate $f_b = 0.17$ (see Chap. 20). Take the cooling function to be $\Lambda \sim 10^{-36} \, \mathrm{kg \, m^5 \, s^{-3}}$ for the hot gas out of which galaxies form [13].

References

1. B.W. Carroll, D.A. Ostlie, *An Introduction to Modern Astrophysics*, 2nd edn. (Addison-Wesley, San Francisco, 2007)
2. D. Maoz, *Astrophysics in a Nutshell* (Princeton University Press, Princeton, 2007)
3. J.H. Jeans, R. Soc. Lon. Philos. Trans. Ser. A **199**, 1 (1902)
4. J. Binney, S. Tremaine, *Galactic Dynamics*, 2nd edn. (Princeton University Press, Princeton, 2008)
5. E.E. Salpeter, Astrophys. J. **121**, 161 (1955)

6. N. Bastian, K.R. Covey, M.R. Meyer, Ann. Rev. Astron. Astrophys. **48**, 339 (2010)

7. D.A. Neufeld, S. Lepp, G.J. Melnick, Astrophys. J. Suppl. Ser. **100**, 132 (1995)

8. A. Izidoro, K. de Souza Torres, O.C. Winter, N. Haghighipour, Astrophys. J. **767**, 54 (2013)

9. D. Charbonneau, Bull. Am. Astron. Soc. **40**, 250 (2008)

10. D.N.C. Lin, P. Bodenheimer, D.C. Richardson, Nature **380**, 606 (1996)

11. K. Heng, Am. Sci. **101**, 184 (2013)

12. D.M. Popper, Ann. Rev. Astron. Astrophys. **18**, 115 (1980)

13. A. Dalgarno, R.A. McCray, Ann. Rev. Astron. Astrophys. **10**, 375 (1972)

Chapter 20
Cosmology: Early Universe

In Chap. 18 we saw how Cepheid stars and type Ia supernovae have been used to measure the expansion of the universe. If we run the clock backward, we deduce that in the past the universe was smaller, denser, and hotter than it is today. In this chapter we use gas physics and particle physics to understand the early, hot phase of the universe, and we discuss observations that probe this phase directly (through the cosmic microwave background radiation) and indirectly (through the abundances of elements created in the first few minutes after the big bang).

20.1 Cosmic Microwave Background Radiation

When the universe was young, it was hot enough for gas to be ionized. The many free electrons were very effective at scattering light, so the gas was optically thick and photons were tightly coupled to matter. As the universe expanded and cooled, the electrons and ions were able to come together to form neutral atoms. Suddenly the photons were liberated, free to travel great distances through the universe. Those photons are still around and visible as the **cosmic microwave background (CMB)** radiation.

These ideas were initially developed in the late 1940s by Ralph Alpher, George Gamow, and Robert Herman (see [1]), and later by Robert Dicke and Jim Peebles (see [2]). In the early 1960s, Arno Penzias and Robert Wilson were working with a new microwave antenna in Holmdel, NJ. They measured a low level of noise no matter what direction they pointed the antenna, which persisted despite all efforts to clean the antenna and eliminate sources of noise (including pigeons). They realized the signal was consistent with blackbody radiation at a temperature ~ 3 K, but did not know how to interpret such a signal. Eventually they heard that Dicke, Peebles, and others at Princeton University had predicted microwave radiation from the early

C. Keeton, *Principles of Astrophysics: Using Gravity and Stellar Physics to Explore the Cosmos*, Undergraduate Lecture Notes in Physics, DOI 10.1007/978-1-4614-9236-8_20,
© Springer Science+Business Media New York 2014

universe and even set out to look for it.[1] The Holmdel and Princeton groups got together, and the rest is history [4,5]. Penzias and Wilson won the 1978 Nobel Prize in Physics for their discovery of the CMB.

20.1.1 Hot Big Bang

We said the universe was hotter when it was younger, but to study the CMB we need to quantify that statement. In Chap. 11 we discussed the theoretical framework for describing the expansion of the universe. For our purposes here, there are two key results. First, we characterize the expansion using the dimensionless scale factor $a(t)$ such that distances scale as a and volumes scale as a^3 relative to today. Second, the expansion causes light waves to stretch, creating the cosmological redshift. From Eqs. (11.19) and (11.20), the ratio of observed and emitted wavelengths is

$$\frac{\lambda_{\text{obs}}}{\lambda_{\text{em}}} = \frac{1}{a}$$

Suppose the universe today is filled with blackbody radiation with some temperature T_0. In the past, the wavelengths were all smaller by the factor a. What was the corresponding temperature? From the Planck spectrum (Eq. 13.8), or equivalently Wien's law (Eq. 13.13), wavelength and temperature are related by $\lambda T = \text{constant}$, which immediately implies

$$T \propto a^{-1} \quad \Rightarrow \quad T = \frac{T_0}{a} \tag{20.1}$$

This allows us to characterize how the universe has cooled as it has expanded.

We also need to specify how the density has changed. Density times volume is mass, so if total mass is conserved[2] then ρa^3 is constant. In other words, conservation of mass implies

$$\rho \propto a^{-3} \quad \Rightarrow \quad \rho = \frac{\rho_0}{a^3} \tag{20.2}$$

Combining Eqs. (20.1) and (20.2), we can write a relation between density and temperature:

$$\frac{\rho}{\rho_0} = \left(\frac{T}{T_0}\right)^3 \tag{20.3}$$

---

[1] See [3] for more of the story.
[2] While some mass is converted to energy via fusion in stars, it is a tiny fraction of the total.

20.1.2   Theory: Recombination Temperature

As the universe expanded, two effects caused photons to **decouple** from matter. As the density decreased, the photon's mean free path increased; at some point this effect alone would have made the gas optically thin. But another process was also underway: as the universe cooled, electrons and ions were able to combine to form neutral atoms in a process known as **recombination**. The gas effectively became transparent, leaving the universe filled with a bath of photons that are free but carry a memory of the physical conditions of the universe at the time they were released.

We can estimate the temperature of the universe at recombination, which helps us understand the state of the universe that we study when we observe the CMB (also see [6]). For simplicity, let's assume the universe was pure hydrogen. Let n_b and ρ_b be the total number and mass density of baryons, which in a hydrogen universe just means protons.[3] Let X be the ionization fraction, so the number densities of free electrons and ions are $n_e = n_i = X n_b$ while the number density of neutral hydrogen atoms is $n_H = (1 - X)n_b$. If the gas was in equilibrium, the Saha equation (14.3) gives the ratio of ionized to neutral atoms to be

$$\frac{n_i}{n_H} = \frac{X}{1 - X} = \frac{2\, Z_{II}}{X\, n_b\, Z_I} \left(\frac{2\pi\, m_e\, k\, T}{h^2}\right)^{3/2} e^{-\chi_I / kT} \qquad (20.4)$$

(Note that X appears on both the left- and right-hand sides.) We can relate the baryon number density, n_b, to its value today using Eq. (20.3):

$$n_b = \frac{\rho_b}{m_p} = \frac{\rho_{b0}}{m_p} \frac{T^3}{T_0^3}$$

Since $n_b \propto T^3$, the net temperature scaling on the right-hand side of Eq. (20.4) is $T^{-3/2} \exp(-\chi_I / kT)$. For hydrogen, the relevant numbers are:

$$Z_I = 2$$
$$Z_{II} = 1$$
$$\chi_I = 13.6\,\text{eV}$$
$$\rho_{b0} = 4.14 \times 10^{-28}\,\text{kg}\,\text{m}^{-3}$$
$$T_0 = 2.725\,\text{K}$$

---

[3]Electrons are leptons, and they contribute so little mass compared to baryons that we neglect them when characterizing the mass density of the universe. They are important when it comes to charge, though.

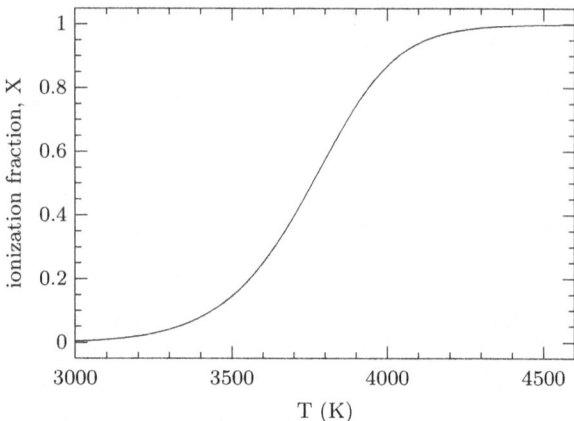

Fig. 20.1 Equilibrium ionization fraction as a function of temperature, for a universe of pure hydrogen

Plugging in the numbers lets us rewrite Eq. (20.4) as

$$\frac{X^2}{1-X} = 1.57 \times 10^{17} \left(\frac{kT}{eV}\right)^{-3/2} e^{-13.6\,eV/kT}$$

This is a quadratic equation that we can solve to plot X as a function of temperature, as shown in Fig. 20.1. According to this simple estimate, recombination should have occurred between about 3,500 and 4,000 K. More detailed analyses account for the fact that a photon released when one atom combined could reionize a nearby neutral atom, and place everything in an expanding universe (e.g., [7]). Those analyses indicate that the temperature had to be a little lower, around 3,000 K, for recombination to be complete.[4]

Incidentally, inverting Eq. (20.1) lets us determine the scale factor at recombination:

$$a_{recomb} = \frac{T_0}{T_{recomb}} \approx \frac{2.7\,K}{3,000\,K} \approx \frac{1}{1,100}$$

At the time we observe with the CMB, distances were more than a 1,000 times smaller, and densities more than a billion times larger, than they are today.

20.1.3 Observations

Mapping the CMB is a vital part of cosmology today. The first detailed maps were obtained by the Cosmic Background Explorer (COBE, launched in 1989). The CMB spectrum measured by COBE (Fig. 20.2) matches a theoretical blackbody spectrum

---

[4]Recall that we made a similar calculation for the Sun's hydrogen in Sect. 14.1.3 and found a higher transition temperature of $\sim 10^4$ K. In the Sun, the higher density facilitates recombination.

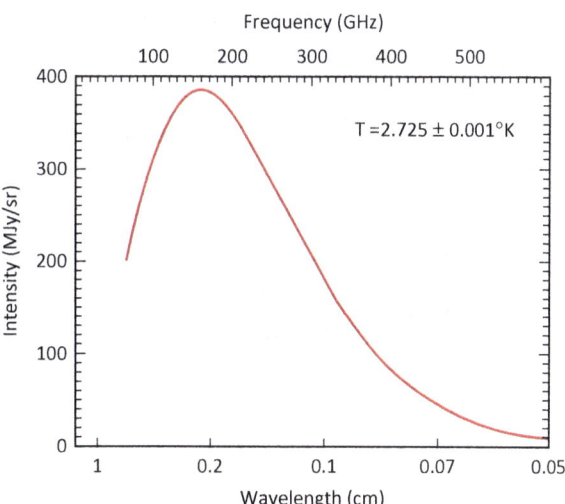

Fig. 20.2 CMB spectrum measured by the FIRAS instrument on COBE. The intensity is plotted as a function of frequency (top axis); the corresponding wavelengths are shown on the bottom axis. The curve shows a theoretical blackbody spectrum with a temperature of 2.725 K. Points with errorbars are shown, but they are actually too small to see (Credit: NASA/WMAP)

phenomenally well; in fact, the CMB is the best blackbody known. This is the first piece of evidence that we truly understand what was happening in the early universe when the CMB was produced. COBE also found that the CMB is not quite uniform: there are small **anisotropies**, or directions where the temperature is slightly warmer or cooler than the average. George Smoot and John Mather won the 2006 Nobel Prize in Physics for their discoveries with COBE.

In the years since COBE, a number of instruments have been used to map the CMB (see Fig. 20.3). The Wilkinson Microwave Anisotropy Probe (WMAP, launched in 2001 [8]) observed the full sky with better resolution than COBE. The ground-based Atacama Cosmology Telescope (ACT [9]) and South Pole Telescope (SPT [10]) have produced maps that cover only a portion of the sky but have even higher resolution than WMAP. The Planck spacecraft (launched in 2009 [11]) recently mapped the full sky at a resolution higher than WMAP but not quite as high as SPT and ACT.

The anisotropies are small—less than 1 part in 10,000—but measurable. In a map like Fig. 20.3, it is apparent that warmer and colder regions tend to have a characteristic angular size. We quantify this effect in terms of the **angular power spectrum**, shown in Fig. 20.4. Roughly speaking, you can think of creating a circle with some particular angular size, computing the average temperature within that circle, and then measuring the variations as you move the circle around the sky. Repeating the process for circles with different sizes yields the power spectrum. The CMB power spectrum shows a prominent peak at about 1°, which is the main scale visible in Fig. 20.3, but there are significant features on other scales as well.

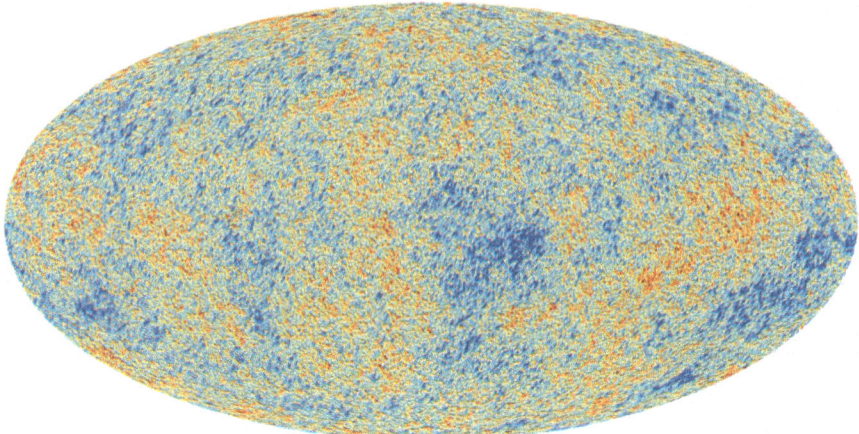

Fig. 20.3 Map of temperature fluctuations in the CMB, from the Planck spacecraft. In this projection, the plane of the Milky Way galaxy runs through the middle of the map from left to right (© ESA and the Planck Collaboration [11])

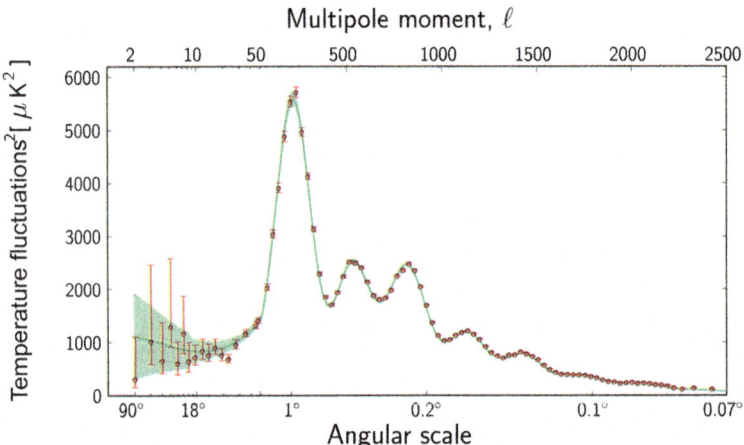

Fig. 20.4 Angular power spectrum of temperature fluctuations in the CMB. The horizontal axis shows the angular scale running from large angles on the left to small angles on the right. (Formally, the power spectrum is computed in terms of the multipole moment ℓ, indicated on the top axis.) The vertical axis quantifies the amplitude of temperature fluctuations. Points with errorbars are Planck measurements; the curve indicates a ΛCDM model fit to the data (© ESA and the Planck Collaboration [11])

20.1.4 Implications

The anisotropies in the CMB are important for two reasons. First, they represent seeds from which galaxies grew. What we see as temperature fluctuations actually correspond to variations in density. Overdensities create cold spots in the temperature map because photons lose energy as they climb out of a gravitational potential well; this is known as the **Sachs-Wolfe effect** [12]. In places where the density was a little higher than average, gravity was a little stronger, which tended to pull in more matter from nearby, causing the overdensity to grow, making gravity even stronger, drawing even more matter, and ... well, you get the picture. There was a runaway process that turned tiny overdensities into the large structures we see today as galaxies and even clusters of galaxies. We know those all-important seeds were there because we observe them in the CMB.

Second, the anisotropies constrain the geometry and composition of the universe. Before recombination, the material in the universe was governed by a fairly straightforward combination of gas physics and gravity. Over- and underdensities acted as sound waves propagating through a nearly-uniform medium in an expanding universe. The sound waves had a characteristic physical scale, which we can compare with the measured angular scale to determine the geometry of the universe. Also, the power spectrum of fluctuations depended on the relative abundances of normal matter, dark matter, and dark energy in the universe. Although we will not delve into the details, it is possible to predict the CMB power spectrum for different assumptions about the composition of the universe. Adjusting the predictions to match the data (as shown by the curve in Fig. 20.4) yields strong constraints on cosmological parameters. (See [13] for a full discussion of cosmological constraints from Planck.) Constraints from the CMB and other datasets are shown in Fig. 11.6. It is striking is that three very different ways of probing the universe yield consistent results. Even if we do not yet know what dark matter and dark energy are, we think we know how much of each substance the universe contains.

20.2 Big Bang Nucleosynthesis

The CMB provides direct access to the physical state of the universe when it was about 380,000 years old. We can reach back even further—to when the universe was only a few minutes old—by using the idea that the very young universe was a nuclear reactor.

20.2.1 Theory: "The First Three Minutes"

When the universe was young and hot, it was filled with a sea of elementary particles: protons, neutrons, electrons, positrons, neutrinos, antineutrinos, and photons. Why not anti-protons and anti-neutrons? This is the unsolved question of **baryogenesis**: why is there a (slight) asymmetry favoring matter over antimatter in our universe? If there were exact symmetry between matter and antimatter, there would have been equal numbers of baryons and anti-baryons in the early universe, and they would have annihilated to leave a universe filled only with photons. Particle physicists are still trying to understand the origin of the asymmetry, but for our purposes we simply accept that there is matter in the universe and attempt to understand how the density of matter affects things we can measure.

Early on, the temperature and density were high enough to allow reactions among the particles[5]:

$$n + e^+ \rightleftharpoons p + \bar{\nu}_e \qquad (20.5a)$$

$$n + \nu_e \rightleftharpoons p + e^- \qquad (20.5b)$$

$$n \rightleftharpoons p + e^- + \bar{\nu}_e \qquad (20.5c)$$

How high did the temperature have to be for such reactions to occur? The neutron weighs a little more than the proton, so there is a Boltzmann factor that describes their relative numbers (in thermodynamic equilibrium):

$$\frac{N_n}{N_p} \sim e^{-(m_n - m_p)c^2/kT}$$

The mass difference corresponds to energy difference

$$\Delta E_{np} = (m_n - m_p)c^2 = 1.29\,\text{MeV}$$

Roughly speaking, then, protons and neutrons could be in thermodynamic equilibrium only when

$$kT \gtrsim 1.29\,\text{MeV} \quad \Rightarrow \quad T \gtrsim 1.5 \times 10^{10}\,\text{K}$$

As the universe cooled, two things happened. First, the "inverse" reactions slowed down and protons began to outnumber neutrons. Second, while the protons and neutrons were doing their thing, electron/positron pairs were forming and annihilating:

$$e^- + e^+ \rightleftharpoons 2\gamma$$

Electrons and positrons each have a mass of 0.51 MeV, so this reaction could occur only when

$$kT \gtrsim 1.02\,\text{MeV} \quad \Rightarrow \quad T \gtrsim 1.2 \times 10^{10}\,\text{K}$$

Once the temperature fell below about 1 MeV, it was no longer possible to form new electron/positron pairs. As existing pairs annihilated, the decreasing number of electrons made it harder to form neutrons (see Eqs. 20.5b and 20.5c). When all the details are taken into account, it turns out that the above reactions ceased when

---

[5]For more discussion of particle physics in the early universe, see [6, 14, 15].

the temperature cooled to about 0.8 MeV. This is called **neutron freezeout**, and it (momentarily) fixed the relative abundance of protons and neutrons to be

$$\frac{N_n}{N_p} \approx e^{-1.3\,\text{MeV}/0.8\,\text{MeV}} \approx 0.20$$

I say "momentarily" because free neutrons spontaneously decay by the process in Eq. (20.5c), with a half-life of just 615 s. The remaining neutrons did begin to decay, and the neutron/proton ratio fell to

$$\frac{N_n}{N_p} \approx \frac{1}{7} \tag{20.6}$$

Another process kicked in once the temperature cooled to about 0.1 MeV: neutrons could combine with protons to form deuterium,

$$n + p \rightarrow {}^2_1\text{H} + \gamma$$

Neutrons that are bound into nuclei are stable, so this was a key step that locked in the (primordial) abundance of neutrons. Once deuterium formed, it could go through various reactions to create helium-4 (see, e.g., [16]). One channel involves hydrogen-3,

$$ {}^2_1\text{H} + {}^2_1\text{H} \rightarrow {}^3_1\text{H} + p $$
$$ {}^3_1\text{H} + {}^2_1\text{H} \rightarrow {}^4_2\text{He} + n $$

while another involves helium-3,

$$ {}^2_1\text{H} + p \rightarrow {}^3_2\text{He} + \gamma $$
$$ {}^2_1\text{H} + {}^2_1\text{H} \rightarrow {}^3_2\text{He} + n $$
$$ {}^3_2\text{He} + {}^2_1\text{H} \rightarrow {}^4_2\text{He} + p $$

The process could even produce a little lithium and beryllium:

$$ {}^4_2\text{He} + {}^3_1\text{H} \rightarrow {}^7_3\text{Li} + \gamma $$
$$ {}^4_2\text{He} + {}^3_2\text{He} \rightarrow {}^7_4\text{Be} + \gamma $$

It is possible to work out the reaction rates in detail, taking into account the expansion and cooling, and thus predict the abundances of different elements as a function of time. Figure 20.5 shows the results and reveals two important points. First, all of the action occurred when the universe was just a few hundred seconds old. This is reflected in the title of a famous book by Steven Weinberg about the formation of the elements: *The First Three Minutes* [18].

Fig. 20.5 Predicted
abundances of light elements
as a function of time in the
early universe ("p" and "n"
indicate protons and neutrons,
respectively). The vertical
axis is mass fraction; note the
enormous range reflected in
the logarithmic scale (Credit:
Coc et al. [17]. Reproduced
by permission of the AAS)

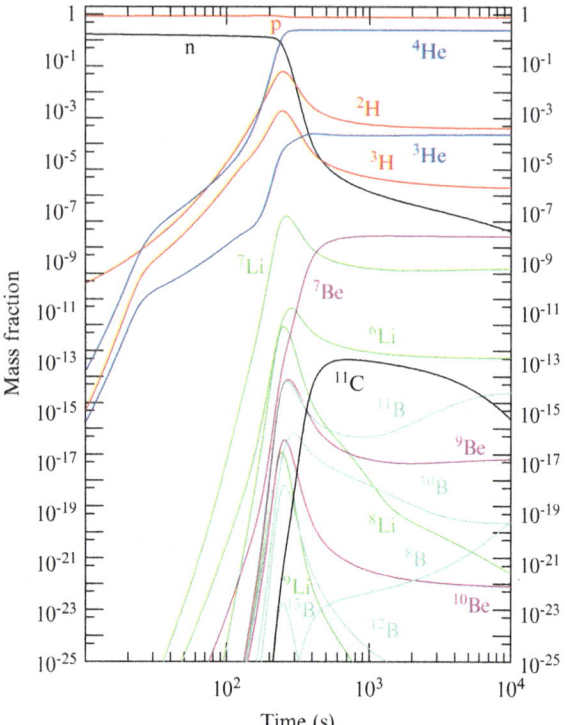

Second, the binding energy of helium-4 makes it very stable (recall our discussion of fusion in Chap. 15), so the vast majority of neutrons wound up in helium-4. We can estimate the mass fraction of helium-4 in the primordial gas as follows. Each helium-4 nucleus has 2 neutrons. From (20.6), there must be about 14 protons for those 2 neutrons. Two of the protons are in the helium-4 with the neutrons, leaving 12 protons left over. Thus, there are about 12 hydrogens for every helium. If we consider the *mass* fraction that is in helium, we have

$$\frac{M_{\text{He}}}{M_{\text{H}} + M_{\text{He}}} \approx \frac{1 \times 4m_p}{12 \times m_p + 1 \times 4m_p} \approx \frac{4}{16} \approx 0.25$$

In other words, we predict that about 25 % of the mass of the primordial gas was helium (and almost all the rest was hydrogen). This follows directly from the neutron/proton ratio and is not very sensitive to other details of the nuclear processes.

While most of the deuterium and helium-3 went into helium-4, small amounts stuck around. It is possible to analyze the reactions and predict the relative abundances of all the different elements. The results depend, not surprisingly, on the total density of baryons in the universe, specifically in the combination $\Omega_b h^2$ where $h = H_0/(100 \, \text{km s}^{-1} \, \text{Mpc}^{-1})$ is a dimensionless version of the Hubble constant.

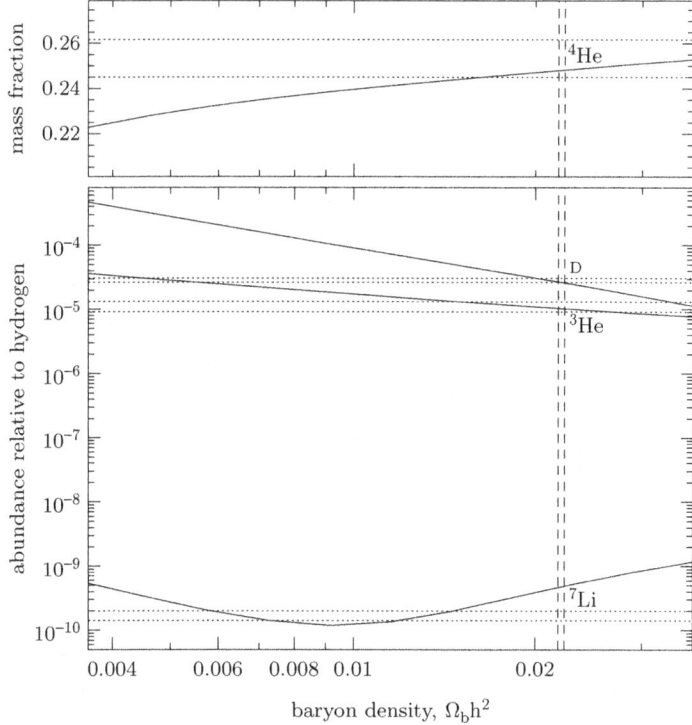

Fig. 20.6 The *solid curves* show theoretical predictions from Burles et al. [19] for the abundance of light elements as a function of the cosmic density of baryons; the natural density parameter is $\Omega_B h^2$ where $h = H_0/(100\,\mathrm{km\,s^{-1}\,Mpc^{-1}})$ is a dimensionless version of the Hubble constant. For the *top panel* the vertical axis is mass fraction, while for the *bottom panel* the vertical axis is the number ratio relative to hydrogen. The horizontal bands with *dotted lines* indicate measurements of primordial abundances (with the band thickness indicating uncertainties) for $^4_2$He [20], D [21], $^3_2$He [22], and $^7_3$Li [23]. The vertical band with *dashed lines* indicates the constraint on $\Omega_b h^2$ from Planck [13]

Figure 20.6 shows the predicted abundances of light elements as a function of this density parameter. The formation of elements in the early universe is known as **big bang nucleosynthesis (BBN)**, and Fig. 20.6 encapsulates the key theoretical results.

20.2.2 Observations: Primordial Abundances

With predictions in hand, we would like to test them observationally. The challenge is figuring out how to uncover the *primordial* abundances of elements, because much of the gas in the universe has been "polluted" by the lives and deaths of stars in the 13.8 billion years since big bang nucleosynthesis.

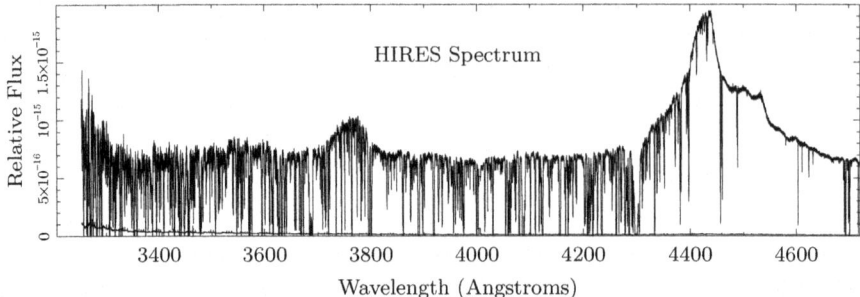

Fig. 20.7 Spectrum of the quasar HS 0105 + 1619 at redshift $z_Q = 2.64$ (when the universe was about 2.5 Gyr old). Lyman-α light from the quasar is emitted with wavelength 1,216 Å but redshifted to the observed wavelength $(1 + z_Q)1,216\,\text{Å} = 4,425\,\text{Å}$. The "forest" of absorption lines at shorter wavelengths correspond to gas clouds between us and the quasar. A cloud at redshift z produces a Ly-α absorption line at wavelength $(1 + z)1,216\,\text{Å}$ (Credit: O'Meara et al. [25]. Reproduced by permission of the AAS)

With a primordial mass fraction of about 25 %, helium-4 would seem to be the easiest element to measure. But helium-4 is also created by fusion in stars. Most stars produce oxygen before they die and release their helium-4 back into the interstellar medium (see Sect. 16.3), so one strategy has been to measure the abundance of helium-4 in gas clouds that contain very little oxygen. Such measurements imply that primordial gas does have an abundance of helium-4 that matches predictions from big bang nucleosynthesis (see [24] for a review).

To probe elements that are more rare, one trick is to look far away, and therefore back in time. If we are hunting for gas that does not contain stars, we cannot rely on *emission* of light from the gas; instead, we search for *absorption* of light that passes through the gas. Quasars provide ideal "flashlights" for this experiment, because they are distant—creating a good chance that the light passed through a gas cloud when the universe was younger—yet bright enough to observe. (The most distant quasars date from when the universe was only ~1 Gyr old.) High-resolution spectra of quasars indeed reveal a "forest" of absorption lines from gas clouds that lie along the intervening line of sight, as shown in Fig. 20.7. Zooming in on different portions of the spectrum, we can identify all the different lines in the Lyman series that are produced by a *single* gas cloud, as shown in Fig. 20.8.

We can use such spectra to search for deuterium. The electron shell structure is similar to that of hydrogen, but the presence of a neutron in the nucleus perturbs the energy levels and shifts the absorption to slightly shorter wavelengths than in hydrogen. With a high-quality spectrum, it is possible to distinguish the deuterium lines from the hydrogen lines (see Fig. 20.8), and thus to measure the abundance of deuterium relative to hydrogen in the distant gas cloud. If we find similar deuterium abundances in a variety of systems, we can infer that it reflects the primordial abundance.

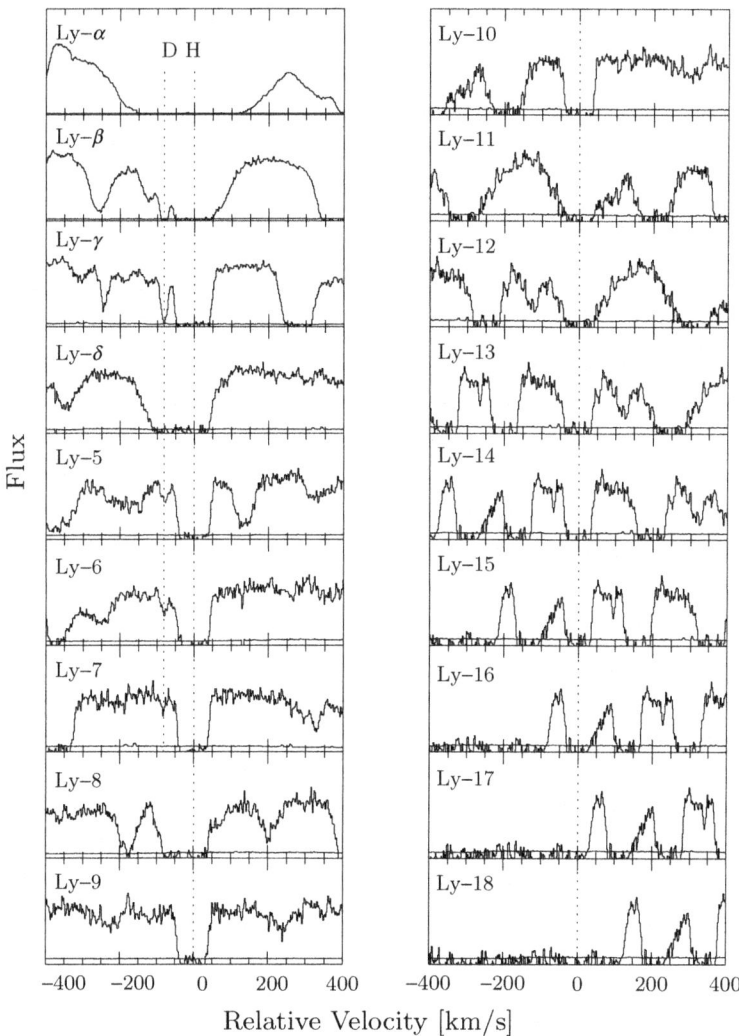

Fig. 20.8 Close-up of various absorption lines in the spectrum from Fig. 20.7. To put the panels on a common scale, wavelength offsets have been converted to velocity offsets using $\Delta v = c \, \Delta \lambda$. The absorption lines shown here are all part of the Lyman series produced by a single gas cloud. In several lines (notably Ly-β through Ly-6) there is clear evidence of absorption by deuterium just blueward of the absorption by hydrogen (Credit: O'Meara et al. [25]. Reproduced by permission of the AAS)

There are ways to recover the primordial abundances of helium-3 and lithium-7 as well, as discussed by Weiss [24]. The measurements are indicated by the horizontal bands in Fig. 20.6. For deuterium, helium-3, and helium-4, the measurements agree well with the theoretical predictions for the value of $\Omega_b h^2$ inferred from Planck observations of the CMB. In other words, for these three elements at least, we seem to have a consistent picture of what the universe was doing when it was just a few minutes old. Curiously, the measured abundance of lithium-7 is a little lower than expected for the relevant value of $\Omega_b h^2$, and we do not yet know what to make of that result. It is thought that lithium-7 might be easier to destroy than other elements, which could explain the low measured abundance. But it is also possible that some "exotic" processes in the early universe could have modified the production of lithium-7 [26]. Either way, we should learn something interesting from the discrepancy between theory and observations.

Notwithstanding the lithium problem, the overall picture suggests that we have a remarkably good understanding of what the universe was like in the first few minutes after the big bang. Together, the CMB and BBN provide exceedingly strong evidence that normal matter makes up a small fraction of the universe, and that the universe is dominated by two exotic substances about which we still have much to learn.

20.3 How Did We Get Here?

This brings to a close our story of how the elements were created: inside stars, and in the early universe. Combining that with our understanding of gas (for the CMB) and stars (especially type Ia supernovae), we can assemble a fairly detailed census of the contents of the universe: today the universe contains about 5 % normal matter, 27 % dark matter, and 68 % dark energy. And with our knowledge of (astro)physics we can reconstruct the history of the universe. To recap: the composition of the primordial gas was determined when the universe was about 3 min old. The gas went through recombination when the universe was around 380,000 years old, and we can observe this with the CMB. In the intervening 13.8 billion years, large gas clouds collapsed under the influence of gravity to become galaxies. Within galaxies, smaller clouds cooled and collapsed to form stars. Inside stars, the density, temperature, and pressure rose to the point that nuclear fusion could begin, releasing energy and creating all the elements heavier than hydrogen, helium, and lithium. At the end of their lives, those stars released heavy elements to the interstellar medium, where they could be incorporated into new stars and planets. Our planet and our bodies are made from elements forged by nuclear fusion in earlier generations of stars; and our lives today are powered by fusion in the Sun. What is even more remarkable than the story itself, I think, is that we can use physical principles to figure it all out. I hope you have enjoyed the journey!

Problems

20.1. Consider the transition from ionized to neutral gas in the early universe. Before recombination, the interaction between photons and ionized hydrogen was dominated by electron scattering characterized by the Thomson cross section, $\sigma_T = 6.65 \times 10^{-29}\,\mathrm{m}^2$.

(a) What was the density of baryons ρ_b just before recombination at redshift $z \approx 1{,}100$? How far (in pc) could an average photon travel before scattering?

(b) The early universe was opaque because of the ionized gas, and then became transparent after recombination. Later on (sometime in the redshift range $z \sim 6$–20), the universe was *reionized* by radiation from stars and quasars. If ionized hydrogen gas pervades the universe today, why isn't the universe opaque now? Be specific and quantitative.

20.2. If the universe were filled with helium rather than hydrogen, when would recombination have occurred? See Problem 14.1 for the ionization stages of helium. Assume the density and temperature today are the same as in our actual universe.

20.3. In Sect. 13.1.4 we thought about blackbody radiation as a gas of photons.

(a) Use dimensional analysis to estimate the energy density in blackbody radiation with temperature T. Hint: photons are both relativistic and quantum entities.

(b) What is the energy density in CMB radiation today? Estimate the corresponding number density of CMB photons, and the ratio of CMB photons to baryons in the universe today.

References

1. R.A. Alpher, R. Herman, Phys. Today **41**, 24 (1988)
2. P.J.E. Peebles, Astrophys. J. **142**, 1317 (1965)
3. P. Peebles, *Principles of Physical Cosmology*. Princeton Series in Physics (Princeton University Press, Princeton, 1993)
4. R.H. Dicke, P.J.E. Peebles, P.G. Roll, D.T. Wilkinson, Astrophys. J. **142**, 414 (1965)
5. A.A. Penzias, R.W. Wilson, Astrophys. J. **142**, 419 (1965)
6. B.W. Carroll, D.A. Ostlie, *An Introduction to Modern Astrophysics*, 2nd edn. (Addison-Wesley, Massachusetts, 2007)
7. E. Kolb, M. Turner, *The Early Universe*. Frontiers in Physics (Addison-Wesley, Massachusetts, 1994)
8. C.L. Bennett et al., Astrophys. J. Suppl. Ser. **208**, 20 (2013)
9. S. Das et al., Phys. Rev. Lett. **107**(2), 021301 (2011)
10. R. Keisler et al., Astrophys. J. **743**, 28 (2011)
11. Planck Collaboration, ArXiv e-prints arXiv:1303.5062 (2013)
12. R.K. Sachs, A.M. Wolfe, Astrophys. J. **147**, 73 (1967)
13. Planck Collaboration, ArXiv e-prints arXiv:1303.5076 (2013)
14. D. Maoz, *Astrophysics in a Nutshell* (Princeton University Press, Princeton, 2007)
15. A. Weiss, Einstein Online **2**, 1018 (2006)
16. K.M. Nollett, S. Burles, Phys. Rev. D **61**(12), 123505 (2000)
17. A. Coc, S. Goriely, Y. Xu, M. Saimpert, E. Vangioni, Astrophys. J. **744**, 158 (2012)
18. S. Weinberg, *The First Three Minutes: A Modern View of the Origin of the Universe* (BasicBooks, New York, 1993)
19. S. Burles, K.M. Nollett, M.S. Turner, Astrophys. J. Lett. **552**, L1 (2001)
20. E. Aver, K.A. Olive, E.D. Skillman, J. Cosmol. Astropart. Phys. **4**, 004 (2012)
21. F. Iocco, G. Mangano, G. Miele, O. Pisanti, P.D. Serpico, Phys. Rep. **472**, 1 (2009)
22. T.M. Bania, R.T. Rood, D.S. Balser, Nature **415**, 54 (2002)
23. C. Charbonnel, F. Primas, Astron. Astrophys. **442**, 961 (2005)

24. A. Weiss, Einstein Online **2**, 1019 (2006)
25. J.M. O'Meara, D. Tytler, D. Kirkman, N. Suzuki, J.X. Prochaska, D. Lubin, A.M. Wolfe, Astrophys. J. **552**, 718 (2001)
26. B.D. Fields, Annu. Rev. Nucl. Part. Sci. **61**, 47 (2011)

Part III
Appendices

Appendix A
Technical Background

A.1 Cartesian and Polar Coordinates

We often study motion that is confined to a plane (e.g., orbits in a spherically symmetric gravitational field), so it is worthwhile to review 2-d coordinate systems. In standard **Cartesian coordinates** the position vector is written as

$$\mathbf{r} = x\,\hat{\mathbf{x}} + y\,\hat{\mathbf{y}}$$

(Alternate notations include $\mathbf{r} = r_x\,\hat{\mathbf{x}} + r_y\,\hat{\mathbf{y}} = x\,\hat{\mathbf{i}} + y\,\hat{\mathbf{j}} = x\,\hat{\mathbf{e}}_x + y\,\hat{\mathbf{e}}_y$.) Here $\hat{\mathbf{x}}$ and $\hat{\mathbf{y}}$ are **unit vectors**, which means their lengths are $|\hat{\mathbf{x}}| = |\hat{\mathbf{y}}| = 1$.

If an object moves in two dimensions as a function of time, we can describe its motion as $\mathbf{r}(t) = x(t)\,\hat{\mathbf{x}} + y(t)\,\hat{\mathbf{y}}$. In Cartesian coordinates, the unit vectors are independent of position and hence independent of time, so we can write the velocity and acceleration vectors as

$$\mathbf{v} \equiv \frac{d\mathbf{r}}{dt} = \frac{dx}{dt}\,\hat{\mathbf{x}} + \frac{dy}{dt}\,\hat{\mathbf{y}}$$

$$\mathbf{a} \equiv \frac{d\mathbf{v}}{dt} = \frac{d^2x}{dt^2}\,\hat{\mathbf{x}} + \frac{d^2y}{dt^2}\,\hat{\mathbf{y}}$$

There are no surprises or complications when the unit vectors are constant.

We can also write the position vector in **polar coordinates**, in which we specify the object's radius r and azimuthal angle ϕ as shown in Fig. A.1. The angle is defined to go counterclockwise by convention. From trigonometry, we have

$$x = r\cos\phi \quad \text{and} \quad y = r\sin\phi$$

C. Keeton, *Principles of Astrophysics: Using Gravity and Stellar Physics to Explore the Cosmos*, Undergraduate Lecture Notes in Physics, DOI 10.1007/978-1-4614-9236-8, © Springer Science+Business Media New York 2014

Fig. A.1 Illustration of
Cartesian and polar
coordinates in two
dimensions. For point A, the
polar coordinates (r, ϕ) are
indicated. The Cartesian unit
vectors $\hat{\mathbf{x}}$ and $\hat{\mathbf{y}}$ are shown
with *solid arrows*, while the
polar unit vectors $\hat{\mathbf{r}}$ and $\hat{\boldsymbol{\phi}}$ are
shown with *dashed arrows*.
For point B, the Cartesian
unit vectors are the same but
the polar unit vectors are
different

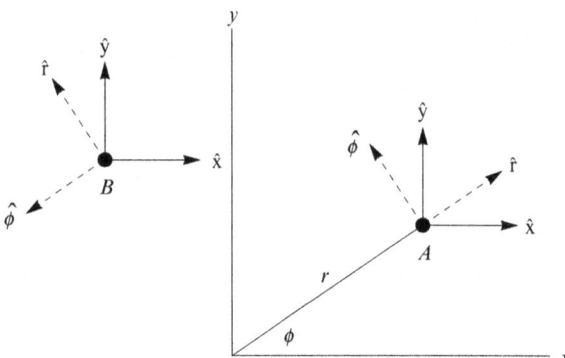

along with the inverse relations

$$r = \left(x^2 + y^2\right)^{1/2} \quad \text{and} \quad \phi = \tan^{-1}\frac{y}{x}$$

We can define unit vectors in polar coordinates, which are also shown in Fig. A.1.
The radial unit vector, $\hat{\mathbf{r}}$, always points away from the origin:

$$\hat{\mathbf{r}} = \frac{\mathbf{r}}{r} \quad \Leftrightarrow \quad \mathbf{r} = r\,\hat{\mathbf{r}}$$

The angular unit vector, $\hat{\boldsymbol{\phi}}$, is perpendicular to $\hat{\mathbf{r}}$ and defined to point in the direction
in which ϕ increases (i.e., counterclockwise). Expressing the polar unit vectors in
Cartesian coordinates gives

$$\hat{\mathbf{r}} = \cos\phi\,\hat{\mathbf{x}} + \sin\phi\,\hat{\mathbf{y}} \quad \text{and} \quad \hat{\boldsymbol{\phi}} = -\sin\phi\,\hat{\mathbf{x}} + \cos\phi\,\hat{\mathbf{y}}$$

Because the polar unit vectors depend on position, they change with time as an
object moves:

$$\frac{d\hat{\mathbf{r}}}{dt} = -\sin\phi\frac{d\phi}{dt}\hat{\mathbf{x}} + \cos\phi\frac{d\phi}{dt}\hat{\mathbf{y}} = \frac{d\phi}{dt}\hat{\boldsymbol{\phi}}$$

$$\frac{d\hat{\boldsymbol{\phi}}}{dt} = -\cos\phi\frac{d\phi}{dt}\hat{\mathbf{x}} - \sin\phi\frac{d\phi}{dt}\hat{\mathbf{y}} = -\frac{d\phi}{dt}\hat{\mathbf{r}}$$

This makes sense: because their lengths are fixed, the unit vectors can only change
direction, so the derivative of each unit vector is perpendicular to that unit vector.
   We can now define the velocity and acceleration vectors in polar coordinates.
The velocity is fairly straightforward:

$$\mathbf{v} \equiv \frac{d\mathbf{r}}{dt} = \frac{dr}{dt}\hat{\mathbf{r}} + r\frac{d\hat{\mathbf{r}}}{dt} = \frac{dr}{dt}\hat{\mathbf{r}} + r\frac{d\phi}{dt}\hat{\boldsymbol{\phi}}$$

The acceleration is a little more complicated:

$$
\mathbf{a} \equiv \frac{d\mathbf{v}}{dt} = \frac{d^2r}{dt^2}\hat{\mathbf{r}} + \frac{dr}{dt}\frac{d\hat{\mathbf{r}}}{dt} + \frac{dr}{dt}\frac{d\phi}{dt}\hat{\boldsymbol{\phi}} + r\frac{d^2\phi}{dt^2}\hat{\boldsymbol{\phi}} + r\frac{d\phi}{dt}\frac{d\hat{\boldsymbol{\phi}}}{dt}
$$

$$
= \frac{d^2r}{dt^2}\hat{\mathbf{r}} + \frac{dr}{dt}\frac{d\phi}{dt}\hat{\boldsymbol{\phi}} + \frac{dr}{dt}\frac{d\phi}{dt}\hat{\boldsymbol{\phi}} + r\frac{d^2\phi}{dt^2}\hat{\boldsymbol{\phi}} - r\left(\frac{d\phi}{dt}\right)^2\hat{\mathbf{r}}
$$

or, after collecting terms,

$$
\mathbf{a} = \left[\frac{d^2r}{dt^2} - r\left(\frac{d\phi}{dt}\right)^2\right]\hat{\mathbf{r}} + \left[r\frac{d^2\phi}{dt^2} + 2\frac{dr}{dt}\frac{d\phi}{dt}\right]\hat{\boldsymbol{\phi}}
$$

$$
= \left[\frac{d^2r}{dt^2} - r\left(\frac{d\phi}{dt}\right)^2\right]\hat{\mathbf{r}} + \frac{1}{r}\frac{d}{dt}\left(r^2\frac{d\phi}{dt}\right)\hat{\boldsymbol{\phi}}
$$

In the second step we rewrite the angular term in a form that is sometimes convenient.

A.2 Cylindrical and Spherical Coordinates

To generalize polar coordinates to three dimensions, one option is to keep polar coordinates in the (x, y) plane and add a simple Cartesian component in the z direction. This leads to **cylindrical coordinates** (R, ϕ, z), as shown in the left panel of Fig. A.2. Here $\hat{\mathbf{R}}$ and $\hat{\boldsymbol{\phi}}$ behave just like the unit vectors in basic polar

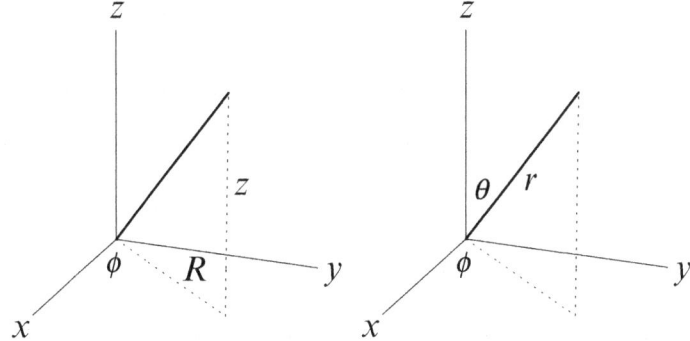

Fig. A.2 The *left panel* illustrates cylindrical coordinates (R, ϕ, z) while the *right panel* illustrates spherical coordinates (r, θ, ϕ). In both cases, the azimuthal angle ϕ is measured in the (x, y) plane. In cylindrical coordinates, R is also measured in the (x, y) plane, and z is perpendicular to the plane. In spherical coordinates, r is the distance from the origin to the point, and θ is the polar angle measured from the z-axis

coordinates, while $\hat{\mathbf{z}}$ is constant. We can therefore write down the position, velocity, and acceleration vectors:

$$\mathbf{r} = R\,\hat{\mathbf{R}} + z\,\hat{\mathbf{z}}$$

$$\mathbf{v} = \frac{dR}{dt}\hat{\mathbf{R}} + R\frac{d\phi}{dt}\hat{\boldsymbol{\phi}} + \frac{dz}{dt}\hat{\mathbf{z}}$$

$$\mathbf{a} = \left[\frac{d^2 R}{dt^2} - R\left(\frac{d\phi}{dt}\right)^2\right]\hat{\mathbf{R}} + \frac{1}{R}\frac{d}{dt}\left(R^2\frac{d\phi}{dt}\right)\hat{\boldsymbol{\phi}} + \frac{d^2 z}{dt^2}\hat{\mathbf{z}}$$

Another option is to keep a single length variable and use two angular variables. This yields **spherical coordinates** (r, θ, ϕ), as shown in the right panel of Fig. A.2. By convention, ϕ still measures the azimuthal angle in the sense of rotation around the z-axis, while θ measures the angle with the z-axis (so it is akin to latitude on Earth's surface). The conversion between spherical and Cartesian coordinates is

$$x = r\sin\theta\cos\phi \qquad y = r\sin\theta\sin\phi \qquad z = r\cos\theta$$

An analysis similar to that in Sect. A.1 yields the following expressions for the position, velocity, and acceleration vectors:

$$\mathbf{r} = r\,\hat{\mathbf{r}}$$

$$\mathbf{v} = \frac{dr}{dt}\hat{\mathbf{r}} + r\frac{d\theta}{dt}\hat{\boldsymbol{\theta}} + r\sin\theta\frac{d\phi}{dt}\hat{\boldsymbol{\phi}}$$

$$\mathbf{a} = \left[\frac{d^2 r}{dt^2} - r\left(\frac{d\theta}{dt}\right)^2 - r\sin^2\theta\left(\frac{d\phi}{dt}\right)^2\right]\hat{\mathbf{r}}$$

$$+ \left[r\frac{d^2\theta}{dt^2} + 2\frac{dr}{dt}\frac{d\theta}{dt} - r\sin\theta\cos\theta\left(\frac{d\phi}{dt}\right)^2\right]\hat{\boldsymbol{\theta}}$$

$$+ \left[r\sin\theta\frac{d^2\phi}{dt^2} + 2\sin\theta\frac{dr}{dt}\frac{d\phi}{dt} + 2r\cos\theta\frac{d\theta}{dt}\frac{d\phi}{dt}\right]\hat{\boldsymbol{\phi}}$$

A.3   Rotating Reference Frame

Consider a coordinate system (x', y', z') that is a rotating version of some stationary reference frame (x, y, z). The rotation axis is fixed, but the rotation rate may be variable. Without loss of generality we can define $\hat{\mathbf{z}} = \hat{\mathbf{z}}'$ to be the axis of rotation. Then we can write the unit vectors in the rotating frame as

$$\hat{\mathbf{x}}' = \cos\phi(t)\,\hat{\mathbf{x}} + \sin\phi(t)\,\hat{\mathbf{y}}$$

$$\hat{\mathbf{y}}' = -\sin\phi(t)\,\hat{\mathbf{x}} + \cos\phi(t)\,\hat{\mathbf{y}}$$

where $\phi(t)$ is the phase angle between the two frames at time t. The derivatives of these are

$$\frac{d\hat{x}'}{dt} = -\omega \sin\phi(t)\,\hat{x} + \omega\cos\phi(t)\,\hat{y} = \omega\,\hat{y}' \tag{A.1}$$

$$\frac{d\hat{y}'}{dt} = -\omega\cos\phi(t)\,\hat{x} - \omega\sin\phi(t)\,\hat{y} = -\omega\,\hat{x}' \tag{A.2}$$

where $\omega = d\phi/dt$ is the angular frequency of rotation. It is useful to define the angular frequency vector Ω so it points along the rotation axis and has a magnitude equal to the angular frequency at which the axes rotate: in our setup, $\Omega = \omega\,\hat{z}$.

Given some general vector $\mathbf{Q}$, we can write its components in both the fixed and rotating frames,

$$\mathbf{Q} = Q_x\,\hat{x} + Q_y\,\hat{y} + Q_z\,\hat{z} = Q_x'\,\hat{x}' + Q_y'\,\hat{y}' + Q_z'\,\hat{z}'$$

(With z as the rotation axis, $Q_z' = Q_z$ and $\hat{z}' = \hat{z}$.) The time derivative can be written in the fixed frame as

$$\frac{d\mathbf{Q}}{dt} = \frac{dQ_x}{dt}\,\hat{x} + \frac{dQ_y}{dt}\,\hat{y} + \frac{dQ_z}{dt}\,\hat{z}$$

Let's identify this as $(d\mathbf{Q}/dt)_{\text{fixed}}$. Now consider the rotating reference frame:

$$\frac{d\mathbf{Q}}{dt} = \left(\frac{dQ_x'}{dt}\,\hat{x}' + \frac{dQ_y'}{dt}\,\hat{y}' + \frac{dQ_z'}{dt}\,\hat{z}'\right) + \left(Q_x'\,\frac{d\hat{x}'}{dt} + Q_y'\,\frac{d\hat{y}'}{dt} + Q_z'\,\frac{d\hat{z}'}{dt}\right)$$

The first set of terms are what we would call $(d\mathbf{Q}/dt)_{\text{rot}}$, the derivative with respect to the rotating coordinates. The second set of terms can be rewritten using Eqs. (A.1) and (A.2) to obtain

$$\frac{d\mathbf{Q}}{dt} = \left(\frac{d\mathbf{Q}}{dt}\right)_{\text{rot}} + \left(\omega Q_x'\,\hat{y}' - \omega Q_y'\,\hat{x}'\right) = \left(\frac{d\mathbf{Q}}{dt}\right)_{\text{rot}} + \Omega \times \mathbf{Q}$$

using the properties of the vector cross product. Since the left-hand side is the same as $(d\mathbf{Q}/dt)_{\text{fixed}}$, we can think of the derivative operator as

$$\left(\frac{d}{dt}\right)_{\text{fixed}} = \left(\frac{d}{dt}\right)_{\text{rot}} + \Omega\times$$

Now if we let $\mathbf{Q}$ be the position vector $\mathbf{r}$, we have

$$\mathbf{v}_{\text{fixed}} = \mathbf{v}_{\text{rot}} + \Omega \times \mathbf{r}$$

Taking the derivative of this yields

$$\left(\frac{d\mathbf{v}}{dt}\right)_{\text{fixed}} = \left(\frac{d\mathbf{v}}{dt}\right)_{\text{rot}} + \Omega \times \mathbf{v}_{\text{rot}} + \frac{d\Omega}{dt} \times \mathbf{r} + \Omega \times \left(\frac{d\mathbf{r}}{dt}\right)_{\text{rot}} + \Omega \times (\Omega \times \mathbf{r})$$

The first term is the acceleration with respect to the rotating frame. Since $(d\mathbf{r}/dt)_{\text{rot}}$ is the velocity in the rotating frame, the fourth term is the same as the second. Collecting and identifying terms gives

$$\mathbf{a}_{\text{fixed}} = \mathbf{a}_{\text{rot}} + \Omega \times (\Omega \times \mathbf{r}) + 2\Omega \times \mathbf{v}_{\text{rot}} + \frac{d\Omega}{dt} \times \mathbf{r} \qquad (A.3)$$

The physical interpretation of the various terms is discussed in Sect. 6.1.1.

A.4   Angular Momentum

At various places in the book we encounter conservation of angular momentum as applied to an individual particle. As a reminder, a particle's angular momentum is

$$\mathbf{L} = \mathbf{r} \times \mathbf{p} = m\,\mathbf{r} \times \mathbf{v}$$

For a circular orbit, $\mathbf{v}$ is perpendicular to $\mathbf{r}$ so this reduces to

$$L = m\,r\,v = m\,r^2\,\omega$$

where $\omega = v/r$ is the angular speed of the orbit.

How do we generalize to an object with a finite size? Consider a sphere with mass M and radius R. Suppose it is rotating in **solid body rotation**, so all parts of the sphere rotate with the same period and hence the same angular speed. A particle with spherical coordinates (r, θ, ϕ) moves in a circle of radius $r \sin \theta$. The total angular momentum is obtained by adding up the contributions of all the individual particles:

$$L = \int_V \rho\,(r \sin \theta)^2\,\omega\,dV = \omega \int_0^R dr\,r^4 \int_0^\pi d\theta\,\sin^3 \theta \int_0^{2\pi} d\phi\,\rho(r, \theta, \phi)$$

where we use the spherical volume element $dV = r^2 \sin \theta\,dr\,d\theta\,d\phi$, and we pull out ω because it is constant. If the sphere has *uniform density* we can also pull out ρ and then evaluate the integrals:

$$L = \omega\rho \times \frac{R^5}{5} \times \frac{4}{3} \times 2\pi = \frac{2}{5} MR^2 \omega$$

where we use $M = (4/3)\pi R^3 \rho$. Finally, we note that the factor of ω just depends on how fast the sphere is spinning, while the factor of $(2/5)MR^2$ depends only on the structure of the sphere. We collect the structure-dependent pieces and write

$$L = I\omega$$

where I is the **moment of inertia**. For a uniform density sphere of mass M and radius R, the moment of inertia is

$$I = \frac{2}{5} MR^2$$

Other geometries would give different values for the moment of inertia.

A.5 Taylor Series Approximation

If we need to study some complicated function $f(x)$ but are mainly interested in its behavior over some "small" region, we can obtain a useful approximation by making a Taylor series expansion. Recall that to expand $f(x)$ around some value $x = a$, we write

$$f(x) \approx f(a) + f'(a)(x - a) + \frac{1}{2} f''(a)(x - a)^2 + \mathcal{O}\left((x - a)^3\right)$$

where $f'(a)$ is the first derivative of the function, evaluated at $x = a$, while $f''(a)$ is the second derivative, and so forth. If we can write the function in a form such that $a = 0$, then we have

$$f(x) \approx f(0) + f'(0)x + \frac{1}{2} f''(0)x^2 + \mathcal{O}(x^3)$$

Example. We often see functions of the form

$$f(x) = (1 + x)^\alpha$$

Let's evaluate the derivatives:

$$f'(x) = \alpha(1 + x)^{\alpha-1}$$
$$f''(x) = \alpha(\alpha - 1)(1 + x)^{\alpha-2}$$

The Taylor series approximation is therefore

$$f(x) \approx 1 + \alpha x + \frac{1}{2}\alpha(\alpha - 1)x^2 + \mathcal{O}(x^3)$$

The approximation $(1 + x)^\alpha \approx 1 + \alpha x$ is one we use a lot.

A.6 Numerical Solution of Differential Equations

If we face a differential equation that is difficult or impossible to solve with pencil and paper, we can turn to numerical techniques. Consider a differential equation of the form

$$\frac{df}{dt} = g(t)$$

We can rewrite the left-hand side as

$$\frac{f(t + \Delta t) - f(t)}{\Delta t} \approx g(t)$$

if Δt is sufficiently small. Rearranging, we can write:

$$f(t + \Delta t) \approx f(t) + g(t)\,\Delta t$$

Suppose as an initial condition we know $f(t_1)$. Then we can take a series of steps:

$$f(t_2) \approx f(t_1) + g(t_1)(t_2 - t_1)$$
$$f(t_3) \approx f(t_2) + g(t_2)(t_3 - t_2)$$
$$f(t_4) \approx f(t_3) + g(t_3)(t_4 - t_3)$$

$$\vdots$$

$$f(t_{i+1}) \approx f(t_i) + g(t_i)(t_{i+1} - t_i)$$

If the steps are small, the approximations are good enough to give something close to the right answer. Taking a long series of small steps is tedious by hand but manageable by computer.

This approach can be extended to handle a system of differential equations. For example, if we have two equations

$$\frac{df_1}{dt} = g_1(t) \quad \text{and} \quad \frac{df_2}{dt} = g_2(t)$$

then we can write the solution as

$$f_1(t_{i+1}) \approx f_1(t_i) + g_1(t_i)(t_{i+1} - t_i) \quad \text{and} \quad f_2(t_{i+1}) \approx f_2(t_i) + g_2(t_i)(t_{i+1} - t_i)$$

and so on for as many time steps as desired. The generalization to three or more equations is straightforward.

So far we have considered first-order differential equations, but we can generalize to second-order equations through a "trick." Consider a second-order equation of the form

$$\frac{d^2 x}{dt^2} = a(t)$$

If we take not only the position x but also the velocity v to be dependent variables, we can write the second-order equation as a pair of coupled first-order equations:

$$\frac{dx}{dt} = v(t) \quad \text{and} \quad \frac{dv}{dt} = a(t)$$

This pair of equations can be solved as above.

There is an important subtlety here: computers have finite precision, and with a large number of steps the numerical error can build up. Keeping the steps small enough to make the approximations valid but large enough to control numerical errors may require a delicate balance. In addition, the equations and/or boundary conditions may have intrinsic difficulties in certain problems. A variety of algorithms have therefore been developed for solving differential equations computationally (see, e.g., [1]).

A.7   Useful Integrals

Here is a compilation of integrals, some of which appear at various places in the text. The integrals can be evaluated by consulting a reference book (e.g., [2]) or using software such as Mathematica [3]. Most of these are expressed as indefinite integrals, which have an arbitrary constant of integration that is omitted here to avoid clutter. First consider integrals of the form:

$$\int \frac{x}{1+x} \, dx = x - \ln(1+x)$$

$$\int \frac{x}{(1+x)^2} \, dx = \frac{1}{1+x} + \ln(1+x)$$

Next consider similar integrals with x^2 instead of x:

$$\int \frac{x^2}{(1+x^2)^{1/2}}\, dx = \frac{1}{2}x\left(1+x^2\right)^{1/2} - \frac{1}{2}\sinh^{-1} x$$

$$\int \frac{x^2}{1+x^2}\, dx = x - \tan^{-1} x$$

$$\int \frac{x^2}{(1+x^2)^{3/2}}\, dx = -\frac{x}{(1+x^2)^{1/2}} + \sinh^{-1} x$$

$$\int \frac{x^2}{(1+x^2)^2}\, dx = -\frac{x}{2(1+x^2)} + \frac{1}{2}\tan^{-1} x$$

$$\int \frac{x^2}{(1+x^2)^{5/2}}\, dx = \frac{x^3}{3(1+x^2)^{3/2}}$$

Here are a few other integrals that may be useful:

$$\int \left(\frac{x}{1-x}\right)^{1/2}\, dx = -[x(1-x)]^{1/2} + \sin^{-1} x^{1/2}$$

$$\int \frac{x^4}{(1+x^2)^4}\, dx = \frac{x(-3-8x^2+3x^4)}{48(1+x^2)^3} + \frac{1}{16}\tan^{-1} x$$

Finally, here are definite integrals with a Gaussian integrand:

$$\int_{-\infty}^{\infty} e^{-x^2/(2\sigma^2)}\, dx = (2\pi)^{1/2}\sigma$$

$$\int_{-\infty}^{\infty} x^2\, e^{-x^2/(2\sigma^2)}\, dx = (2\pi)^{1/2}\sigma^3$$

$$\int_{-\infty}^{\infty} x^4\, e^{-x^2/(2\sigma^2)}\, dx = 3(2\pi)^{1/2}\sigma^5$$

References

1. A. Iserles, *A First Course in the Numerical Analysis of Differential Equations*, 2nd edn. (Cambridge University Press, Cambridge/New York, 2009)
2. A. Jeffrey, D. Zwillinger, *Table of Integrals, Series, and Products*. Table of Integrals, Series, and Products Series, 7th edn. (Elsevier, Amsterdam, 2007)
3. Wolfram Research, Inc., *Mathematica*, 8th edn. (Wolfram Research, Champaign, 2010)

Appendix B
Solutions

This Appendix provides partial solutions for some of the end-of-chapter problems. The answers are intended to help you check your work without revealing too much about the analysis itself.

Chapter 1

1.1 (a) $v \sim 30\,\mathrm{km\,s^{-1}}$

1.2 (b) The amount of mass required is $<0.001\,M_\odot$.

1.3 (b) $m \sim 0.3\,\mathrm{g}$

1.4 <100 hydrogen atoms per cubic meter.

1.5 (b) Hydrogen nuclei have a typical speed of $\sim 4 \times 10^5\,\mathrm{m\,s^{-1}}$.

1.6 (b) $F \sim 0.6\,\mathrm{kg\,m\,s^{-2}}$

Chapter 2

2.5 If $\Phi(0) = 0$ then $\Phi(R) = 3GM/2R$.

2.7 For mass $M_\odot$, the Schwarzschild radius is 3 km.

2.8 (b) I estimate that I could jump off an asteroid smaller than a few kilometers in radius.

C. Keeton, *Principles of Astrophysics: Using Gravity and Stellar Physics to Explore the Cosmos*, Undergraduate Lecture Notes in Physics, DOI 10.1007/978-1-4614-9236-8,
© Springer Science+Business Media New York 2014

Chapter 3

3.6 Star #16 moves faster than 10^4 km s^{-1} at pericenter.

Chapter 4

4.4 (b) The uncertainty in the mass of Sirius A associated with the uncertainty in the distance is about $0.02\,M_\odot$.

4.5 $a \approx 0.006$ AU.

4.8 (a) Roughly 500.

Chapter 5

5.3 (b) Triton's orbit must be shrinking by $\sim$15 cm yr^{-1}.

5.4 (d) For $M \approx 16\,M_\odot$, the tidal acceleration would be in excess of 10^7 m s^{-2}.

5.6 If we catch the asteroid when it is 1 AU from Earth, the amount of tangential velocity we need to impart is <1 m s^{-1}. (Depending on the asteroid's mass, that may still translate into a lot of kinetic energy.)

Chapter 6

6.1 The space station would have to spin once every 20 s.

6.2 (a) Dimensional analysis is sufficient here, but if you want full details the mathematical analysis is similar to Problem 16.9. (b) By dimensional analysis, v must scale as $(GM/d)^{1/2}$ where M is the mass of each body and d is the distance between them. Here you should be able to find the exact multiplicative constant.

Chapter 7

7.2 (a) $M \approx 9 \times 10^{10} M_\odot$

7.4 (b) The rotation curve scales as $r^{1/2}$ for $r \ll a$ and as $r^{-1/2}$ for $r \gg a$.

7.6 (b) $\rho \sim 0.4\,M_\odot$ pc^{-3} using the thick disk parameters.

Chapter 8

8.2 (d) $\sigma \approx 20\,\mathrm{km\,s^{-1}}$

8.3 (d) $R_f \approx 250\,\mathrm{kpc}$

8.4 (b), (d) The probability of a direct hit is less than 10^{-13}, but the probability of a perturbation to the motion is ~ 0.1.

Chapter 9

9.3 (c) The Einstein crossing time is more than a century.

9.4 For the second point, $\mu_{\mathrm{tot}} \approx 2.2$.

9.5 (d) $\tau \sim 3 \times 10^{-7}$.

9.7 (a) $M \approx 6 \times 10^{10}\,M_\odot$.

9.8 (d) The estimated number of lenses is $<10^{-5}$.

Chapter 10

10.2 (a) $v_x = c$ (independent of u).

10.4 (b) $\lambda_{\mathrm{obs}} = 487.0\,\mathrm{nm}$.

10.6 (a) From $2R_S$ to R_S, the time elapsed in the probe's reference frame is a little over 143 s.

10.7 (a) The innermost stable circular orbit lies at $r = 3R_S$.

10.8 (b) If the stars have the same mass M, the time to merge is $t = 5c^5 a_0^4 / 512 G^3 M^3$.

Chapter 11

11.5 (b) One case has a power law dependence on t, while the other has an exponential dependence.

11.6 The lookback time to $z = 1$ is a little less than 6 Gyr.

11.7 Around 80 Gyr in the future.

Chapter 12

12.3 (b) $H \approx 10$ km.

12.5 For Jupiter, $T_{esc} = 3{,}960$ K for atomic hydrogen (H), and twice that for molecular hydrogen (H_2).

12.6 (b) At the center, $T \approx 5 \times 10^4$ K and $v_{rms} \approx 4 \times 10^4$ m s^{-1} for protons.

12.7 (c) $T(0) > 2 \times 10^6$ K.

Chapter 13

13.4 The effective temperature of Sirius B is about 2.5×10^4 K.

13.6 At $2.2\,\mu$m, the ratio of emitted brightnesses is planet/star $= 3.3 \times 10^{-4}$ if the albedo is $a = 0$.

13.8 (d) Using the numbers given in the problem, $R_{Eris} \approx 1{,}800$ km. (Other recent measurements and analyses have led to somewhat different results.)

Chapter 14

14.1 (b) $N_4/N_1 = 5.4 \times 10^{-4}$ for the O star.

14.3 (b) $X = 0.975$ at the center of a uniform star the same size and mass as the Sun.

14.5 $\rho \approx 9 \times 10^{-4}$ kg m^{-3}

Chapter 15

15.5 (a) $\alpha \approx 4$

15.7 The mean free path is more than 10 light-years.

Chapter 16

16.5 The lower limit to the main sequence is actually around $0.08\,M_\odot$, and our estimate is fairly close.

16.7 Assuming 10% of the mass is involved in fusion, the core helium burning phase lasts $\sim 3 \times 10^6$ yr.

16.8 A little less than a decade.

Chapter 17

17.3 (d) $I_3/I_1^{5/3} = 1.14$ for $kT/E_F = 0.2$.

17.5 (b) $p_F/(m_e c) \approx 1$ for $M = 0.7\,M_\odot$, so the gas is not truly non-relativistic (although it is not highly-relativistic either, for that would require $p_F/(m_e c) \gg 1$).

17.6 (c) A neutron star with $\eta \sim 10$ has $R \sim 3$ km. (This is surprisingly similar to our dimensional analysis estimate in Sect. 1.3.2.)

Chapter 19

19.2 The rule of thumb is roughly one supernova per galaxy per century.

19.5 I get a final radius of ~ 0.2 AU.

19.6 (e) The largest object that could cool within the age of the universe is a few times $10^{13}\,M_\odot$.

Chapter 20

20.1 (a) $\ell \sim 1{,}500$ pc

20.3 (b) The photon/baryon ratio in the universe is $\sim 10^9$.

Index

C. Keeton, *Principles of Astrophysics: Using Gravity and Stellar Physics to Explore the Cosmos*, Undergraduate Lecture Notes in Physics, DOI 10.1007/978-1-4614-9236-8,
© Springer Science+Business Media New York 2014

Lightning Source UK Ltd.
Milton Keynes UK
UKOW06f1143040917
308549UK00002B/13/P